The International Correspondence School's

STEAM LOCOMOTIVE STUDY COURSE

VOLUME 2

Stephenson Valve Gear

Walschaert Valve Gear
Part 1 and Part 2

Baker Locomotive Valve Gear

Southern Locomotive Valve Gear

Locomotive Boilers
Part 1 and Part 2

Locomotive Injectors

Edited by William C. Fitt

Wildwood Publications — Traverse City, Michigan

Reprinted by permission of:
International Correspondence School
Division of INTEXT
Scranton, Pennsylvania 18503

Library of Congress Catalog Card No. 79-65782
ISBN No. 0-914104-05-5

International Correspondence School's

STEAM LOCOMOTIVE STUDY COURSE

The Set Includes:

VOLUME 1

The Steam Locomotive - Part 1 • The Steam Locomotive - Part 2 • Locomotive Valves • Locomotive Valve Setting

VOLUME 2

Stephenson Valve Gear • Walschaert Valve Gear - Part 1 • Walschaert Valve Gear - Part 2 • Baker Locomotive Valve Gear • Southern Valve Gear • Locomotive Boilers - Part 1 • Locomotive Boilers - Part 2 • Locomotive Injectors

VOLUME 3

Locomotive Feedwater Heating Equipment • Heat and Superheaters • Locomotive Appliances - Part 1 • Locomotive Appliances - Part 2 • Locomotive Lubricators • Locomotive Headlights

VOLUME 4

Hand Firing of Locomotives • Locomotive Stokers • Type D Duplex Stokers • Locomotive Management • Locomotive Breakdowns • Foundation Brake Rigging - Part 1 • Foundation Brake Rigging - Part 2 • "AB" Freight Brake Equipment • Type K Freight Car Brake Equipment

Volume Two
Table of Contents

Stephenson Valve Gear

1968 Printed in U.S.A. Edition 3

1944 Edition

STEPHENSON VALVE GEAR

ARRANGEMENT AND OPERATION

DETAILS AND ACTION OF GEAR

INTRODUCTION

1. Definition of Valve Gear.—A *valve gear*, commonly but incorrectly referred to as a valve motion, is the mechanism which is employed to move the valves of a locomotive. The movement of the valve gear causes the valves to admit the steam to, and exhaust the steam from, the cylinders in the proper manner to make the locomotive run either forwards or backwards. Also the valve gear permits of the cut-off being varied, that is, the travel of the valves across the steam ports can be either shortened or lengthened, and the admission of steam to the cylinders can be cut off at different points in the stroke of the piston. This allows the steam pressure in the cylinders and on the pistons, and therefore the power of the locomotive to be changed to meet such variable conditions as speed, weight of train, grade, etc., under which the locomotive may be working. Briefly, a valve gear permits the locomotive to be reversed, and the cut-off to be changed.

Valve gears usually obtain their movement from a form of crank on either the main driving axle or on the ends of the main crankpins. The locomotive valve gears in general use are the Walschaert, the Baker, and the Stephenson, although the Southern valve gear is also used to some extent.

2. Historical.—The Stephenson valve gear derives its name from the fact that it was first applied to a locomotive by George Stephenson in 1842. The invention of the gear has been the subject of much controversy, but the credit is usually given to a draftsman named Williams who was employed by the Stephenson locomotive works.

The gear was universally used in America for many years, but with the introduction of heavy locomotives, it was found that other types of valve gears could be employed to a better advantage. Therefore the Stephenson valve gear is no longer applied to new locomotives and its use is confined to those that were built before the application of other gears became general.

3. The Stephenson valve gear is an inside gear, that is, one which is placed between the driving wheels. A gear of this type obtains its movement from eccentrics which are keyed to the main driving axle. The axles of modern locomotives finally became so large as to require eccentrics and straps of such a size as to be impractical. Also inside gears, especially with large locomotives, are difficult to inspect and maintain. The Stephenson gear was, therefore, supplanted by outside gears, or by gears which are placed on the outside of the driving wheels.

Outside valve gears do not require eccentrics, but obtain their movement from small crank-arms on the main pins. Also they are readily accessible for inspection, oiling, and repairs.

4. Valve Gears Required.—The ordinary steam locomotive consists of two engines which are supplied with steam from one boiler, with the driving wheels of the engines connected by axles. Each engine has a cylinder, a piston, a valve, and the parts necessary to convert the reciprocating, or to-and-fro, movement of the piston to a turning movement at the main driving wheel on that side. Each engine requires a valve gear and, therefore, the common type of locomotive has two valve gears. Mallet locomotives require four valve gears, as this type of locomotive usually consists of four engines or two complete locomotives under one boiler. If the two valve gears were separate and distinct units, each would require a lever in order to operate them. As such an arrangement would not be practicable, the two valve gears are so connected that both are under the control of a single reverse lever.

UNDERLYING PRINCIPLES OF GEAR

5. Principle on Which Gear Operates.—The principle on which the Stephenson valve gear operates will be explained by tracing its development from a simple form of gear such as is shown in Fig. 1 *(a)*. The valve used with this gear has neither lap nor lead; that is, the width of the valve at each end is the same as the width of the steam ports.

6. Arrangement of Simple Gear.—The arrangement of the gear shown in Fig. 1 *(a)*, so that the piston will be made to travel back and forth in the cylinder and thereby transmit a turning movement to the driving wheel through the medium of the main crank *a c* and the main rod *b c*, is as follows: With the piston at the beginning of the backward stroke as shown in *(a)*, the gear must be so arranged that the valve will be on the point of admitting steam to the cylinder in front of the piston, and the same condition must exist when the piston is at the beginning of the forward stroke. To meet this requirement, the valve with the piston at the beginning of the stroke must be in mid-position or mid-stroke as shown; that is, a line *d d* drawn through the center of the valve (half way between the ends) should come midway between the steam ports. The valve when so placed is a half stroke in advance of the piston, provided that the movement of both is to the right, and if this difference of positions is maintained the engine will keep moving as long as steam is supplied.

7. Position of Cranks.—As the main crank *a c* moves with the piston, and the valve moves with the valve crank *a e*, the cranks must be set with respect to each other so as to maintain the valve in the proper position or one-half stroke in advance of the piston.

To place the valve a half stroke ahead of the piston, and to maintain this difference in position between them, requires that the valve crank *a e*, or eccentric, as it will be called, must be placed a quarter turn ahead of the main crank, *a c*. This position places the valve a half stroke in advance of the piston, because, like the main crank and the piston, the eccentric moves the valve a half stroke when it makes a quarter turn. The eccentric is accordingly placed at *a e*, a quarter turn ahead of the main crank *a c*, as the wheel is assumed to be turning forwards.

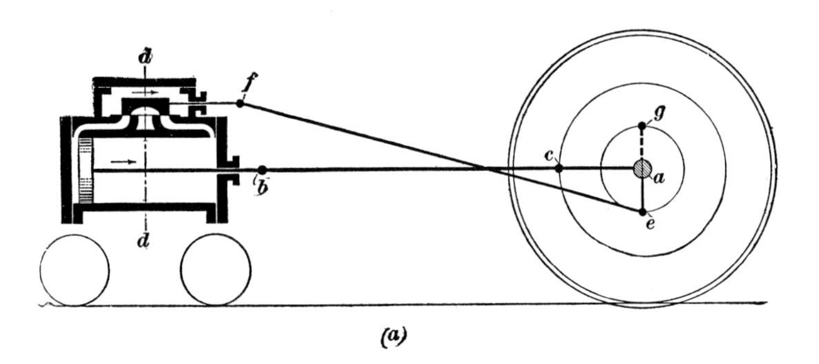

(a)

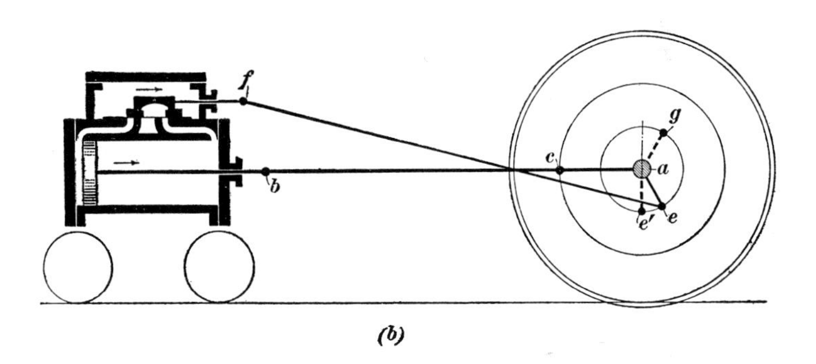

(b)

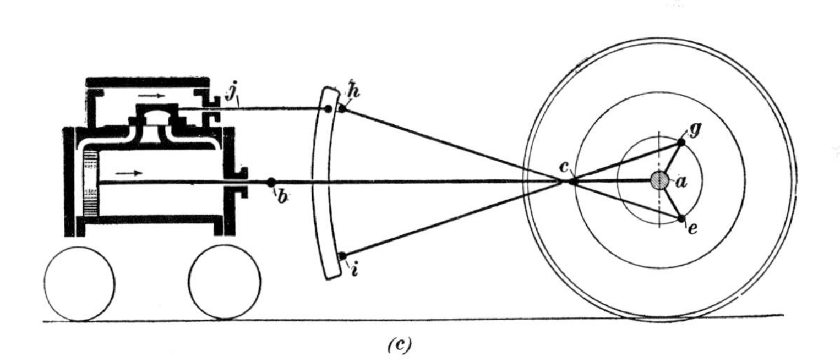

(c)

Fig. 1

8. Operation.—If the driving wheel, Fig. 1 *(a)* is turned in the forward direction, the eccentric, the valve rod *e f*, and the valve move to the right, and steam is admitted through the front steam port to the cylinder. When the piston is at the beginning of the other stroke, the eccentric brings the valve to admission at the back steam port, and a further rotation of the wheel causes the valve to open the port. With the piston at mid-stroke, the valve has one port wide open for the admission of steam, and the other port wide open to the exhaust. The valve cuts off the admission of steam to the cylinder just as the piston completes a stroke. If it is desired to have the wheel turn in the reverse direction, the eccentric *a e* has only to be changed to the position *a g*, which brings it a quarter turn ahead of the main crank for this direction of rotation, and places the valve a half stroke ahead of the piston. Hence, to reverse merely requires an eccentric the same distance from the main pin when running backwards as when running forwards.

9. Addition of Steam Lap.—With the valve shown in Fig. 1 *(a)*, no benefit is obtained from the expansive force of steam for the steam is admitted to the cylinder during the complete stroke of the piston. Therefore, the valve in Fig. 1 *(b)* has been given steam lap, that is, the ends of the valve have been made wider than the steam ports. This change in the valve keeps the steam in the cylinder for a time after the supply has been cut off, and the steam expands and moves the piston.

When the valve has steam lap, the eccentric, in order to bring the valve to admission with the piston at the beginning of a stroke, must be turned away from its quarter position *a e'*, Fig. 1 *(b)*, until the valve is displaced from mid-position by an amount equal to the steam lap, and also the lead if any, and this brings the eccentric to *a e*. To reverse the engine, the eccentric has only to be placed at *a g* or at the same distance from *c* on the other side of the axle. The eccentric *a e*, as it is set to cause the forward movement of the engine, will be called the go-ahead eccentric, and the one *a g*, set to cause the backward movement, the back-up eccentric.

10. Reversing Arrangement.—It will be evident, if both a go-ahead and a back-up eccentric are fixed in their correct positions on the axle, that the engine can be made to run in either direction if a means is provided to throw the valve under the influence of either one.

An arrangement for this purpose is shown in Fig. 1 *(c)* consisting of a slotted link *h i* which can be raised and lowered, and in which the end of the valve rod *j* can slide, and eccentric rods *e h* and *g i* which connect the eccentrics to the link. When the link is all the way down as shown, the go-ahead eccentric rod is directly in line with the valve rod *j*, the valve receiving its entire movement from the go-ahead eccentric *a e*, and the engine moves forwards. With the link all the way up, the back-up eccentric rod *g i* is in line with the valve rod, the valve being placed under the control of the back-up eccentric *a g*, and the engine will run backwards. The link merely provides a means of throwing the valve under the control of the eccentric desired.

As will be explained later, the Stephenson valve gear uses an eccentric for each gear, and a link so that the valve can be thrown under the control of either eccentric and the locomotive thereby reversed. Therefore with the addition of another eccentric, an eccentric rod, and a link, the simple form of valve gear shown in Fig. 1 *(a)* is converted into a gear which permits the engine to be run in either direction.

11. Views of Valve Gear.—As a certain portion of each valve gear consists of the same parts, it is not necessary when identifying the parts to show both valve gears in the same view. However, it is necessary to show both gears in order to understand how they are operated by one reverse lever. Therefore, in Fig. 2 are shown three side views of the valve gear on the right side of the locomotive as viewed from the left side with all the parts on the left side removed which would obstruct the view. Fig. 3 is a view of the front end of both gears and shows how the two gears are connected as well as some details which cannot be clearly shown in Fig. 2. Fig. 2 also shows the position of the parts of the valve gear when the main crankpin is at *P*. In these views the reverse lever *16* and the valve gear are shown in three different positions. The main driving axle is marked *A*.

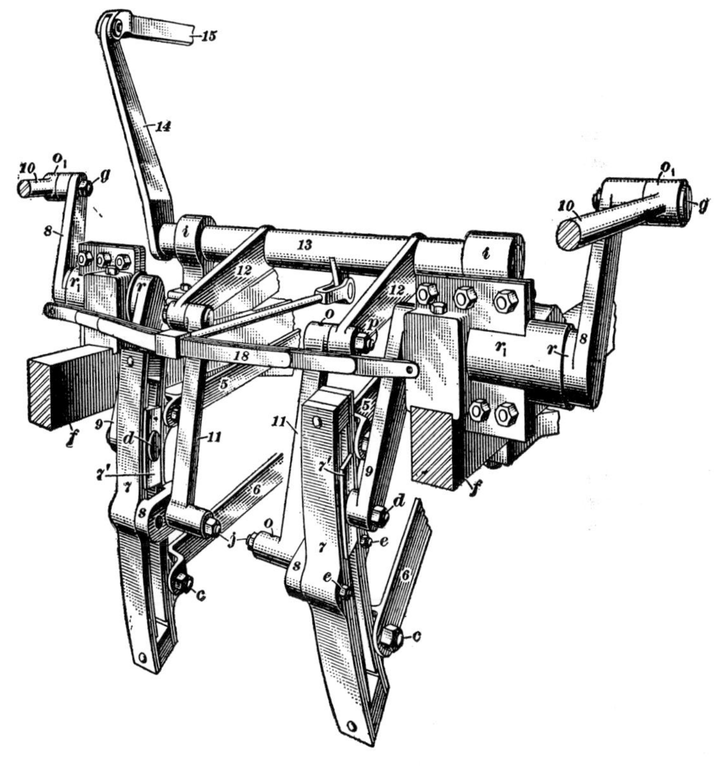

Fig. 3

12. Names of the Parts.—Referring to Fig. 2 *(a),* the names of the principal parts of a single valve gear, it being understood that similar parts comprise the other valve gear, are as follows: *1,* the forward-motion, or go-ahead, eccentric; *2,* the backward-motion, or back-up, eccentric; *3,* the go-ahead eccentric strap; *4,* the back-up eccentric strap; *5,* also see Fig. 3, the go-ahead eccentric rod or blade; *6,* the back-up eccentric rod or blade; *7,* the link, with a link saddle *s*; *7′*, the link block; *r*, the rocker with two arms, the upper rocker-arm *8,* and the lower rocker-arm *9; 10,* the valve rod; *11,* the link hanger; and *12,* the link lifter or the reverse shaft arm. Duplicates of all the mentioned parts are found in the other gear. The parts which are not duplicated but which are common to both gears are shown in Figs. 2 and 3 and are as follows: *13,* the reverse shaft, sometimes called the tumbling shaft; *14,* the upper reverse-shaft arm; *15,* the reach rod; *16,* Fig. 2, the reverse lever; *17,* the quadrant; *18,* the counter-balance spring.

13. Arrangement of Parts.—The eccentrics *1* and *2,* Fig. 2, are secured to the axle *A* by keys *t,* and setscrews, not shown, and rotate within the eccentric straps *3* and *4* when the axle *A* turns. The eccentric marked *1* is called the go-ahead eccentric because its movement when transmitted to the valve causes the locomotive to run forwards. The eccentric marked *2* is called the back-up eccentric because it gives the required movement to the valve to cause the locomotive to run backwards. The eccentric straps *3* and *4* encircle the eccentrics and are made in halves which are held together by the bolts and nuts *a.* The back ends of the eccentric rods *5* and *6* are connected to the eccentric straps by the bolts and nuts *b.* The front ends of the eccentric rods are forked, and are connected to the links *7* by the eccentric-rod pins *c.* The connection at these points is flexible; that is, the link turns freely on the pins *c* when the locomotive is moving.

The link saddle *s,* Figs. 2 and 3, which is secured to the link by the two bolts and nuts *e,* is connected to the link lifter *12* by the link hanger *11.* The link hanger swings freely on the link-saddle pin *j,* which is cast with the saddle, and also on the link-hanger pin *p,* Fig. 2 *(c)* which connects it to the link lifter *12.* The link block *7′* which is placed within the slotted link *7,* is connected to the lower rocker-arm *9* by the link-block pin *d.* The block turns freely on the pin but the pin is fixed with respect to the rocker-arm. The rocker *r* turns freely in the rocker box r^1, Fig. 3, which is bolted to the frame *f* of the locomotive. The valve rod *19,* Figs. 2 and 3, is flexibly connected to the upper rocker-arm *8* by the rocker-arm pin *g,* which is oiled through the oil hole o^1, Fig. 3. The valve rod *10* and the valve stem are held together by the valve-stem key *h,* Fig. 2

14. The link lifter or lifting arms *12,* Fig. 3, and the upper reverse-shaft arm *14* are secured rigidly to the reverse shaft *13,* the ends of which are carried in brackets *i,* bolted to the frames. The front end of the reach rod *15,* Fig. 2, is pivoted to the top of the upper reverse-shaft arm *14,* and the other end is pivoted to the reverse lever *16.* The lower end of the reverse lever is pivoted at i^1 to the frame or deck plate. The quadrant *17* with notches cut in its upper face is bolted at each end to the quadrant stands j^1, which are usually bolted to brackets on the boiler. The reverse-lever latch box *k* serves to hold the latch *l* in place. The latch *l,* which is a bolt or tooth, engages with the teeth in the quadrant and thereby holds the reverse lever in position in any part of the quadrant. The latch is raised when it is desired to move the lever by the latch handle *m,* Fig. 2 *(a),* through the medium of the latch links *n.* A spring returns the bolt *l* into the teeth of the quadrant when the latch handle *m* is released. The inner end of the stud shown on the front of the latch box extends through an elongated slot in the bolt *l* and serves to guide it.

The valve gears of modern locomotives are usually power operated, the reach rod being connected to a power device instead of to the reverse lever. The device is generally operated by compressed air, and is controlled by a small lever.

15. A complete locomotive valve gear, by which is meant the two gears as shown in Fig. 3, may be considered as being made up of two parts, one part being directly concerned with valve movement, and in operation when the locomotive is running, and a second part, the purpose of which is to transmit movement to the first part through the reverse lever. The eccentrics, the eccentric straps, the eccentric rods, (there are four of each) the links, the link blocks, the rockers, and the valve rods (there are two of each), transmit movement to the valves when the locomotive is working, and may be considered as one part of the gear.

The other or second part, which comprises the two link hangers, the two link lifters, the reverse shaft, the upper reverse-shaft arm, and the reach rod, may be considered as the second part of the gear, and is used to transmit the movement of the reverse lever to the first part. With the exception of the link hangers, the second named part receives its movement from the reverse lever only.

16. Use of Transmission Bar.—In Fig. 2 is shown the arrangement of the valve gear as applied to an eight-wheel or 4-4-0 type of locomotive. With this type the main driving axle *A* on which the eccentrics are placed is quite close to the valves and the cylinders. In Fig. 4 *(a)* the main driving axle is the second one and is some distance from the valves. In order to avoid using very long eccentric rods, which would be the case were the link block *7′* pivoted to the lower rocker arm *9*, or a very long valve rod *10* were the rocker *r* placed opposite the link, a transmission bar *16′* is used, which serves to connect the link block to the lower rocker-arm. Both ends of the bar are forked and the rear end straddles the link and the link block, and is secured to the latter by the link-block pin *d*. The transmission bar is suspended from the frame by the transmission-bar hanger *17′*, on which the rear end of the bar swings. As shown in view *(b)*, the rocker differs from the one shown in Figs. 2 and 3, as in this view both the rocker-arms *8* and *9* point downwards.

17. The rocker shown in Fig. 4 *(b)* is referred to as a direct rocker because, when in operation, both arms move in the same direction. The rocker shown in Figs. 2 and 3 is an indirect rocker because the movement of the two arms is always in opposite directions. For example, if the lower arm *9* is moved ahead, the upper arm *8* moves back, and this arm moves ahead if the lower arm is moved back.

The valve gear in Fig. 4 *(a)* operates an inside-admission piston valve *1′* instead of a slide valve as in Fig. 2. Also, the arrangement of the counterbalance *18* differs from that shown in Fig. 2, as the casing shown in Fig. 4 *(a)* contains a coil spring.

GENERAL OPERATION

18. Positions of Reverse Lever.—The valve gear is said to be in full forward-gear position with the reverse lever *16* in the position shown in Fig. 2 *(a)*, in mid-gear position with the lever in the center of the quadrant *17*, as in Fig. 2 *(b)*, and in full back-gear position with the lever in the back corner of the quadrant, Fig. 2 *(c)*. The locomotive will run forwards with the reverse lever in full forward gear position, or in any

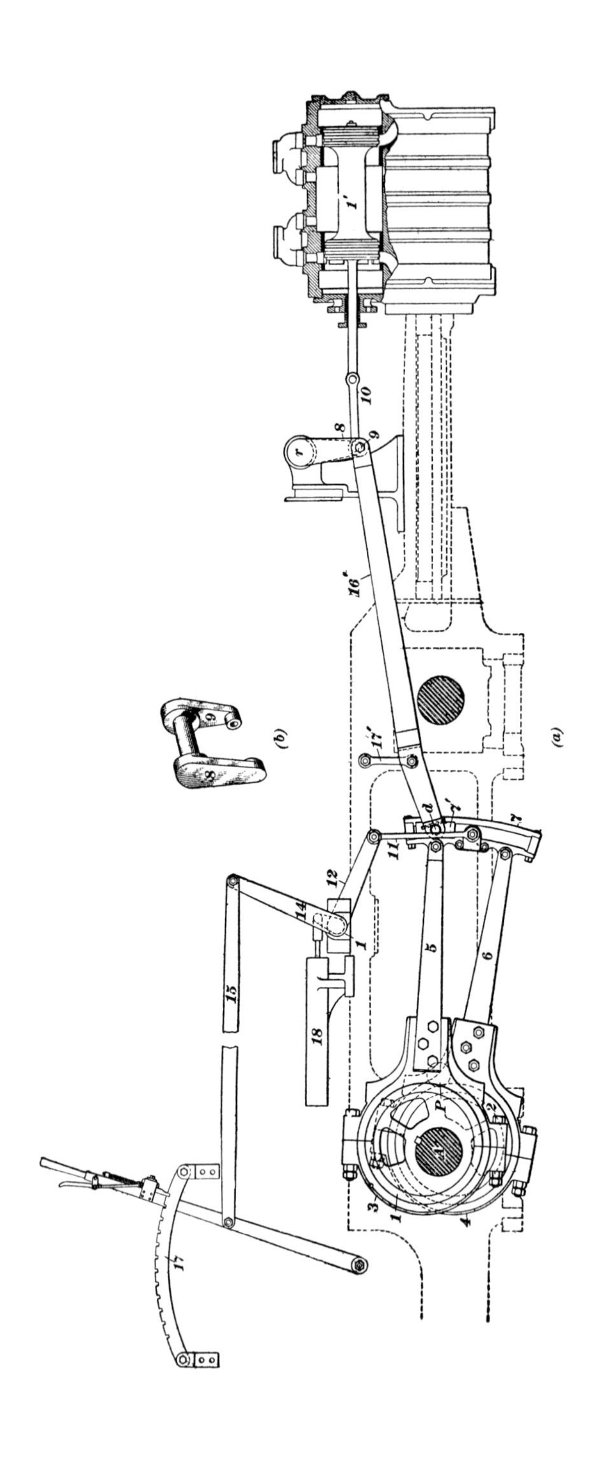

Fig. 4

intermediate position between it and mid-gear. The locomotive will run backwards with the reverse lever in full back-gear position, or in any intermediate position between it and mid-gear position. In mid-gear position, a locomotive will not move if the valve gear is accurately designed and installed, but usually such inaccuracies exist that the locomotive may start either forwards or backwards, and then continue to move in the direction in which it started.

The longest cut-off possible in forward and backward gears is obtained with the reverse lever in the front and back corners of the quadrant, as shown in Fig. 2 *(a)* and *(c)*. The cut-off decreases from the maximum to the minimum as the lever is moved from the corners to the notch on either side of the center notch. With the reverse lever in either corner, the steam will be admitted to the cylinder for about 80 or 85 per cent of the piston stroke before it is cut off by the valves.

19. Movement Imparted by Reverse Lever.—When the reverse lever *16,* Fig. 2 *(a),* is drawn back, the following movement will be imparted to the valve gear: The reach rod *15,* Fig. 3, and the upper end of the reverse-shaft arm *14* move back with the reverse lever, and turn the reverse shaft *13.* When the shaft turns, the link lifters *12* and the link-hangers *11* pull the links *7* and the forward ends of the eccentric rods upwards, and the eccentric straps *3* and *4,* Fig. 2, turn on the eccentrics *1* and *2.* The links *7,* Fig. 3, slide freely past the link blocks *7′* on account of the blocks having a working fit in the slots in each link. When the reverse lever is mid-gear, Fig. 2 *(b),* the links will be raised half way up, and this brings the link blocks in the center of the links. Therefore the link blocks are in the upper half of the links when the reverse lever is between full forward-gear and mid-gear. If the reverse lever is moved toward the back corner of the quadrant, the links, Fig. 3, will continue to move upwards until, with the lever in the back corner, Fig. 2 *(c),* the link blocks will be near the lower end of the slots in the links. The link blocks are therefore in the lower half of the links when the reverse lever is between mid-gear and full back-gear positions.

20. With the link blocks near the upper ends of the links, as shown in Fig. 3 or in any intermediate position between that and mid-gear, the valves receive their movement from the go-ahead eccentrics *1* and the go-ahead eccentric rods *5,* Fig. 2 *(a),* and the locomotive moves forwards. When the links are moved upwards until the link blocks are in the lower half of the links, Fig. 2 *(c),* the valves are under the control of the back-up eccentrics *2* and the back-up eccentric rods *6,* and the locomotive moves backwards. The cut-off is longest with the link blocks near the ends of the links and is shortened by bringing the blocks nearer to the center of the links.

The foregoing shows that the movement of the reverse lever merely changes the position of the link blocks in the links, and thereby lengthens or shortens the cut-off, or brings the link blocks and the valves under the control of the go-ahead or the back-up eccentrics and thereby reverses the locomotive. The movement imparted to the valves when the reverse lever is moved from one gear to the other depends upon the position of the main crankpins. The movement of the valves will be greatest with the pins on the quarters. The movement decreases according as the pins are turned nearer to the centers in which latter position the minimum valve movement is obtained.

21. Movement Imparted by Eccentrics.—The manner in which the go-ahead eccentric *1*, Fig. 2 *(a)*, transmits movement to the valve is as follows, it being understood that the other go-ahead eccentric is imparting movement to its valve in the same way. When the axle *A* turns, the go-ahead eccentric *1* turns with it within the eccentric strap *3*. As the eccentric serves the same purpose as a crank the eccentric strap is given a circular motion, and a to-and-fro movement is imparted to the front end of the eccentric rod *5*, and the link *7*. A similar movement is transmitted to the lower end of the link by the back-up eccentric *2*, and the rod *6*, although this latter movement does not affect the valve with the gear in the position shown. The link also swings on the lower end of the link hanger *11*. The to-and-fro movement of the link transmits the same movement to the link block *7′* and to the lower rocker-arm *9*. The upper rocker-arm *8* accordingly moves back and forth and imparts a similar movement to the valve which admits steam to and exhausts steam from the cylinder in the proper manner to cause the locomotive to run forwards.

22. With the link block in the lower half of the link, as shown in Fig. 2 *(c)*, the back-up eccentric causes the link, the link block, and the rocker-arm to move in the same way as the go-ahead eccentric. The valve is also moved to and fro, and the steam is admitted to and exhausted from the cylinder so as to cause the locomotive to run backwards.

In other words, the back-up eccentric moves the valve gear in the same way when the locomotive is running backwards, as the go-ahead eccentric does when the locomotive is running forwards. With the link blocks in either end of the links, the valves receive their entire movement from the eccentrics which are in control. However, with the link blocks in any other position the valves are also influenced to some extent by the eccentrics which are not in control.

DETAILS OF THE PARTS

23. The Eccentric.—The valve gear is driven from the axle which has a circular movement, and the valve which is operated by the valve gear has a to-and-fro movement. A circular movement at one point can be transformed into a to-and-fro movement at another point only by the use of a crank or some device which gives the same action as a crank. Therefore, in order for the valve gear to give the required movement to the valve, the axle where the eccentric strap is connected must be either shaped like a crank, or some device must be placed on the axle that will produce the same movement as a crank. The device used is called an eccentric because it is impracticable to make the axle of the required shape.

24. One of the eccentrics in Fig. 2 is shown detached from the axle in Fig. 5 *(a)*. A locomotive equipped with the Stephenson valve gear has four eccentrics which are all alike, two being used to impart valve movement when going ahead, and two when backing up. The go-ahead eccentrics can be identified because the go-ahead eccentric rods are always connected to the tops of the links.

The eccentric is a disk which is attached to the axle with a center with the axle. Thus the center *C* of the eccentric in Fig. 5 *(a)* is outside the center *S* of the circular opening provided for the axle. The eccentric

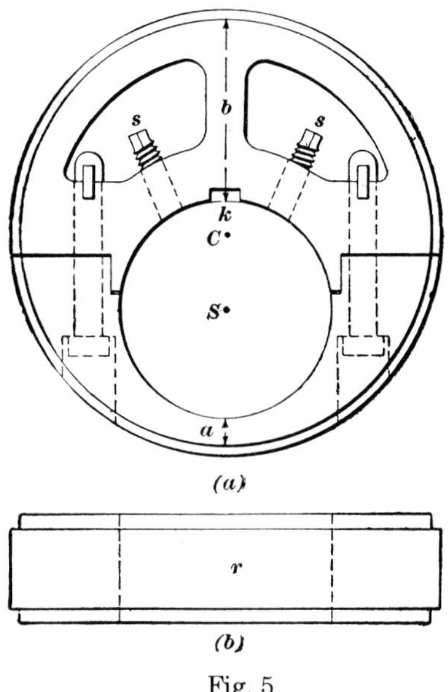

Fig. 5

rotates within the eccentric strap when the locomotive is moving, and at this time it performs the same function as a crank because it transmits a to-and-fro movement to the front end of the eccentric rod.

25. An eccentric is made in two parts so that it can be conveniently applied to the axle, but the manner in which the parts are held together varies. In some cases, studs are screwed into one-half of the eccentric and these project and pass through holes in the other half and the parts are held together by applying nuts to the studs.

In Fig. 5 *(a)* the two halves of the eccentric are held together by the two bolts and keys shown. As shown in the plan view of the eccentric in *(b)*, the outer surface has a raised rib *r*, which serves to hold the eccentric strap in place on the eccentric because the rib sets in a groove in the wearing surface of the eccentric strap. Eccentrics are fastened to the axle in different ways, but usually by a combination of set-screw and keys. In *(a)*, the setscrews are marked *s* and the keyway *k*. A keyway is cut in the axle and, with the eccentric in the proper position, the two keyways come opposite each other. A key is then driven in and prevents the eccentric from turning on the axle. The setscrews *s* prevent movement of the eccentrics lengthwise of the axle.

26. Action of Eccentric.—In Fig. 6 *(a)*, *(b)*, and *(c)* is shown how an eccentric produces the same movement as a crank. View *(a)* shows an axle bent in the shape of a crank with the center marked *A*, and view *(b)* shows the axle straight, with a disk placed on it, the center *a* of which coincides with the center of the imaginary crank on the axle shown by dotted lines. Therefore the disk serves the same purpose as the crank in view *(a)* and the axle can be made straight. The center *a* of the eccentric is exterior to or out of center with the center *A* of the axle, hence the term eccentric as applied to this device. When a rod is connected to the eccentric by a strap, the strap will have a circular movement like the eccentric, but the front end of the rod will have a to-and-fro movement the same as if it were connected to the crank shown in view *(a)*. The center *a* of the eccentric need not be outside of the axle as shown in view *(b)*. The eccentric *b* may be made so that its center *a* comes within the axle as

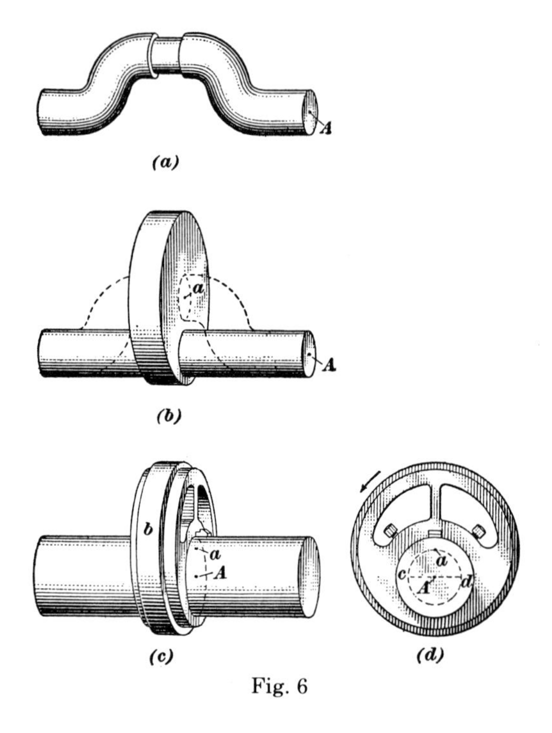

Fig. 6

shown in view *(c)*, and the crank action will still be obtained so long as the center *a* of the eccentric is outside of the center *A* of the axle.

27. The length of the crank-arm formed by the eccentric, Fig. 6 *(d)*, is equal to the distance between the center of the axle *A* and the center of the eccentric *a*. This distance is called the *eccentricity* of the eccentric or the amount it is out of center. The *throw* of the eccentric is the maximum movement it imparts to the front end of the eccentric rod, and is equal to twice the eccentricity *A a*, just as the throw of a crank is twice the length of the crank-arm. Thus, in view *(d)*, the center of the eccentric *a* describes the circle *a c d* when it turns. When the center *a* is at *c*, the end of the eccentric rod is ahead as far as it can be moved. When the center of the eccentric is at *d*, the rod is at the end of its backward movement. The maximum movement of the end of the eccentric rod, when the center of the eccentric turns from *c* to *d*, is equal to the line *c d* or twice the eccentricity *A a*. The distance *A a* is sometimes called the throw of the eccentric and the distance *c d* the full throw. The throw of the eccentric, when on the axle, may be determined by subtracting the least distance *a*, Fig. 5 *(a)*, from the greatest distance *b*.

28. Eccentric Strap.—An eccentric strap *3*, Fig. 2, is a ring of cast iron or steel which fits on the eccentric *1*. Its purpose is to transform the

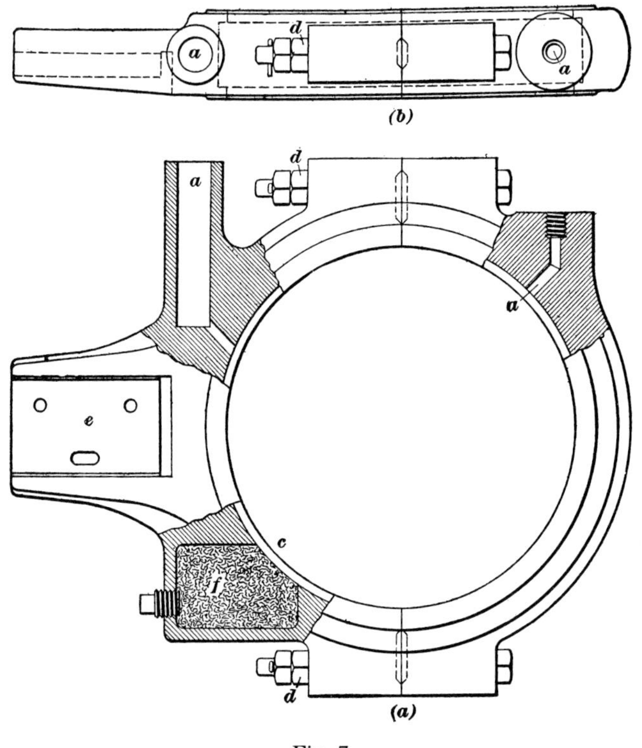

Fig. 7

rotary motion of the eccentric into a reciprocating or a to-and-fro motion at the front end of the eccentric rod. As shown in Fig. 7, the eccentric strap is cast in two parts which are held together on the eccentric by bolts and nuts *d*. The strap is held on the eccentric by making the inner surface *c* grooved or channeled to fit the raised rib *r*, Fig. 5 *(b)*, on the eccentric. Oil is applied to the wearing surface of the strap and the eccentric through the two openings *a*, Fig. 7, the strap being cut away at these points to show the oil passages. An oil cellar *f* shown in section is filled with oily waste by removing the plug shown. Part *e* of the strap is grooved to receive the back end of the eccentric rod.

29. The bolt holes in part *e* of the strap are bored in the same position for straps of engines of the same class. When setting the valves, the length of the eccentric rod is adjusted by the elongated hole. If necessary, the rod is then lengthened or shortened the required amount to make the other two holes in the strap and the rod come in line. If the holes are only slightly out, smaller bolts are used.

The part of the eccentric strap that wears on the eccentric is lined with brass or bronze. Long service can therefore be obtained from the straps as the liners have only to be renewed when worn. Slight wear in the liners is taken up by shims which are placed in the straps at the junction of the two halves.

30. The Link.—The link *7*, Figs. 2 and 3, two views of which are shown in Fig. 8, is a slightly curved slotted piece of steel. The links provide a means of readily reversing the locomotive since by moving the reverse lever the valve can be readily thrown under the control of either the go-ahead or back-up eccentrics. The links also provide a means whereby the admission of steam to the cylinders can be cut off at different points in the stroke of the piston, because the cut-off is shortened by moving the links and bringing the link blocks nearer to the center of the links, and lengthened with the blocks nearer to the ends of the links.

The link shown in Fig. 8 *(a)* is of the solid, or one-piece type, and view *(b)* shows a link of the built-up type, the parts being held together by a bolt and nut at each end. The solid type is in more general use because the parts of the built-up type are liable to becomes loose. The eccentric rods are pinned to the link at the holes *a* and case-hardened bushings are pressed into the link at these points as shown. The parts of the links adjacent to the upper holes are in section so as to show the oil passages through which the eccentric-rod pins are lubricated.

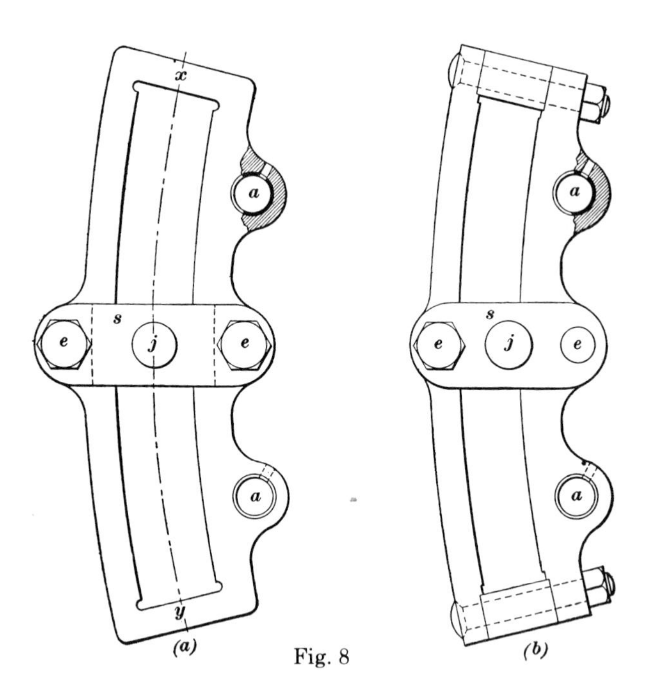

Fig. 8

31. Link Saddle.—The link saddle *s*, Figs. 2 and 3, as shown in Fig. 8, is a plate or holder which is connected to the link midway between the ends by the bolts and nuts *e*. The link-saddle pin *j* is made in one piece with the link saddle. The link is raised and lowered by the link-saddle pin through the medium of the link lifter.

32. Link Block.—The purpose of the link block *7′*, Fig. 3, is to transmit the movement of the link to the lower rocker-arm *9* or to the transmission bar depending on the connection.

A side view of the link block, shown partly in section, is given in Fig. 9 *(a),* and view *(b)* shows the block as seen from the rear. The link block is a steel block fitted on two faces to the curvature of the link, and this permits the link to be moved up and down on the block. The link block shown in *(b)* is made in three pieces, a centerpiece *2* which fits in the slot in the link and two side plates *1* and *3,* which, as they are wider than the link slot, serve to hold the centerpiece in the link.

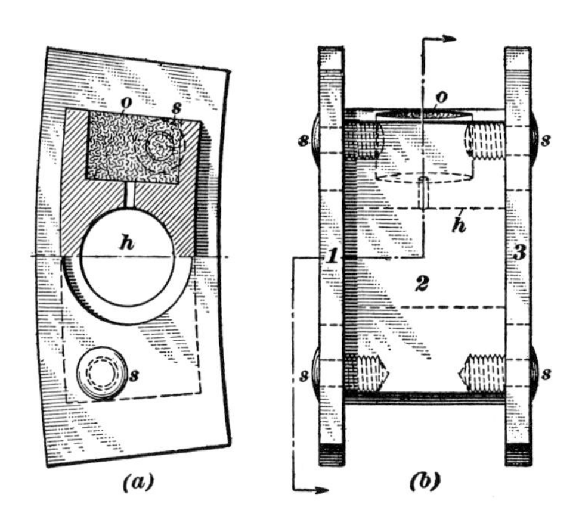

Fig. 9

The side plates *1* and *3*, Fig. 9 *(b),* are connected to the centerpieces *2* by four studs *s*, which are screwed in the centerpiece. As the inner ends of the studs cannot be seen, they are here shown by dotted lines. When assembling the link block in the link, the centerpiece is placed in the link slot, the side plates are placed on each side of the centerpiece and the ends of the studs are riveted over. Usually the side plates are riveted to the centerpiece by four rivets which go all the way through and are riveted over at the ends.

The link-block pin which connects the link block to the lower rocker-arm passes through the large hole *h,* as in *(a).* The opening *o* is provided for holding oily waste. The oil passes through the oil passage in the bottom of the opening *o* to the hole *h* and lubricates the link-block pin.

33. A built-up link block is sometimes made in two parts, in which case one of the side plates only is made separate from the remainder of the block. The block is assembled in the link by riveting this plate to the other part of the block. However, the three-piece construction is preferable because link blocks wear more on the side plates, and they have to be removed oftener than the centerpiece. The side plates are renewed by chipping off the ends of the rivets and driving them out. Built-up link blocks have to be used with one-piece or solid links, as shown in Fig. 8 *(a),* so that they can be assembled in the links. With a built-up link, the block is usually made in one piece and the link is assembled about the block. However, built-up link blocks are sometimes used with built-up links.

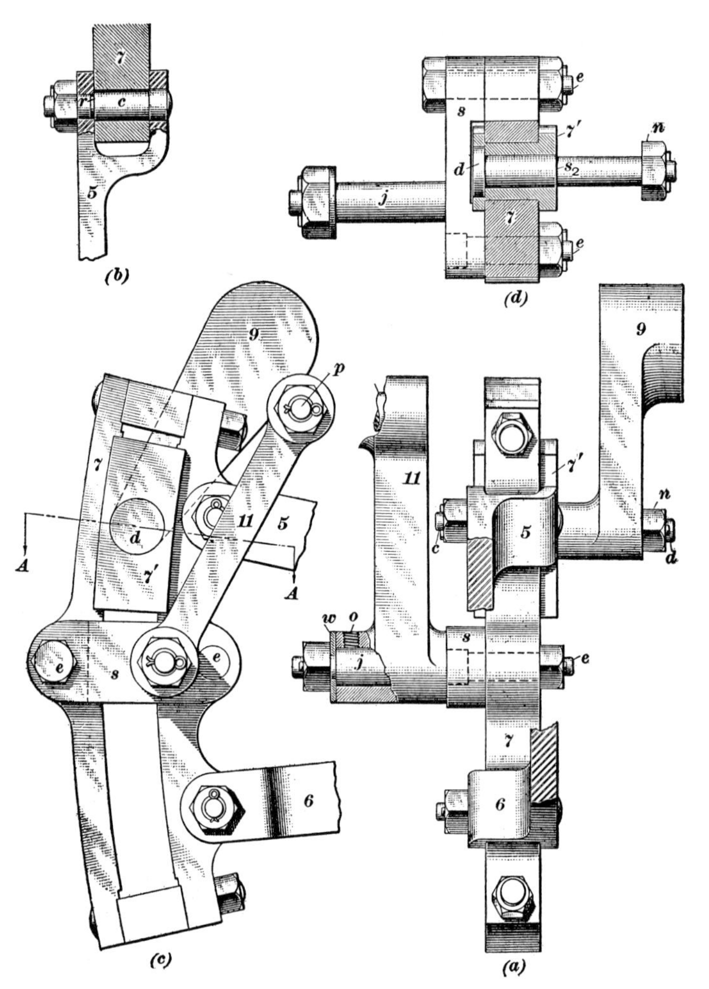

Fig. 10

34. Link and Link Block Assembled.—A rear view of the link *7*, with the link block assembled in it is shown in Fig. 10 *(a)*, a side view of these parts is given in *(c)*, and view *(d)* is a section taken on the line *AA*, view *(c)*. In view *(b)* the forked end of the eccentric rod *5*, and the link *7* is shown in section. As shown in view *(a)*, the link block *7′* is connected to the lower rocker-arm *9* by the link-block pin *d* and the nut *n*. This pin should not be confused with the eccentric-rod pin *c*. An end view of pin *d* is shown in view *(c)*. The link-block pin *d*, view *(d)*, has a shoulder s^2 which bears against the side of the lower rocker-arm in contact with the link block. This construction permits the link block to turn freely on the link-block pin, because the pin cannot draw the block up against the side of the rocker-arm when the nut *n* is tightened up.

The part of the link-block pin between the shoulder s_2 and the nut *n*, Fig. 10 *(d)*, is held in the lower rocker-arm, and this portion of the pin is tapered slightly so that it can be readily driven out. View *(d)* also shows that the side of the link saddle *s* next to the link *7* is cut away so as to permit the link to be moved without the saddle striking the link block. The head of the link-block pin *d* is countersunk in the link block so as not to strike the saddle. View *(c)* shows that the side plate of the link block *7* is wider than the slot in the link. Therefore this plate, in combination with the one on the other side, keeps the link block in position in the link.

35. In Fig. 10 *(c)* is shown that the top of the link block clears the top of the slot in the link when the reverse lever is all the way forwards. The reason is that the link, due to being suspended on a hanger, swings in an arc and therefore has a slight up-and-down movement or slip on the block. With a space above the block, the link will not strike the block when it is moving.

The link hanger *11*, Fig. 10 *(a)*, is held on the link-saddle pin *j* by a washer *w*, a nut, and a cotter key. The washer bears against a shoulder on the pin, hence the link hanger cannot be cramped on the pin when the nut is tightened up. The link-saddle pin is oiled through the oil hole *o* in the link hanger, which is shown broken away at this point. The link saddle is shown at *s*, and *e* is one of the link-saddle bolts. The link-hanger pin *p* is oiled through the oil hole *o*, in the hanger, Fig. 3.

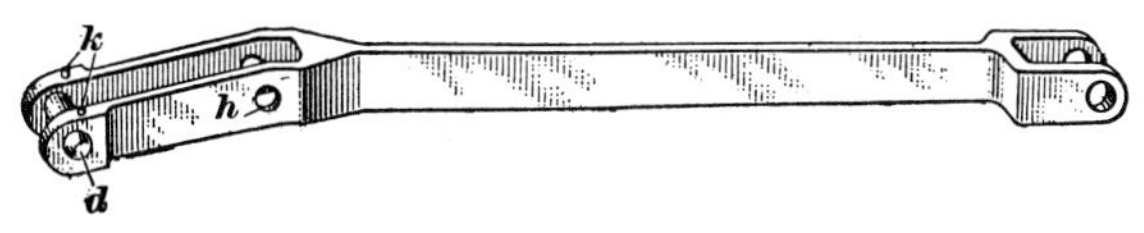

Fig. 11

36. The front ends of the eccentric rods *5* and *6*, Fig. 3, are forked, and Fig. 10 *(b)* shows how the forked end of the eccentric rod *5* is connected to the link *7*. The eccentric-rod pin *c* passes through the forked end and the hole in the link *7*, and is held in place by the nut shown. The ends of the pin *c* fit tightly in the forks of the rod. The reason for this arrangement is to confine the wear on the eccentric-rod pin to the link bushing and the part of the pin that passes through it. The eccentric-rod pin *c* has a shoulder *r* where it comes against one of the forks of the rod. This construction permits the link to work freely on the eccentric-rod pin *c* and prevents the forked end of the rod from being cramped on the link when the nut on the pin is tightened up. Each end of the pin *c* is tapered where it fits in the rod, therefore, when the pin is being driven out, the first movement loosens it. The eccentric-rod pins are oiled through the holes that lead into the pin holes *a* in the link, Fig. 8.

37. Transmission Bar.—A view of the transmission bar *16'*, Fig. 4 *(a)*, is given in Fig. 11. Each end of the bar is forked, the lower rocker-arm is placed in the front fork and the link block and link in the other fork. The link-block pin *a* is held in position by two split keys *k*. The transmission-bar hanger *17'*, Fig. 4 *(a)*, is pinned at *h*, Fig. 11.

38. Reverse Shaft.—The reverse shaft *13*, Fig. 12, is a transverse shaft which rests at each end in bearings or boxes *i* bolted to the frames *f*. The reverse shaft carries the two lifting arms *12*, to which the link hangers are pinned, and the upper reverse-shaft arm *14* to which the

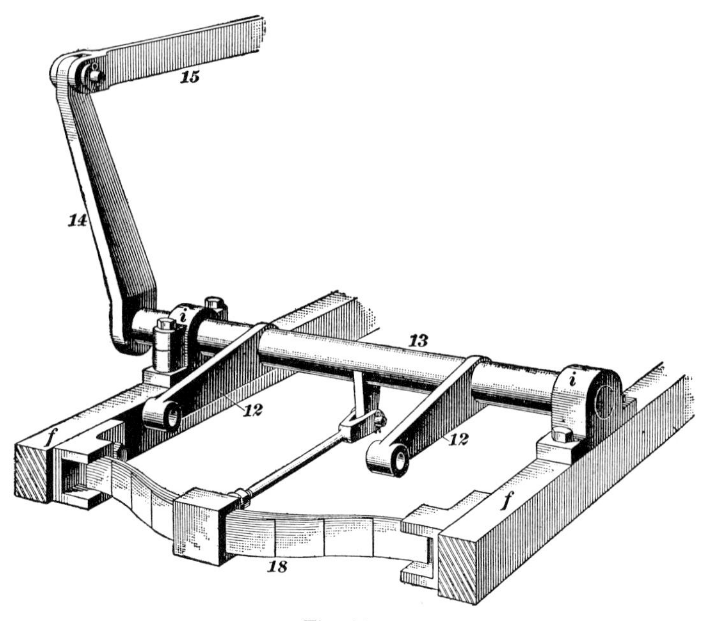

Fig. 12

reach rod *15* is connected. The lifting arms are forged to the reverse shaft. The purpose of the reverse shaft with the arm *14,* and arms *12,* is to transmit the movement of the reverse lever and the reach rod simultaneously to both valve gears, through the medium of the link hangers. The purpose of the counterbalance spring *18,* which is connected to the reverse shaft *13* by the rod shown, is to make it easier to draw back the reverse lever against the weight and friction of the valve gear when the lever is changed. The spring *18* is compressed when the lever is moved forwards, and the tension of the spring therefore assists the movement of the lever when it is drawn back. Different types of counterbalance springs are used but they all serve the same purpose.

39. Link Hanger.—The link hanger *11*, Fig. 3, serves to connect the lifting arms *12* to the link-saddle pins *j*. The link hanger swings freely on the link-hanger pin *p* at the upper end and on the link-saddle pin at the lower end. The pin *p* is a taper fit in the lifting arm and has a shoulder where it comes against the side of the arm. Hence the nut can be tightened up, and the pin prevented from turning without drawing the pin through the arm and clamping the hanger. The link hanger is provided with hardened bushings where it turns on the link-saddle pin and the link-hanger pin.

40. Rocker and Rocker Box.—The rocker *r* and the rocker box r_1, Fig. 3, are shown in Fig. 13 *(a)* and *(b)*; *(a)* is an end view with a part of the box broken away to show the rocker-shaft *r*, and *(b)* is a side view. The rocker consists of a rocker-shaft *r*, with two arms, the upper rocker-arm *8*, and the lower rocker-arm *9*. A rocker is necessary with an inside valve gear in order to make the connection between the link block, which is on the inside of the frame, and the valve rod, which is on the outside. The

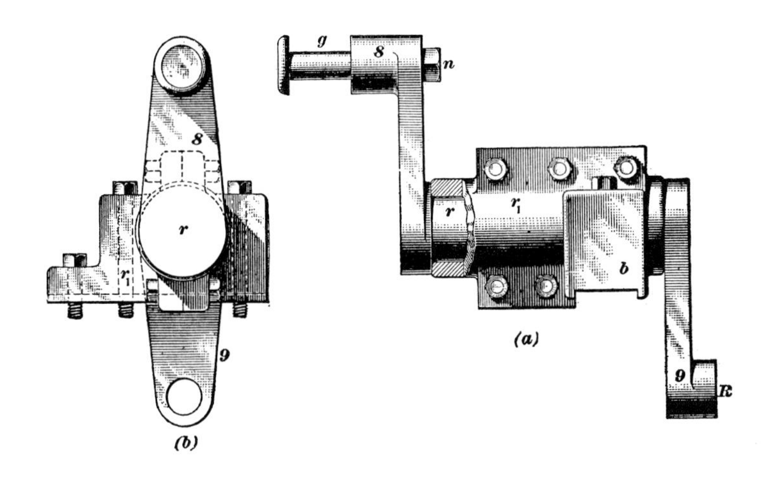

Fig. 13

link block is pivoted to the side *R* of the lower rocker-arm *9*, view *(a)* and the valve rod is connected to the upper rocker-arm by the valve-rod pin *g*. The end of the pin in the rocker-arm is tapered and has a shoulder which prevents the pin from being pulled through the rocker-arm and cramping the valve rod when the nut *n* is drawn up. Various types of rockers are used. The rocker-shaft is frequently made hollow so as to lighten it, and the valve rod may be connected to the upper rocker-arm by a stud, a washer, and a nut. The rocker-shaft *r*, view *(a)*, turns within the rocker box r_1, which is made in two parts, held together by the five bolts and nuts shown. The part *b* of the rocker box, view *(a)*, is connected to the frame by the vertical studs shown in view *(b)*. Usually the rocker box is secured to the frame by bolts which pass all the way through it. The part of the rocker box next to the frame is shaped to fit over the edges of the frame. The rocker *r* is oiled through oil cups, not shown, but which are placed on the rocker box.

41. Oiling Points in Valve Gear.—The points in the valve gear at which oil holes are provided, and the parts which are lubricated through them are as follows: In the eccentric straps, to lubricate the wearing surfaces of the eccentrics and the straps; in the links for the purpose of oiling the eccentric-rod pins; in the link blocks to lubricate the link-block pins; in the ends of the link hangers to oil the link-saddle pins, and the link-hanger pins; in the rocker boxes to oil the wearing surfaces of the rockers and the rocker boxes; in the valve rod to oil the valve-rod pins, and in the reverse-shaft brackets.

42. Indirect and Direct Motion.—The motion of the valve gear is said to be indirect when the eccentric rod and the valve move in opposite directions and the valve gear is direct motion when the eccentric rod and the valve move in the same direction. The type of rocker used determines whether the valve gear is indirect or direct motion. With the rocker shown in Fig. 13 *(a)*, the arms *8* and *9* point in opposite directions and the gear is indirect motion because the upper arm will always move in the opposite direction to the lower arm. The gear is direct motion with the type of rocker shown in Fig. 4 *(a)*, because both arms move in the same direction.

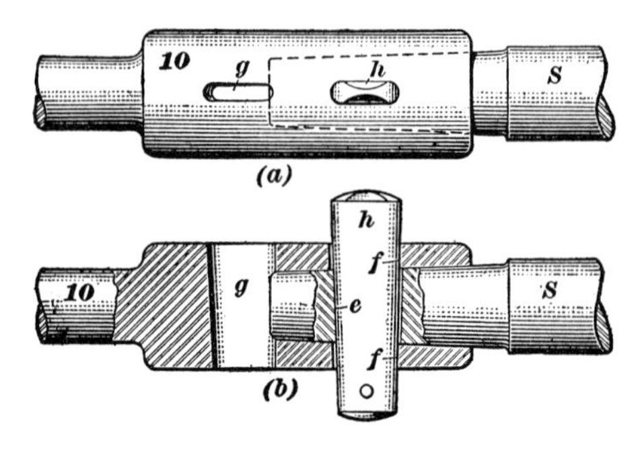

Fig. 14

43. Valve-Rod and Valve-Stem Connection.—The connection between the valve rod *10*, Fig. 2, and the valve stem is shown in Fig. 14 *(a)* and *(b)*, the latter view being a section. The end of the valve stem *S* is tapered and fits into the end of the valve rod *10*. The valve-stem key *h* is driven through the slots in the rod and the stem and holds the parts together. The slots in the rod and stem are longer than the width of the key, and the back edge of the key bears against the end of the slot in the valve stem at *e*, and the front edge against the slot in the valve rod at *f*. Therefore, the valve stem *S* can be drawn back tight in the rod by driving down the key. The slot *g* is provided so that a wedge can be inserted when it is desired to drive the valve stem out of the valve rod. Due to the end of the valve stem being tapered, the first movement loosens the stem when it is being driven out of the rod.

POSITION OF ECCENTRICS

44. The valve gear outside of the eccentrics is merely an arrangement of parts whereby the valves can be thrown under the full or partial throw of either the go-ahead or back-up eccentrics. The fact that the locomotive runs either forwards or backwards depends solely on the position of the eccentrics on the axle. Unless they are correctly placed, the proper movement will not be imparted to the valves through the other parts of the valve gear, and an improper distribution of steam will result. It is therefore essential that the eccentrics should be placed on the axle in the proper positions and the reasons for placing them in such positions should be carefully studied.

45. Valve Movement Required.—In order for the locomotive to run forwards, the go-ahead eccentric must be set on the axle to bring about the following valve movement, it being understood that the correspond- ing eccentric performs a similar function for the other valve: *(a)* The eccentric must move the valve so that the steam port to the cylinder is opened slightly when the piston is at the beginning of either stroke, that is, the valve must have the steam port open the lead; *(b)* when the piston starts, the eccentric must move the valve to open the steam port wider. The two eccentrics that impart the proper movement to the valves to make the locomotive run forwards are called the forward-motion, or go-ahead, eccentrics.

To make the locomotive run backwards, the two other eccentrics must be set on the axle so as to bring about the same valve movement when the locomotive is running backwards as is produced by the go-ahead eccentrics when the locomotive is running forwards. The two eccentrics that give the proper valve movement when the locomotive is running backwards are called the backward-motion, or back-up, eccentrics.

The back-up eccentric is set the same distance from the main pin on one side of the axle as the go-ahead eccentric on the other side, as otherwise the locomotive could not be reversed.

46. Determining Position of Eccentrics.—The fact that the valve must be in a certain position when the piston is at the beginning of either stroke implies that the position of the eccentric on the axle must bear a certain definite relation to the position of the main crankpin, because the position of the main pin determines the position of the piston.

The position of the eccentrics on the axle to cover all conditions is shown in Figs. 15, 16, 17, and 18, in which the principal parts of the valve gear are shown by lines. The axle is marked *A*, the main pin *P*, the go-ahead eccentric *1*, with the center marked *a*, and the back-up eccentric *2*, with the center marked *b*. The line *c d* is an imaginary vertical line drawn through the center of the axle and is used to identify the positions of the eccentrics.

47. There are four conditions to be taken into account when the position of the eccentrics on the axle is considered, namely an indirect-motion valve gear and an outside-admission valve, Fig. 15, a direct-motion valve gear and an outside-admission valve, Fig. 16, an indirect-motion valve gear and an inside-admission valve, Fig. 17, and a direct-motion valve gear and an inside-admission valve, Fig. 18.

However, the four mentioned conditions do not require four different positions of the eccentrics as two different positions cover all four of the conditions, that is, the eccentrics are set alike in Fig. 15 and 18 and in Figs. 16 and 17. The usual condition met with is an indirect-motion valve gear and outside-admission valves, and a direct-motion valve gear and inside-admission valves.

48. Indirect-Motion and Outside-Admission.—Suppose with an indirect-motion valve gear and an outside-admission valve, Fig. 15, that it is required to place an eccentric on the axle in the proper position to cause the locomotive to move forwards, or in other words, to locate the position of the go-ahead eccentric *1*.

The general position of this eccentric can be more readily understood by considering the movement which must be imparted to the valve in order to cause the locomotive to move forwards. With the main crankpin *P*, Fig. 15 *(a)*, turning in the direction of the arrow, and with the piston *e* moving back, the valve to admit steam to the cylinder must move back or to the right. Therefore, the go-ahead eccentric *1* must be somewhere behind the main pin *P* so that when the pin turns forwards, the eccentric will move toward the valve, push the lower rocker-arm *9* ahead, and the upper arm *8* and the valve back.

49. The position of the go-ahead eccentric is found thus: In Fig. 15 *(a)*, with the center of the eccentric at a_1, one-quarter of a turn behind the main pin, the valve would be in mid-position and the steam lap of the valve would cover both steam ports. However, it is necessary to have the front steam port open the lead, and this requires that the eccentric be turned in the proper direction until the valve is moved the lap and lead from mid-position. The center of the eccentric is then turned from a_1, toward the main pin and this pushes the eccentric rod *5* and the lower

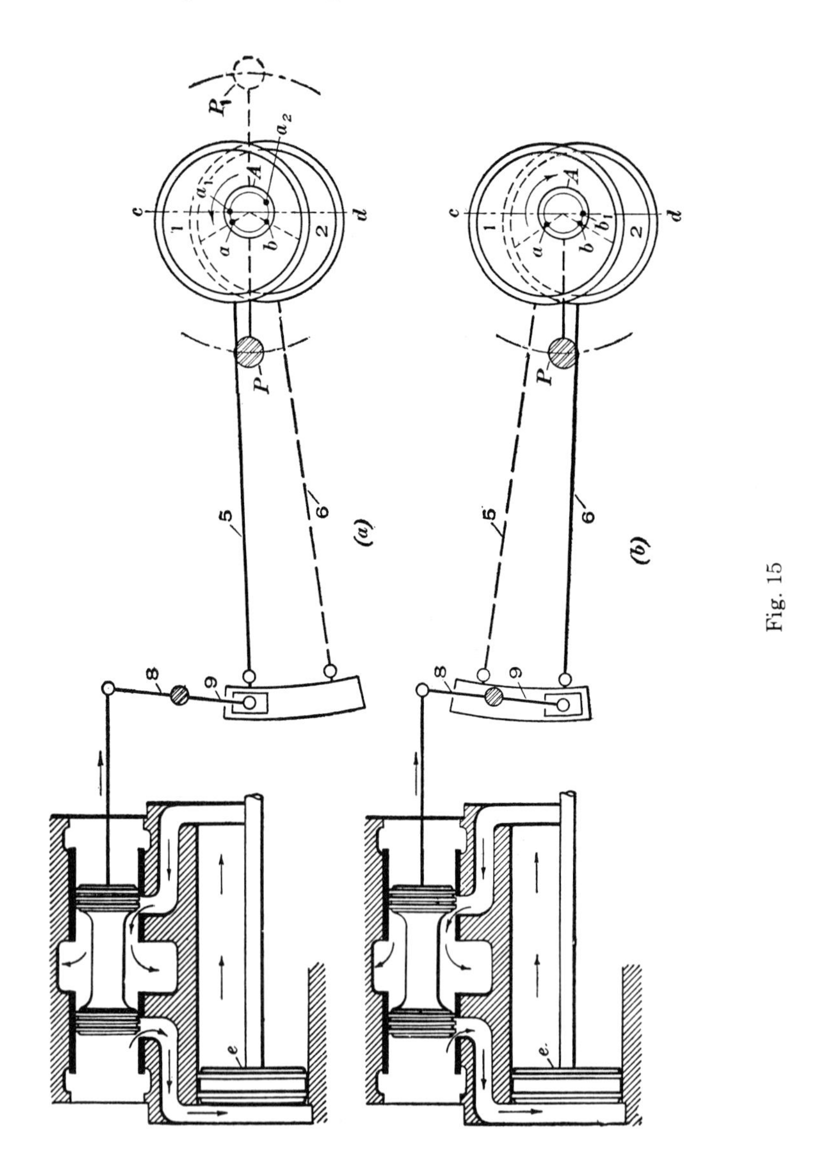

Fig. 15

rocker-arm *9* ahead, and the upper rocker-arm *8* and the valve back until, with the center at a, the valve has been moved to the rear far enough to open the front steam port the lead.

50. The position of the go-ahead eccentric is identified by saying that its center is behind the main pin one-quarter of a turn, less the amount the eccentric has been turned to move the valve the lap and the lead, or from a_1 to a, Fig. 15. Ordinarily the eccentric is located by stating that it is one-quarter of a turn less the lap and the lead behind the main pin. The position of the eccentric just noted causes the locomotive to move forwards because the valve has opened the steam port the lead, and the valve moves back thereby opening the steam port wider when the main pin turns in the direction of the arrow.

When the main pin is on the other center, as P_1, Fig. 15, the center of the go-ahead eccentric *1* will be at a_2, a half turn from its other position. In this position the back steam port is open the lead, and the eccentric will cause the valve to open this port wider as the pin turns. Therefore, the go-ahead eccentric *1* is properly set to cause the locomotive to move forwards as long as steam is supplied.

51. The following explains the setting of the eccentric to make the locomotive run backwards. In Fig. 15 *(b)*, the valve gear is in backward gear. With the wheel turning in the direction of the arrow, the valve must be moved in the same direction as before or backwards. Such a movement requires that the back-up eccentric *2* be placed somewhere behind the main pin so that the valve will be moved back with the locomotive running backwards. The exact position of the eccentric is found by turning its center from the quarter position at b_1, where the valve is in mid-position, in a direction that causes the valve to open the front steam port the lead. The eccentric must then be turned toward the main pin, thereby moving the eccentric rod *6* and the lower rocker-arm *9* ahead and the upper rocker-arm *8* and the valve back until it is displaced from mid-position the lap and the lead. This brings the center b_1 of the back-up eccentric at b or one-quarter of a turn behind the main pin with the locomotive running backwards, less the amount b_1 b that the eccentric has been turned to move the valve the lap and the lead. Both eccentrics are now set so as to cause the engine to move in either direction desired. The eccentrics are always set with the main pin on the exact dead center because the eccentrics can then be set to give the required lead that is always known.

52. Direct-Motion and Outside-Admission.—With a direct-motion valve gear and an outside-admission valve, Fig. 16 *(a)*, the position of the eccentrics can be determined by following the reasoning already given. With the main pin P on the forward dead center and turning in the direction of the arrow, the valve must be moved to open the front steam port, and the go-ahead eccentric *1* must be in a position to pull the valve backwards or its general position must be ahead of the main pin. With its center at a^1 ahead of the main pin, the valve would be in mid-position. Therefore, the eccentric must be turned in the proper direction, or away from the main pin more than one-quarter of a turn, so as to move the valve from mid-position the lap and the lead. This brings the center of the eccentric at a or one-quarter of a turn plus the movement $a_1 a$ required for the lap and the lead. In this position the go-ahead eccentric *1* has the valve open the lead and, if the locomotive is started forwards, the eccentric *1* will move the valve to open the port.

The gear is shown in backward motion in Fig. 16 *(b)*. In order for the main pin *P* to move in the direction of the arrow or backwards, the back-up eccentric *2* must be set on the axle one-quarter of a turn plus the lap and lead behind the main pin with the locomotive running forwards, or ahead of the main pin if the locomotive is running backwards. This brings the center of the eccentric at *b*.

53. Indirect-Motion and Inside-Admission.—With an indirect-motion valve gear and an inside-admission valve, Fig. 17, the eccentrics

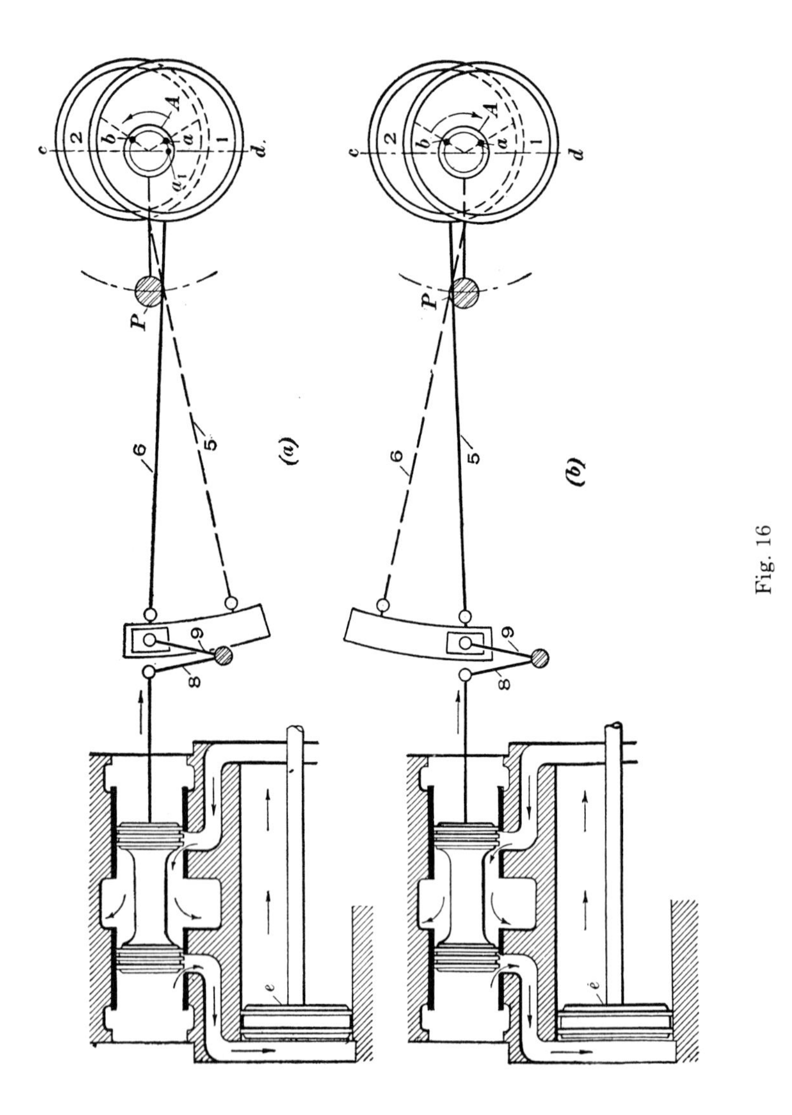

Fig. 16

are set in the same position as in Fig. 16. That is, the center *a* of the go-ahead eccentric *1* is one-quarter of a turn plus the lap and lead ahead of the main pin, and the center *b* of the back-up eccentric *2* is one-quarter of a turn behind the main pin plus the lap and lead with the locomotive running forwards, or ahead of the main pin with the locomotive running backwards. In Fig. 17 the valve gear is shown in forward gear only because it is unnecessary to explain the reason for the position of the back-up eccentric in detail as the explanation regarding Fig. 16 *(b)* also applies to Fig. 17.

54. Direct-Motion and Inside-Admission.—In Fig. 18 is shown a

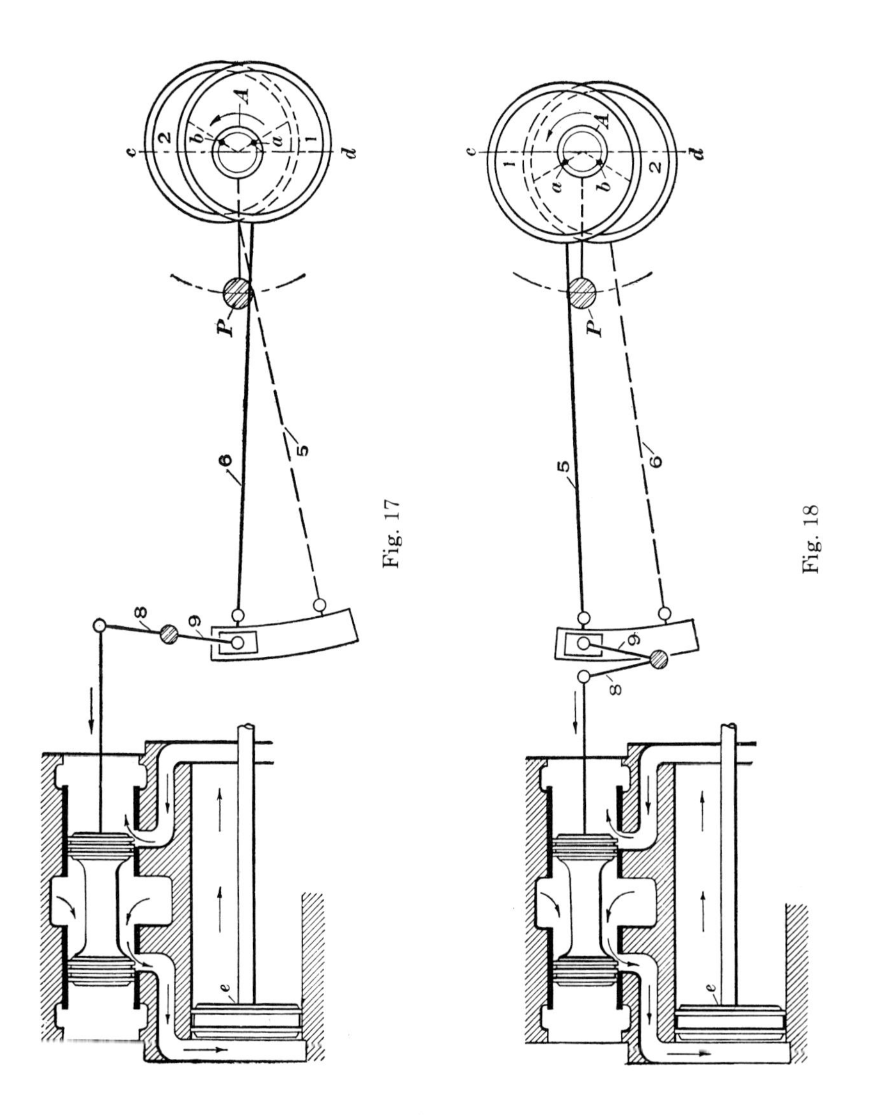

Fig. 17

Fig. 18

direct-motion valve gear and an inside-admission valve. The gear is shown in forward gear only because the reason for the position of the eccentrics is the same as has already been given. The center *a* of the go-ahead eccentric *1* is one-quarter of a turn less the lap and lead behind the main pin when the locomotive is running forwards, and the center *b* of the back-up eccentric *2* is the same distance behind the main pin with the locomotive running backwards. The position of the eccentrics is then the same as with indirect-motion and outside-admission valves.

It will be noted in all cases that the back-up eccentric is set the same distance from the main pin on one side of the axle as the go-ahead eccentric is on the other side. This is necessary because, as explained in Art. **8**, reversal of motion requires an eccentric the same distance from the main pin when running backwards as when running forwards.

55. Sometimes the position of the eccentrics is stated in angles instead of considering the centers of the eccentrics as in Figs. 15, 16, 17, and 18. When the measurement is made in angles, the eccentric cranks and the main crank must be considered as in Fig. 19. Thus the main crank is represented by the line *oz* drawn from the center of the axle to the center of the main crankpin and the eccentric cranks are represented by the lines *oax* and *oby* drawn through the center of the axle and the center of the eccentrics. The angles *Aox* and *Boy* through which the eccentric cranks have been turned from their right-angle positions to their true positions with respect to the main crank are known as the angles of advance. Therefore the go-ahead eccentric *c* is at the right angle *Aoz* less the angle of advance *Aox* behind the main crank, and the back-up eccentric *d* is at the right angle *Boz* less the angle of advance *Boy* ahead of the main crank. The correct position of the eccentrics for indirect-motion and outside-admission valves and for direct-motion and inside-admission valves are illustrated in Fig. 19.

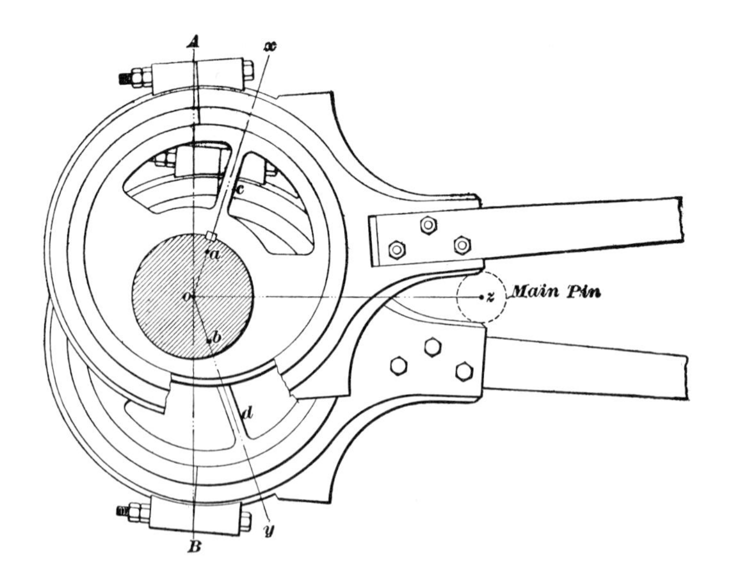

Fig. 19

56. It will be generally found more convenient to identify the position of the eccentrics by their centers rather than by their crank-arms. However, when studying the action of valve gears, the position of the eccentrics is always considered in angles. Although the eccentrics are usually said to be set in relation to the main pin, they are actually set in relation to the center line of motion. This is a line drawn through the center of the main axle and the center of the link-block pin when the valve is in mid-position. The main crankpin, when on the dead center, falls on or comes near to this line; hence, the eccentrics, when set in relation to the main pin, are in practically their correct positions.

57. Relative Positions of All Eccentrics.—The main crankpins are one-quarter of a turn apart, the right pin leading when the locomotive is running forwards. Therefore, the eccentrics on the right side are one-quarter of a turn in advance of the eccentrics on the left side. This brings the go-ahead eccentric on the right side one-quarter of a turn ahead of the go-ahead eccentric on the left side, and the back-up eccentric on the right side the same distance ahead of the back-up eccentric on the left side, when the locomotive is moving ahead.

INCREASE IN LEAD

58. Reason for Increase.—With the Stephenson valve gear the lead increases when the reverse lever is moved from the forward corner of the quadrant to the center. The lead then decreases as the lever is moved to the back corner of the quadrant until it is the same as existed with the lever in the forward corner. A similar action occurs when the reverse lever is moved from the back corner of the quadrant to the forward corner. The lead which is given the valve when the eccentric is set on the axle is called the *fixed lead,* and the gain in lead when the reverse lever is moved is called the *increased lead.*

59. The increase in the lead with the Stephenson valve gear will be explained by referring to Fig. 20. In this illustration the gear is shown by lines. The center of the axle is marked A, P is the main pin, a is the go-ahead eccentric, b the back-up eccentric, and $a\ o$ and $b\ d$ the go-ahead and back-up eccentric rods. The link block is marked o. The line $o\ d$ shows the link in full forward gear, $c_1 d_1$ shows the link in mid-gear, and its position in backward gear is shown by $c\ o$.

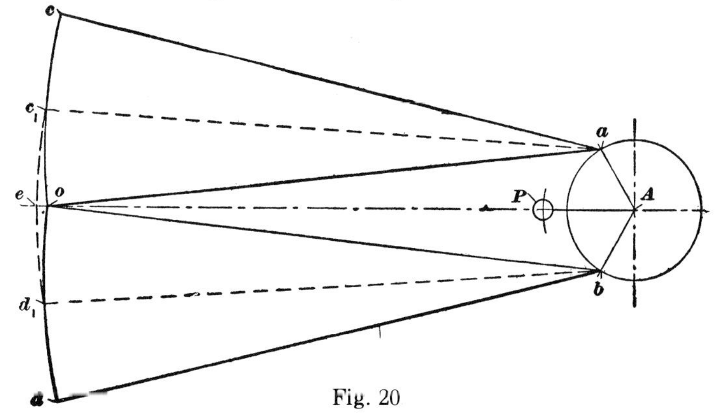

Fig. 20

60. When the link is being raised from full-gear position *o d*, Fig. 20, the forward end of the go-ahead eccentric rod and the top of the link moves in the arc *o c*, and the end *d* of the back-up eccentric rod and the bottom of the link moves in the arc *o d* because the rods are swinging from different centers *a* and *b*. The effect on the link with the eccentric rods describing separate arcs is to move it forwards, and when the link is in mid-gear, as shown by the line c^1d^1, the ends c^1 and d^1 and the center of the link are considerably in advance of the corresponding parts when in full-gear position. The outward movement of the link carries the link block from *o* to *e* and the valve (not shown) is drawn back an equal amount. The fixed lead which existed with the link in full gear will be increased by the amount of the increased lead *o e*. As the link is moved from mid-gear position c^1d^1 to the full back-gear position *c o*, the lower end d^1 of the link follows the arc d^1o and the upper end the arc c^1c. As a result the lower end d^1 of the link is moved back to the point *o*, the position from which the top of the link started, and the upper end of the link moves back to the point *c*. The link block, while the link is being moved from c^1 d^1 to *c o*, moves from *e* to *o*. This moves the valve ahead, until the steam port is again opened the fixed lead, the same as in full forward gear. The lead in this case then increases from full gear to mid-gear when it is at the maximum, and then decreases an equal amount as the link is moved to full backward gear.

61. The reason that the lead increases and decreases is, briefly, as follows: Due to the fact that the rear ends of the eccentric rods are connected at different points *a* and *b*, Fig. 20, the front ends *o* and *d* of the rods move in different arcs *d o* and *o c*. One end of the link follows an arc described by the front end of the rod and the other end of the link an arc described by the front end of the other rod. Therefore, the increase in the lead is due to the eccentrics not having a common or the same center, which causes the ends of the eccentric rods to follow separate arcs when the link is raised or lowered. If the points *a* and *b* were together, the ends of both rods would describe arcs which would coincide and there would be no change in the lead. The increase in the lead is greater with short rods than with long rods because the arcs described by short rods have a greater curvature than when the rods are longer.

62. Crossed and Open Rods.—If the go-ahead eccentric rod were connected to the bottom of the link and the back-up rod connected to the top of the link, the lead would decrease when the reverse lever is drawn up. The rods are said to be *crossed* when arranged in this manner, and with the rods as ordinarily connected they are said to be *open*. The foregoing explains why the go-ahead eccentric rod is always connected to the top of the link and the back-up eccentric rod to the bottom of the link. With the rods connected up in this manner, the lead increases when the lever is hooked up, and this is less objectionable than to have the lead decrease which would occur if the position of the rods at the link were reversed.

VALVE MOVEMENT IN MID-GEAR

63. The valve movement in mid-gear for one-half a turn of the drivers is shown by the outline of the valve gear given in Fig. 21. The position of the link in mid-gear with the main pin at *P* on the forward dead center is shown by $c\ d$, and $c_1\ d_1$ shows the position of the link with the main pin on the back dead center at P_1. The lines $a\ f$ and $b\ g$ are equal to the lap and the lead, because they indicate the amount measured in a straight line that the eccentrics have been turned toward the main pin to displace the valve, the lap, and the lead. when the main pin moves from P to P_1, or makes one-half a turn, the go-ahead eccentric moves the same amount or it moves from a to a_1. This movement of the eccentric moves the top c of the link backwards an amount equal to $a\ f$ plus $g\ a_1$, or twice the lap and the fixed lead. Similarly, a half-turn of the back-up eccentric from b to b_1 moves the bottom of the link backwards an amount equal to $b\ g$ plus $f\ b_1$, or twice the lap and the fixed lead. The position of the link with the main pin on the back dead center is shown by $c_1\ d_1$.

Therefore, for each half-turn of the drivers, with the valve gear in mid-gear, the valve moves twice the steam lap and twice the fixed lead. To this must be added twice the increased lead, due to moving the reverse lever to the center of the quadrant. With valve gears having a constant lead, by which is meant that the lead is not affected by the movement of the reverse lever, the valve moves twice the lap and lead for each half-turn of the wheels with the reverse lever in mid-gear.

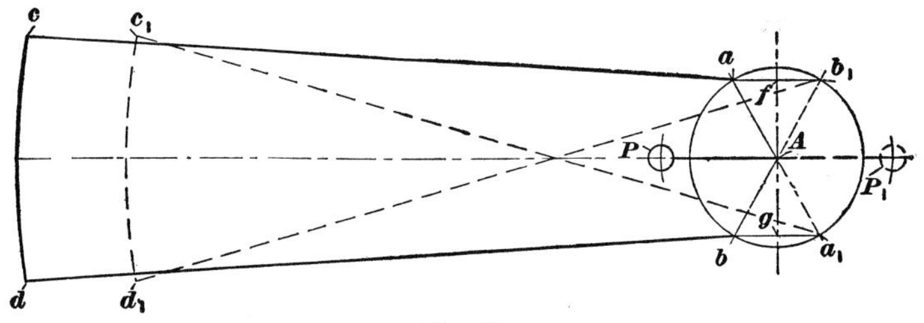

Fig. 21

64. A quarter turn of the drivers from the dead center, with the reverse lever in mid-gear, moves the valve the lap and the lead and brings it to mid-position. The maximum port opening for the admission of steam, obtainable when a locomotive is drifting with the lever in the center of the quadrant, is equal to the fixed lead and the increased lead. As soon as the main pin moves from the dead centers, the valve begins to close the steam port.

REVERSING

65. General Explanation.—A locomotive valve gear will not positively reverse one of the engines of a locomotive in all positions. For example, it is well known that a locomotive operating on one side cannot be reversed if near, or on, the dead centers although it can in all other positions. In order for a locomotive to be reversed in any position the main pins must be set one-quarter turn, or 90° apart, because with the pins more or less than one-quarter of a turn apart there would be positions in which reversal would not take place. The reverse lever must be moved all the way in the quadrant in order to insure reversal, and even then the starting impulse is frequently imparted to one side of the locomotive. The starting impulse varies with the position of the main pins. In certain positions, the slack has to be taken so as to obtain a more advantageous position from which to start the train.

66. Principle Involved in Reversing.—The principle involved in the reversing of a locomotive will be explained by referring to Fig. 1 *(a)*. It was explained in Art. **8** that the engine shown in Fig. 1 *(a)* will run forwards with the eccentric *a e* one-quarter turn in advance of the main pin *c*, and will run backwards with the eccentric at *a g*, one-quarter turn in advance of the main pin for this direction of rotation.

Hence the engine is reversed by placing the eccentric the same amount in advance of the main pin when running backwards as when running forwards.

67. The following fundamental principle can be derived from the foregoing: Reversal of motion requires eccentrics set at the same distance from the main pins when the locomotive is moving backwards as when moving forwards.

With the Stephenson valve gear, the back-up eccentrics meet the requirement of eccentrics the same distance from the main crankpins when running backwards as when running forwards and to reverse merely requires a movement of the links sufficient to throw the valves under the control of these eccentrics.

68. Effect on Valves.—The effect on the valves, when reversal occurs, of eccentrics set at the same distance from the main pins when moving back as when moving ahead, is as follows: The valves are so moved that steam is admitted to the opposite end of one or both cylinders depending on the position of the main pins, and the movement of the valves is such when the engine starts as to keep it moving in the reverse direction.

On account of practically the infinite number of positions in which the main pins can be placed, it is impossible to study the valve movement that occurs when the locomotive is reversed in all positions. However, in Figs. 22, 23, 24, and 25 is shown the change in the position of the valves when the locomotive is reversed with the main pins in certain representative positions.

The movement of the valves when reversing is due to the movement that is imparted to the link blocks by the links when they are raised or lowered. The extent of the movement is dependent on the inclination of the links, and this in turn depends on the position of the eccentrics.

With the gear in the position shown in Fig. 2 *(a)* the top of the link is farther ahead than the bottom, hence when the link is raised to the position shown in Fig. 2 *(c)* the link block and the lower rocker arm will be drawn back and the valve will open the back steam port.

69. Movement of Valves When Reversing.—The movement of inside-admission valves when a locomotive is reversed, with the main crankpins in different positions, is shown by Figs. 22, 23, 24, and 25. The positions of the main crankpins are shown in view *(a)*, the position of the valves before reversal in views *(b)* and *(c)*, and their position after reversal in views *(d)* and *(e)*. The crankpins and the cylinders on the left side are marked *l*, and on the right side they are marked *r*. It is assumed that the reverse lever is in full forward gear before reversal, and that it is in full backward gear after reversal. The lever is usually moved all the way back when reversing so that the maximum valve movement and the greatest port opening possible will be obtained.

70. With the main crankpins in the positions shown in Fig. 22 *(a)*, the

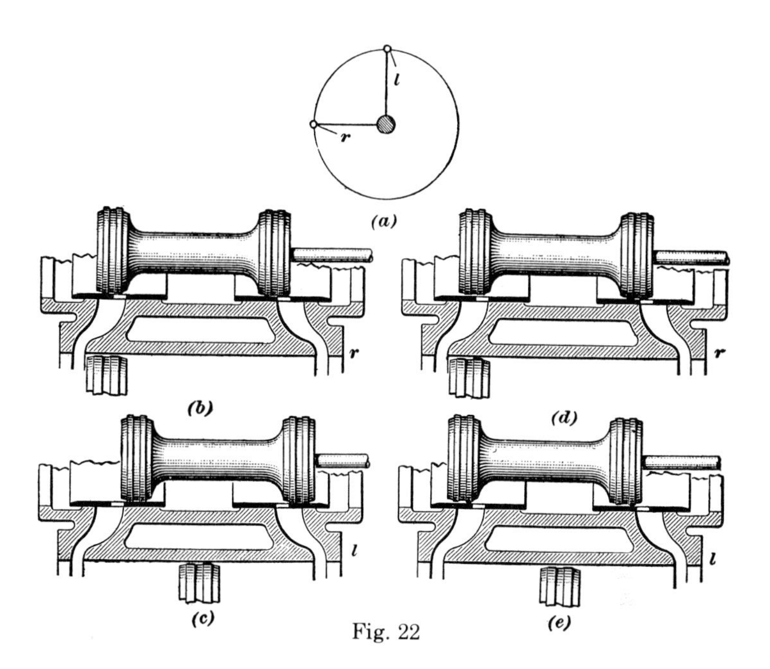

Fig. 22

valve on the right side, view *(b)*, has the front steam port open the lead, and the valve on the left side, view *(c)*, has the back steam port about wide open.

When the locomotive is reversed, the valves move to the same positions as if the locomotive had been running backwards and had

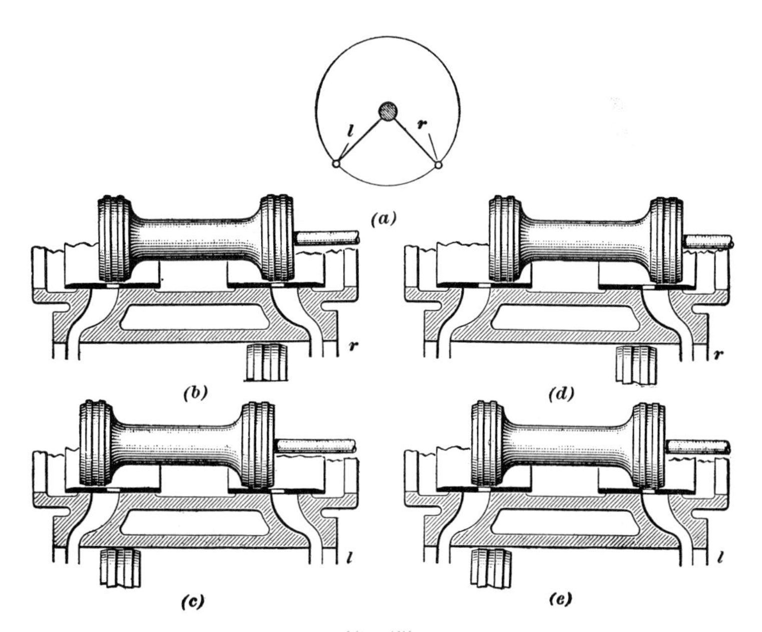

Fig. 23

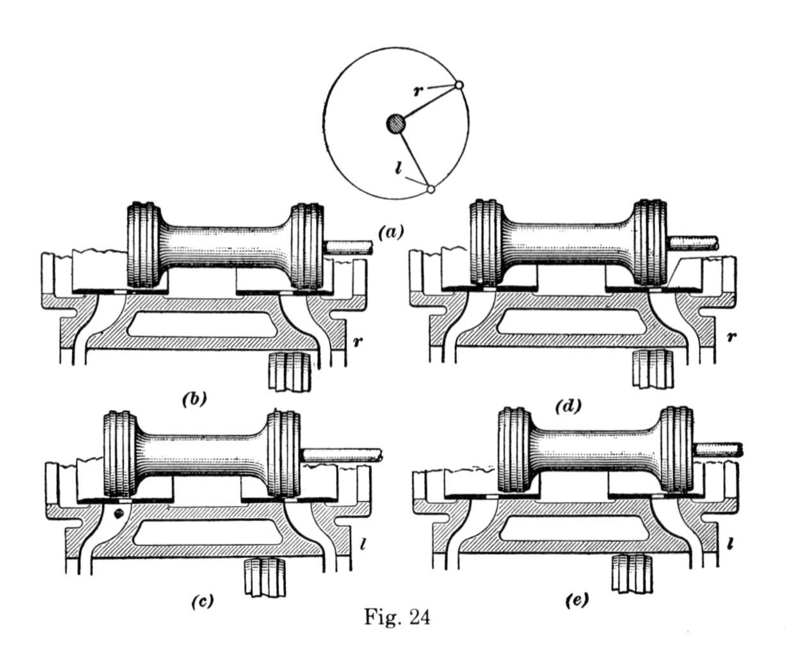

Fig. 24

stopped in the position shown in view *(a),* Fig. 22. The valve on the right side, view *(d)*, has not changed its position; this valve merely moves twice the increased lead when the reverse lever is shifted. The valve on the left side, view *(e)*, opens the front steam port the same amount as the back port before reversal, view *(c)*. In this case, the locomotive for the instant starts backwards from the left side.

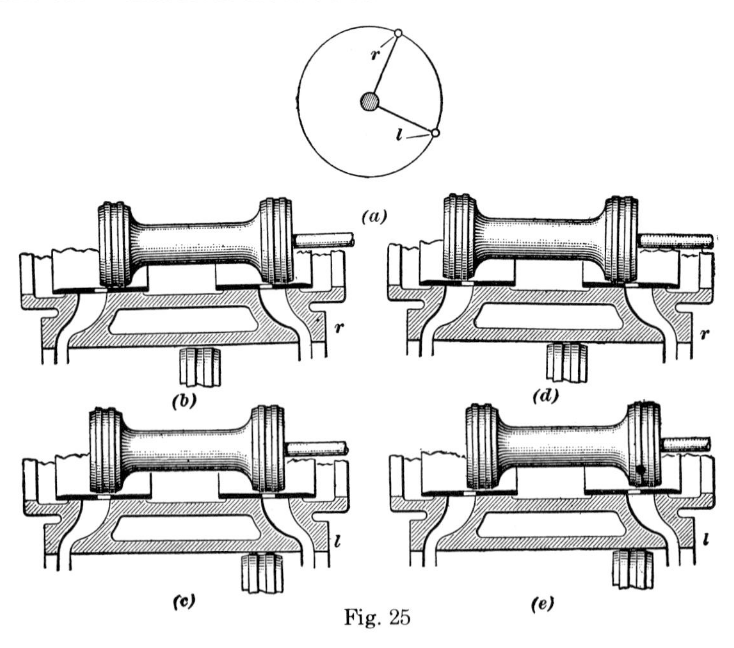

Fig. 25

71. In Fig. 23 is shown the valve movement during reversal with the main pins in the positions shown in view *(a)*. The valve on the right side, view *(b)*, has moved a little beyond cut-off at the front steam port, and the front steam port on the left side, view *(c)*, is fully open. The positions assumed by the valves after the locomotive is reversed will be the same as if the locomotive had been running backwards, and had stopped in the position shown in *(a)*. The valve on the right side, view *(d)*, has the back steam port fully opened and the valve on the left side, view *(e)*, has moved slightly past cut-off at the back steam port. In this instance, the locomotive starts backwards from the right side.

72. With the crankpins in the position shown in Fig. 24 *(a)*, and the locomotive moving forwards, the valve on the right side, view *(b)*, has the back steam port wide open, and the valve on the left side, view *(c)*, is at cut-off at the front steam port. When the locomotive is reversed, the valve on the right side, view *(d)*, moves to cut-off at the front steam port, and the valve on the left side, view *(e)*, fully opens the back steam port. Therefore, the locomotive starts backwards from the left side.

With the crankpins in the positions shown in Fig. 25 *(a)*, the valve on the right side, view *(b)*, has the back steam port almost wide open, and the valve on the left side is about in mid-position and both steam ports are closed. When the locomotive is reversed, the valve on the right side, view *(d)*, moves so as to open partly the front steam port, and the valve on the left side, view *(e)*, also partly opens the back steam port. Therefore, the locomotive starts backwards from both sides.

73. The foregoing shows that the locomotive, when reversed, may receive a starting impulse backwards from one or both sides, depending on the positions of the main crankpins. Generally a locomotive starts backwards from both sides when the crankpins are above and below the dead center, with one of them near the dead center, as in Fig. 25. With both the pins above or below the dead-center positions, Fig. 23 *(a)*, the starting movement in the reverse direction is generally received on one side.

LINK ARC AND LINK RADIUS

74. The link arc is a curved line $x\ y$, Fig. 8 *(a)*, drawn through the center of the link. The link arc is a part of a circle and is struck with a radius equal to the distance between the center of the eccentric and the center of the link-block pin. The link arc or the link curvature, which results when the radius of the link is taken between the points named, gives equal mid-gear leads at each end of the cylinder.

If the same length of eccentric rods were used and the link were struck with a shorter radius, the curvature would be too great, and the mid-gear lead at the front steam port, depending on the type of valve and motion, would be more or less than the lead at the back port. The link, if struck with a longer radius, would be too straight and the lead at the front port would be less or more, depending on conditions, than the lead at the back port.

REASON FOR VARIATION IN CUT-OFF

75. Certain actions of a valve gear are difficult to understand unless a comparison is made between it and some device that operates the valve in the same way but with a simpler arrangement of parts. For this reason the study of the following explanation of the variation in cut-off is optional and reference need not be made to it unless it is desired.

76. Definition of Cut-Off.—Cut-off is said to occur when the steam edge of the valve comes in line with the steam edge of the port, and thereby cuts off the admission of steam to the cylinder. A characteristic of all locomotive valve gears is that their action permits of the cut-off being varied or changed, that is, the valves can be made to cut off the steam to the cylinders at different points in the stroke of the piston. The shortest or earliest working cut-off obtainable is secured with any valve gear with the reverse lever in the notch next to the center notch of the quadrant, and with the longest or latest cut-off with the reverse lever in either corner of the quadrant. A variable cut-off permits the power of the locomotive to be changed in accordance with the work that is being performed.

77. Single Shifting Eccentric.—A single shifting eccentric, Fig. 26, may be considered as a simplified form of valve gear and reference is commonly made to it when the action of a gear is investigated. An analysis of a valve gear from the gear itself is very difficult on account of the movement of so many parts having to be taken into account. The operation of a shifting eccentric is easily understood because its movement is not complicated by the addition of other parts. Moreover its action is practically identical with that of a valve gear. The reason for the term shifting eccentric is that its center can be moved across the axle in a

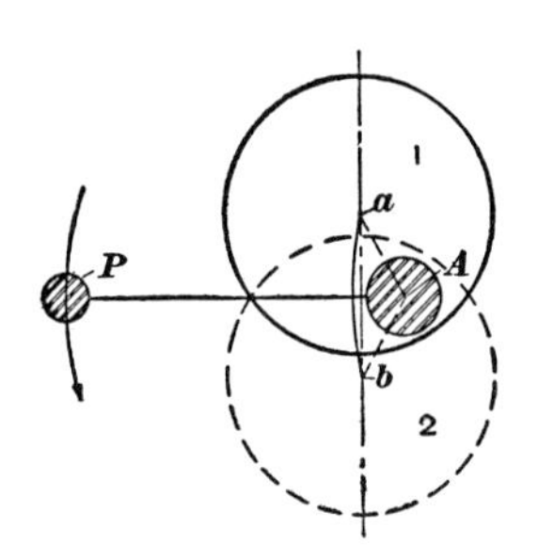

Fig. 26

slightly curved or a straight line. Thus, in this figure, the center *a* of the shifting eccentric *1* follows the slightly curved line *a b* when moved across the axle by an arrangement of parts not shown. If the movement imparted by the eccentric at *1* is made indirect by a rocker and an outside-admission valve is used, the engine will run forwards or with the main pin *P* turning in the direction of the arrow. The engine will run backwards by moving the eccentric from position *1* to position *2*, the center then moving from *a* to *b*. A shifting eccentric is merely a crank for driving a valve and is so constructed that it can be shifted from one side of the axle to the other.

78. Shifting Eccentric and Valve Gear Compared.—The go-ahead eccentric *1* of the Stephenson valve gear, Fig. 2 *(a)*, can be compared to the eccentric in position *1*, Fig. 26, and the back-up eccentric to the one shown at *2*. Raising the link its full travel is equivalent to moving the eccentric from *1* to *2*, because the locomotive is reversed in both cases. When the link is lowered far enough, or when the eccentric is moved from *2* to *1*, the locomotive runs forwards.

It is interesting to note the evolution of the valve gear from the shifting eccentric. It would be impracticable to use one eccentric and move it across the axle when reversing the locomotive. However, by securing two eccentrics to the axle, one in the proper position to run forwards, and the other to run backwards, and using a link to throw the valve under the control of either one, the locomotive can be quickly reversed.

79. Change in Cut-Off.—The reason for the change in the point of cut-off each time the reverse lever is moved can be easily explained by considering the shifting eccentric, but to do so with the valve gear would be very difficult. Therefore reference will be made to Fig. 27. In this illustration, A is the center of the axle, a the center of the eccentric, and P the main pin. The angle of advance or the lap angle, because it is more convenient in this case to consider the valve as having no lead, is $a_1\ A\ a$. That is, the eccentric crank $a_1\ A$ is moved to $a\ A$ in order to bring the edge of the valve in line with the edge of the front steam port when the piston is at the beginning of the stroke. If the eccentric when at a is on the point of opening the port, it follows that the valve will be in exactly the same position, or at cut-off, when the center a of the eccentric has turned through the arc $a\ c\ a_2$ to a_2, because a_2 is exactly opposite a. That is, the eccentric crank $A\ a$, when the main pin P is turned, moves the valve back and opens the front steam port until it comes to $A\ c$ and the crank then moves the valve ahead until at $A\ a_2$ the valve is at cut-off.

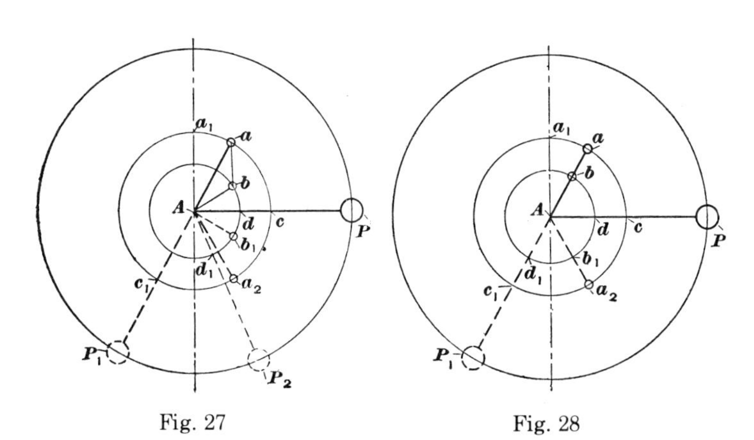

Fig. 27

Fig. 28

80. The point c, Fig. 27, where the main crank cuts the path of the eccentric, moves the same distance on the arc $a\ c\ a_2$ as does the eccentric a because the eccentric is secured rigidly to the axle. Therefore, the position of the main crankpin, when the eccentric is at a_2, or at cut-off, is obtained by using dividers and measuring off from c an arc $c\ c_1$ equal to the arc $a\ a_2$. By drawing a line through A and c_1 the position of the main pin is found to be at P_1 with the eccentric at a_2 and the valve at cut-off. Let it be assumed that the eccentric a is moved across the axle A to b, this action being equivalent to pulling up the reverse lever and raising the link of the valve gear. By using the same reasoning as before, the position of the eccentric with the valve at cut-off is at b_1. By measuring off an arc d d_1, the position of the main crank on the arc described by the eccentric b is located at d_1. The main pin is therefore on the line drawn through A and d_1 or at P_2.

81. The reason the cut-off occurs with the main pin at P_1, in Fig. 27, in one instance, and with the pin at P_2 or earlier in the stroke in the other is as follows: With the center of the eccentric at a, the angle of advance is a_1 $A\ a$ and the eccentricity of the eccentric is $A\ a$. With the center at b, the angle of advance is $a_1\ A\ b$, which exceeds by the addition of the angle $a\ A$ b, the angle $a_1\ A\ a$, and the eccentricity is decreased to $A\ b$. Therefore, moving the center a across the axle decreases the throw and increases the angle of advance. As a result, the center b of the eccentric moves through a shorter arc when it turns to b_1, than does the center a when it turns to a_2. The main pin P as it turns with the eccentric will also move a lesser distance before cut-off takes place when the center of the eccentric is at b than at a, and the piston will not move so far in the cylinder.

Therefore, the shorter cut-off is due to a decrease in the throw of the eccentric and an increase in the angle of advance. The reason for a shorter cut-off with the valve gear is that drawing up the reverse lever produces an effect which is equivalent to throwing the valves under the control of eccentrics of a decreased throw and an increased angular advance. This also applies to types of locomotive valve gears in which the valve crank has no angular advance, the valve displacement for the lap and the lead being accomplished by a lap-and-lead lever

82. Cut-Off Not Affected by Decreased Throw.—In Fig. 28 is shown that decreasing the throw of the eccentric without increasing the angle of advance does not shorten the cut-off. With the center of the eccentric at a, and the main pin at P, the valve is at admission, and it is therefore at cut-off with the eccentric center at a_2, directly opposite its position at a. The position of the main crankpin is located at P_1 by laying off the arc $c\ c_1$ equal to the arc $a\ a_2$ and drawing in the line $A\ c_1\ P_1$.

If the throw of the eccentric is decreased by moving the center from a to b, and the eccentric rod is lengthened so that the valve will be in the same position at b as at a, Fig. 28, the valve will be at cut-off with the eccentric at b_1. Measuring off the arc $d\ d_1$, equal to the arc $b\ b_1$ brings the main pin P_1 to the same position as before. The foregoing is evident because the movement of the center of the eccentric from a to b does not change the angle $a\ A\ P$ between the eccentric crank and the main crank. Therefore, the main crankpin will always move through the same arc while the valve crank is moving the valve to cut-off. The valve does not open the ports so wide with an eccentric of a shorter throw, but the piston is in the same position in the cylinder when cut-off takes place.

OFFSET OF LINK SADDLE PIN

83. The ultimate aim in the design of a valve gear is to obtain equal cut-offs and this implies an equal movement of the valve each way from mid-position. However, in the design of a valve gear, certain factors inherent in mechanisms in which a rotating movement is transformed into a reciprocating or to-and-fro movement must be taken into consideration. The principal one of these factors is the angularity of the eccentric rod; the angularity of the main rod is a minor condition that can be easily corrected during the valve-setting operation. With the Walschaert and Baker valve gears, the angularity of the eccentric rod, unless corrected, causes the link or the gear connecting rod to swing too far forwards and not far enough backwards. However, by introducing a backset in these parts, thereby increasing the backward swing without affecting materially the forward swing, the angularity of the eccentric rod can be compensated for and the cut-offs equalized. The effect of the angularity of the main rod is to cause a greater movement of the piston on the backward stroke than on the forward stroke for equal crankpin arcs. That is, the piston will move farther back for a certain movement of the crankpin from the front dead center than it will move forwards while the crankpin is traversing arcs of the same length from the back dead center. Such a condition is not considered when designing the gear, as it can be easily corrected by making a slight change in the length of the eccentric rod when setting the valves. Such a change will hasten the cut-off on the backward or long portion of the stroke and delay it the same amount on the short portion, hence making them equal. This, however, will result in a lesser port opening at the front port and a wider port opening at the back port, whereas were it not for the main-rod error the port openings would be equal. The main-rod error appears only in the valve-setting operation, where it is corrected without its significance being generally recognized.

84. With the Stephenson valve gear two factors contribute to inequality of cut-off, namely, the angularity of the eccentric rod, which introduces the lesser error, and the location of the eccentric rod pins to the rear of the link arc, which introduces the greater one.

This last error is caused by what is virtually a knuckle-joint action between the eccentric rod and the link in any other than a right-angle position. That is, the eccentric rod may be considered as being in two parts, the rod proper, and an additional length equal to the distance between the center of the pin hole and the link arc. The closing of the joint, which occurs while the link is moving to the rear, gives the same effect as shortening the eccentric rod for such a movement, and a lengthening of the rod for a movement forwards, at which time the joint is opening up. The angularity of the eccentric rod causes the valve to be moved too far forwards and not far enough back, and can be corrected by locating the link saddle pin in front of the link arc. The second condition is the reverse of the first and causes the valve to be carried too far back and not far enough ahead, which requires the saddle pin to be placed behind the link arc. These irregularities then act to counteract or neutralize each other, and, if of the same extent, would require the link saddle pin to be located on the link arc. But, as already pointed out, the location of the eccentric-rod pins introduces the greater inequality, so that the link saddle pin is located the amount of the difference between the above two positions, which places it to the rear of the link arc.

85. Why Back-Setting Link-Saddle Pin Equalizes Cut-Offs.— The irregularities introduced by the angularity of the eccentric rods and

	Backward Swing	Forward Swing
Error Introduced by Angularity of Eccentric Rod	Backward Swing Decreased	Forward Swing Increased
Error Introduced by Location of Eccentric Rod Pins Behind Link Arc	Backward Swing Increased	Forward Swing Decreased

(a)

Increased Backward Swing	Decreased Forward Swing

Net Effect of The Two Errors

(b)

Fig. 29

the location of the eccentric-rod pins is shown graphically in Fig. 29 *(a)*. It can readily be seen by reading vertically that one error acts to neutralize the other, but, as the error due to the location of the pins is the greater one, the net error will be the difference between the two and the final effect will be as shown in *(b)*, that is, the link will still swing too far back and not far enough ahead. This condition can be corrected by causing the link to move farther down on the block at the time the link swing is being decreased and to operate higher up on the block when the link swing is being increased. Accordingly, the cut-off will be lengthened at the time it otherwise would be too short and shortened at the time it would otherwise be too long. Thus, although the swing of the link is still unequal, the action is the same as if the forward swing of the link was increased and the backward swing decreased. It is precisely this action as shown in Fig. 30 that is accomplished by back-setting the link-saddle pin. In this illustration, the backset is exaggerated in order to make clear the action that occurs. The link-saddle pin is assumed to be pivoted at *a*, *b* is the outline of the link when it is at its extreme movement forward in full gear; and the dash outline *c* is its position at its extreme backward movement or just before it begins to move forwards.

86. The effect of locating the point *a*, Fig. 30, behind the link arc and then swinging the link about this point is to cause a point *d* on the link to follow the arc *e*. The ends of this arc are not level, the left end *d′* being the higher, so that the link will rise as it swings to the rear, and lower as it swings to the front. The reduced swing of the link forwards owing to the errors mentioned above is then compensated for by the fact that the link is operating nearer to full gear, whereas the increased swing backwards is taken care of by causing the link to operate nearer to mid-gear. This is shown by the fact that the center of the link-block pin *f* is nearer to the point *d* in the full-line position than the point *f′* is to the point *d′* in the dotted-line position, this showing that the block is lower in the link in one position than in the other. The link will have a greater influence on the valve in the full-line position, as the block is nearer to the top of the link or nearer to full gear; hence the cut-off will be delayed. The reverse will hold true in the dotted-line position, so that the unequal swing of the link will be compensated for. The extent of the rise and fall of the link can be best illustrated by cutting out a cardboard outline of the link and pinning it at *a*. In the foregoing it has been assumed that *a* is stationary, whereas it actually swings almost in a straight line, but this does not have any

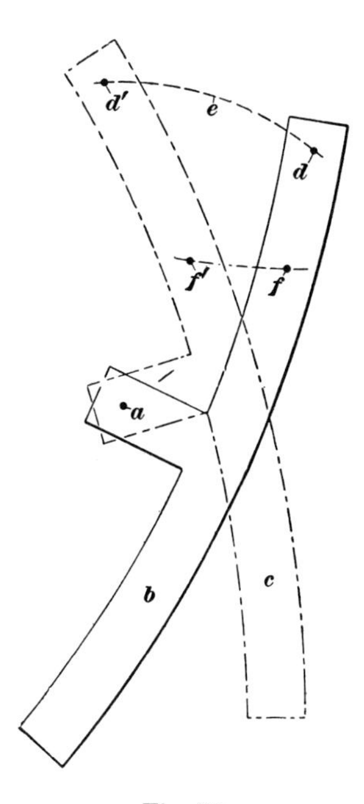

Fig. 30

bearing on the principle given. Considered as a whole, the movement of the link is extremely complex.

87. It was formerly thought that the reason for the back-set of the link-saddle pin was to compensate for the angularity of the main rod, which causes unequal spaces to be traversed by the piston while the main crankpin is passing through equal arcs. This belief was an error, because the assumption that the saddle pin was backset for this reason implied that the valves are set relative to crankpin movement, whereas such is never the case. The valves are always set with respect to crosshead movement, that is, they are so set as to cause cut-off to occur when the piston moves practically the same amount from each end of the cylinder and not while the crankpin is traversing equal arcs. Hence, the pin is backset to compensate for or to offset the errors introduced in the cut-offs by the angularity of the eccentric rods and the location of the eccentric-rod pins back of the link arc.

As already pointed out, allowance for the angularity of the main rod is automatically taken care of during the valve-setting operation. Therefore in the design of the valve gear no consideration is given to the angularity of the main rod.

For a main rod 10 feet long and with a 32-inch stroke, the error in the movement of the crosshead due to the angularity of the rod is less than 1 inch, and this condition can be easily corrected at the expense of equal port openings by a slight adjustment of the eccentric rod.

88. In Fig. 31 is shown the rise and fall of the link due to the offset of the link-saddle pin for the different positions of the main pin *P* in the small views at the left. The go-ahead eccentric is marked *f* and the back-up eccentric *b*. The positions of the link when the main pin is in positions (1), (2), (3), etc., in the small views is indicated by the same figures in the large views. It will be noted, as the pin *P* turns from position *(1)* to position *(4)*, that the link lowers on the block, while the link rises on the block as the pin turns from *(5)* to *(8)*. Therefore, the effect of the offset of the link-saddle pin is to equalize the cut-offs, by delaying or causing a longer cut-off at the front steam port and hastening or causing a shorter cut-off at the back steam port because the front port is supplying steam to the cylinder in positions *(1)* to *(4)* and the back steam port in positions *(5)* to *(8)*.

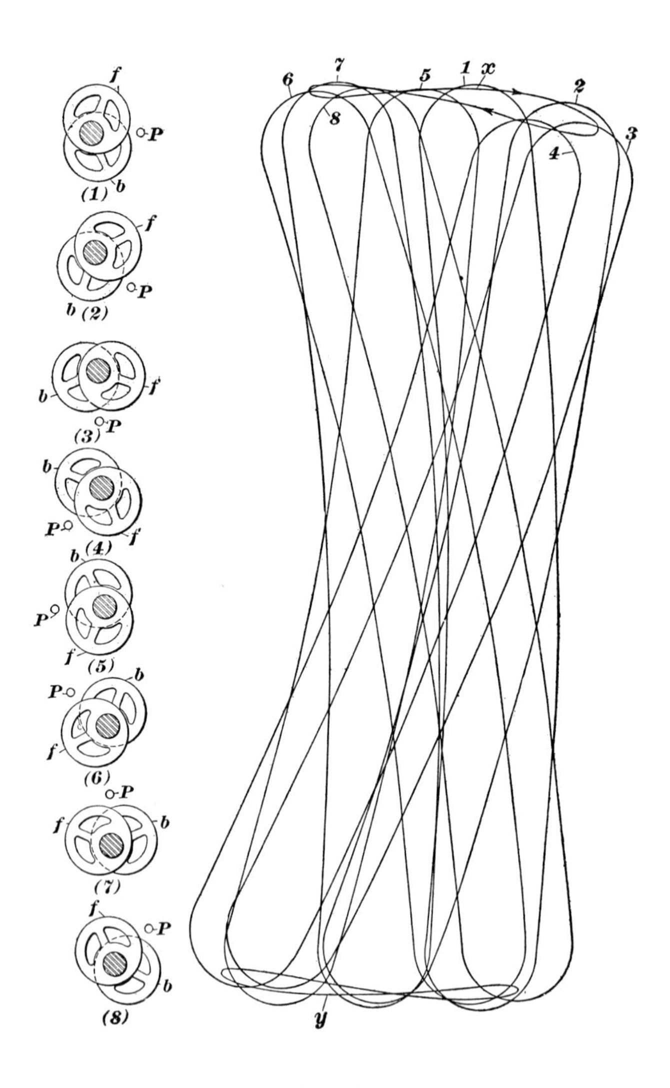

Fig. 31

89. Slipped Eccentric.—An eccentric is said to slip when it moves out of the correct position owing to the key which secures it to the axle being sheared off or lost. If the locomotive is moving forwards at the time, a go-ahead eccentric is the one most liable to slip, and it will move backwards or contrary to the direction of motion. The same will apply when a back-up eccentric slips with the locomotive running backwards. When an eccentric slips, the valve affected will be thrown out of its proper position with respect to the piston and the locomotive will go lame, that is, the exhausts will not be of the same intensity, the exhaust from one end of the cylinder being heavy, and from the other end, light.

90. Setting a Slipped Eccentric.—The eccentric that has slipped can be readily located by noting whether to go-ahead eccentrics, if the locomotive has been running forwards, or the back-up eccentrics, if moving in the reverse direction, are in their proper relation to the main crankpins. After the eccentric has been located, it can be reset by turning it until the keyways come opposite each other, and then driving in a substitute for the key. If the eccentric cannot be turned against the friction of the valve gear, it will be necessary to disconnect the front end of the eccentric rod from the link.

If there is more than one keyway in the eccentric and axle (sometimes there are as many as five in each), probably the best way to set the eccentric is by means of an improvised plumb bob. First place the engine on the forward dead center because this position gives more room in which to work, also place the reverse lever in mid-gear. Next, place the string of the plumb bob on the top back end of the link, and have the eccentric turned until the bottom back end comes vertical with the string, then secure the eccentric. This method is given because, with several keyways, it is possible that they may come in line, with the eccentric in an incorrect position.

The go-ahead eccentric can be set more easily than the back-up eccentric because the latter is next to the wheel. On some types of locomotives it is impossible to work on the back-up eccentric until the go-ahead eccentric and strap are removed. In this event a back-up eccentric should be treated in the same manner as a broken eccentric, strap, or rod.

91. Broken Forward-Motion Eccentric, Strap, or Rod.—The failure of a forward-motion eccentric, strap, or rod, although the strap usually breaks, prevents the valve gear on the side affected from moving the valve when running forwards and the locomotive must be run in on one side. If the eccentric breaks, remove the strap and rod, if the strap or rod breaks, the eccentric may be left up. Also remove the back-up strap and rod. The link should be tied to the link hanger in a vertical position; otherwise, the link will turn at right angles to the hanger and thereby prevent the link on the good side from being moved by the reverse lever.

The valve should be disconnected from the valve rod, and clamped as nearly as possible in mid-position. Lubrication may be applied through the indicator-plug openings, although a cylinder that previously has been receiving ample lubrication may be run a long distance before becoming dry.

Owing to the width of the steam lap, it is not difficult to secure a valve so as to exclude steam from the cylinders; however, it is difficult to place it accurately enough to shut off the cylinder from the exhaust passages.

If the valve has exhaust clearance, the cylinder is open to the exhaust passages when the valve is exactly central. Therefore, more or less smoke and cinders will always be drawn into the cylinder by the suction of the moving piston. This condition may be helped somewhat by leaving the indicator plugs out. Opening the cylinder cocks will cause too much dirt to be drawn in with the air.

The valve can be held in position by cocking the valve-stem gland. This can be done by loosening the gland nuts and prying one side of the gland away from the steam chest. A small wedge should then be applied on that side behind the gland and the nuts tightened on the opposite side. The locomotive will not steam freely, owing to the exhausts being reduced by one-half.

The locomotive can be prevented from stopping on the dead center on the good side by setting the brake hard as soon as this crankpin begins to leave the center.

92. Broken Back-Motion Eccentric, Strap, or Rod.—If on a busy road when a back-motion eccentric, strap, or rod breaks, the main track should be cleared by taking down the broken parts and blocking tightly in full forward gear the link affected. Then the engine should be moved slowly to the nearest siding. If not run slowly, or if the engine is slipped, the link will invariably turn over and cause more delay and damage. After the engine is on the siding, the forward-motion parts should be removed, the link tied vertical, and the valve clamped in mid-position.

93. Broken Link Hanger, Saddle Pin, or Lifting Arm.—A broken link hanger, saddle pin, or lifting arm prevents both valve gears from being moved at the same time. When it is desired to reverse, the uninjured gear will move to back gear and the disabled gear will remain in forward gear. The result is that the locomotive will stop when attempting to reverse because one side tends to move ahead and the other side back. The remedy when any of these parts break is to remove them and place the reverse lever in a position in the quadrant where the train can be started. Then block the disabled link, Fig. 32, at the same height as the good one by placing a block of wood of the required length on top of the link block and tying it in position. Also, place a block in the slot below the link block and allow enough space for the movement of the link on the block. The purpose of this block is to prevent any tendency for the link to work up into back gear.

The reverse lever should be secured in case an attempt is made to reverse. If necessary to run backwards, a block of sufficient length should be used to raise the link into back gear. If the time is too limited to permit of blocking up the link, the broken parts can be removed, and the locomotive run without attempting any repairs with the exception of blocking the reverse lever.

94. Broken Link-Block Pin or Lower Rocker-Arm.—A broken link-block pin or lower rocker-arm renders this valve gear useless because the motion of the eccentrics cannot be transmitted to the rocker-arm and the locomotive has to be run on one side. The link-block pin is the part that usually fails and it may break or it may be twisted off on account of the link block seizing on the pin because of lack of oil. The valve rod should be disconnected from the rocker-arm, and the valve secured in the proper position. Then the rocker should be turned until the arms are in about a horizontal position, or until there is no danger of the link striking the lower rocker-arm and the rocker should be tied in this position.

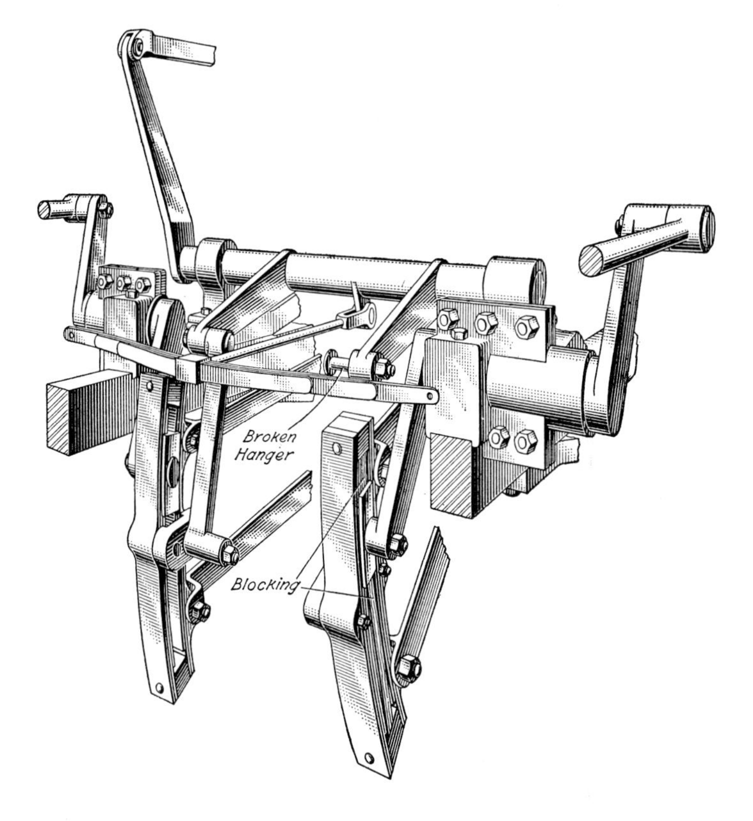

Fig. 32

If the lower rocker-arm breaks, the link block and the portion of the lower rocker-arm connected to it may be allowed to remain in the link. The valve stem should be disconnected from the valve rod and the valve secured in mid-position. If a transmission bar is used, and the link-block pin breaks, the bar should be removed and the valve secured as already described.

95. Broken Transmission Bar.—Failure of the transmission bar usually occurs at the link-block pin, the transmission bar hanger, or at the connection to the rocker-arm. The link block drops into back gear when the hanger breaks. The transmission bar should be disconnected at the rocker-arm and tied up should the hanger fail; the valve should be secured in mid-position.

96. Broken Top Rocker-Arm or Valve Rod.—If the top rocker-arm or valve rod breaks, the broken parts should be removed and the valve secured in mid-position.

97. Failure of Reverse Lever.—The reverse lever may fail to move the valve gear, owing to some of the parts breaking or to the loss or breaking of the pins that connect the lever to the reach rod or the reach rod to the arm of the tumbling shaft. In this event the weight of the links and of the eccentric rods causes the valve gears to drop into forward motion.

The locomotive can be run at full stroke if desired; if so, a block should be placed in each link under its link block, otherwise the links will work into back gear. It may be advisable to block under both links; then they will still be kept in the proper position should one block be lost out.

If the links are blocked at less than full stroke by placing blocking in the tops of the links, blocking must also be placed in the bottom of one or both links. If necessary to run backwards, the blocking should be removed, the links pried up, and the blocking reversed. The length of the blocking should be such as to permit of a little slack between the link block and the blocking.

With some types of locomotives, the breaking of a spring or spring hanger prevents the reverse lever from being moved, owing to the boiler cramping the reach rod when the frames settle down on the boxes. If the reverse lever is caught at a cut-off too short to start the train, the pin which connects the reach rod to the tumbling-shaft arm should be knocked out. This allows the links to fall into full forward gear.

STEPHENSON VALVE GEAR

EXAMINATION QUESTIONS

Notice to Students.—*Study the Instruction Paper thoroughly before you attempt to answer these questions.* **Read each question carefully and be sure you understand it;** *then write the best answer you can. When your answers are completed, examine them closely, correct all the errors you can find, and* **see that every question is answered.**

(1) Define a valve gear.

(2) Name the valve gears in general use.

(3) Why is the Stephenson valve gear no longer applied to modern locomotives?

(4) Name the principal parts of the Stephenson valve gear.

(5) Why is one eccentric called the go-ahead and the other the back-up?

(6) In what positions of the reverse lever is: *(a)* the longest cut-off obtained; *(b)* the shortest cut-off?

(7) What pins are used to connect the eccentric rods to the links?

(8) What pin is used to connect the link block to the lower rocker arm, or to the transmission bar?

(9) What is the purpose of the reverse shaft?

(10) What is the difference between a direct-motion and an indirect-motion valve gear?

(11) *(a)* What is the position of the go-ahead eccentric relative to the main pin with an indirect-motion valve gear and outside-admission valves, with the locomotive running forwards? *(b)* What is the position of the back-up eccentric with the locomotive running backwards?

(12) How much do the valves move with the locomotive drifting and the reverse lever in mid-gear?

(13) Generally speaking, how is the steam pressure shifted in the cylinders when a locomotive is reversed?

(14) Explain the valve movement when a locomotive is reversed with the main pin on the right side on the forward dead center and the pin on the left side on the top quarter.

(15) What should be done when a go-ahead eccentric, strap, or rod breaks?

(16) What should be done when a link-saddle pin breaks?

(17) What should be done when a link-block pin breaks?

(18) What should be done when a back-up eccentric, strap, or rod breaks?

(19) What should be done if a transmission bar breaks?

(20) What should be done if the reverse lever is caught at a short cut-off?

Walschaert Valve Gear
Part 1

By
J. W. Harding

4003A-2 Printed in U.S.A. Edition 2

1946 Edition

WALSCHAERT VALVE GEAR
(PART 1)

ARRANGEMENT AND OPERATION

ACTION OF GEAR

INTRODUCTION

1. The Walschaert valve gear was invented by Edige Walschaerts in 1844. In naming the gear, the last letter of the name has been dropped. The gear, although used to a large extent in Continental Europe, was not applied to locomotives in America to any extent until about 1900, when locomotives had increased in size to such an extent as to make it impracticable to use a valve gear like the Stephenson, which is placed between the wheels. At the present time the Walschaert and other gears have entirely superseded the Stephenson gear on new locomotives.

The only valve gear that can be successfully applied to a modern locomotive is one that, like the Walschaert or some other gear similarly located, is placed outside of the frames. The reason an inside valve gear cannot be used on a modern locomotive is that the axles are so large as to require eccentrics and eccentric straps of too great a size. In addition, large locomotives require cross-bracing between the main frames and also between the frames and the boiler, so as to maintain alignment and stability and to reduce frame failures, and the bracing interferes with the application of an inside valve gear.

The advantages of the Walschaert valve gear over the Stephenson are that it is light, accessible for inspection and maintenance, and low in maintenance cost, and that it will keep square from shopping to shopping when once correctly set. The gear does not interfere with the proper bracing of the frames, which is most essential with heavy locomotives.

PRINCIPLES UNDERLYING GEAR

2. Characteristics of Walschaert Gear.—A characteristic of the Walschaert valve gear as compared with the Stephenson gear is the former uses but one eccentric crank for both the forward and backward motions. The eccentric crank also has no angular advance, which means that it is set at approximately 90 degrees, or a quarter of a turn from the main crankpin. With the Stephenson gear the eccentrics have to be set to displace the valve the amount of the lap plus the lead from mid-position when the piston is at the beginning of the stroke.

3. Investigation of Principles.—An investigation of the principles underlying the arrangement and operation of the Walschaert valve gear will show why one eccentric crank can be used for both forward and backward motions, as well as the reason why the eccentric crank requires no angular advance.

Like any other device, the arrangement of the Walschaert valve gear can be traced back to some elementary type of gear of very simple design. Therefore, it is desirable when beginning an investigation of the principles on which the gear is based, to start with a simple gear and trace its development to one of the Walschaert type.

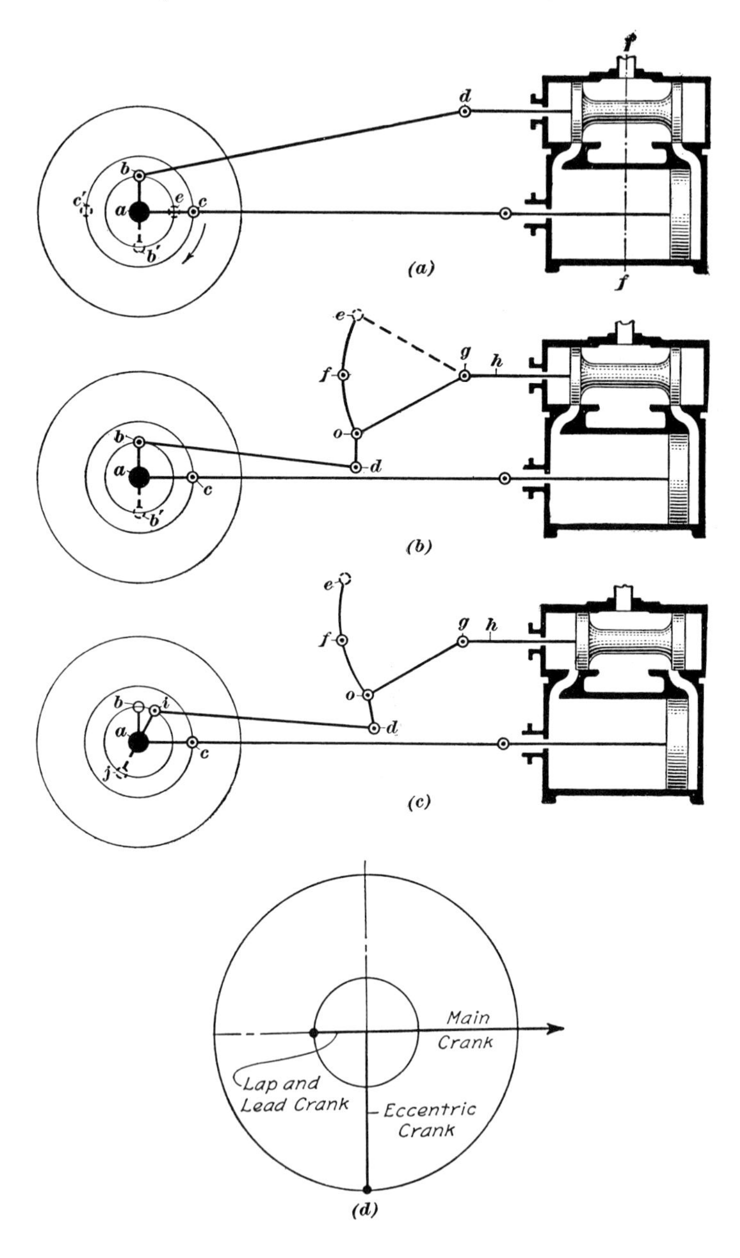

Fig. 1

A simple form of valve gear is shown in Fig. 1 *(a)*, and it will be explained how a gear can be evolved that will use but one eccentric crank *ab* with no angular advance for both forward and backward motions; *a* is the main axle, *ab* the eccentric crank for moving the valve through the eccentric rod *bd*, and *ac* is the main crank with the main pin *c*. The valve, that is inside admission, has no lap or lead; that is, the width of the valve between the steam and the exhaust edges is exactly the same as the width of the steam ports.

4. Position of Valve With Respect to Piston.—To start the piston moving when the wheel is given a slight turn in the direction of rotation, which, in this case, is forwards, the valve with the piston at the beginning of the stroke must be at the point of admitting steam to the cylinder in front of the piston, or it must be in the position shown in Fig. 1 *(a)*. In this position the valve is at mid-stroke or in mid-position, because the line *f* drawn through the center of the valve (half-way between the outer ends) comes midway between the steam ports in the valve seat. The valve when in mid-position is one-half of a stroke behind the piston, and if this difference in position is maintained, the steam will be admitted to and exhausted from the cylinder in such a manner as to keep the piston, and, therefore, the driving wheel in motion as long as steam is supplied.

5. Position of Valve Crank With Respect to Main Crank.—The position of the valve crank *ab* in Fig. 1, in relation to the main crank *ac*, in order to keep the valve one-half of a stroke behind the piston, will be next considered. To keep the valve one-half a stroke behind the piston, assuming that the driving wheel is turning forwards, the eccentric crank must be placed one-quarter of a turn behind the main crank, because the eccentric crank in moving the valve is similar to the main crank in moving the piston where a one-quarter turn of the crank moves the piston one-half stroke.

6. Operation.—The operation of the valve gear in keeping the engine in motion is as follows: When the main crankpin turns in the direction of the arrow, the rotation of the axle *a*, Fig. 1, imparts to the eccentric crank *b* a circular motion that draws the eccentric rod *bd* and the valve to the right. When the main pin *c* arrives at the bottom-quarter position, the piston is at practically half stroke, and the eccentric crank is then at *e*. The valve has now moved the limit of its travel to the right, and will have the front port wide open for the admission of steam, and the back port wide open to the exhaust. When the main pin reaches the back dead center at *c'*, the eccentric crank is at *ab'*, and the valve is in mid-position. A further movement of the main pin causes the eccentric crank to move the valve and open the back steam port for steam and the front port to exhaust. Accordingly, the valve gear shown in Fig. 1 *(a)* will keep the piston in motion and the driving wheel turning forwards so long as the steam supply is maintained.

7. Gear Not Reversible.—While the valve gear shown in Fig. 1 *(a)* imparts the proper movement to the valve to keep the engine running forwards, yet it is deficient to the extent that the engine cannot be reversed and run backwards. For example, if the main crankpin *c* is turned backwards, the eccentric crank will draw the eccentric rod *bd* backwards and the valve will open the back steam port instead of the front port. To give the valve the proper movement for a backward rotation of the driving wheel, it is necessary merely to set an eccentric crank on the axle the same distance from the main pin with the engine running backwards as when running forwards. If the eccentric crank *ab* is given one-half turn and set at *ab'*, the crank will be the same amount behind the main pin *c* when the engine is moving back as when moving ahead. When the wheel is turned backwards, the eccentric crank *ab'* will push the valve ahead and the engine will continue to move in the reverse direction. It would be impracticable to give the eccentric crank one-half a turn each time it was desired to reverse, and more convenient means must be devised to give the same effect as moving the eccentric when it is desired to run backwards.

8. Arrangement for Reversing.—An arrangement that produces the same effect as if the eccentric crank were moved one-half turn on the axle when it is desired to reverse the engine is shown in Fig. 1 *(b)*. The arrangement is very simple and requires merely the addition of a slotted link *ed,* and a rod *go*, to the parts shown in view *(a)*. The link is arranged to swing on a pivot *f,* and the eccentric rod *bd* is connected to the lower end. The front end of the rod *og* is flexibly connected to the valve rod at *g* and the back end is free to slide in the slot *oe* in the link *de*. The rod *og* will be referred to as the *radius rod,* because it is the radius of a circle of which the link slot *oe* is a part. If the wheel is turned forwards, the arrangement shown in view *(b)* is in reality the same as in view *(a)* because the eccentric rod *bd* and the radius rod *og* may be considered as one continuous rod that connects the eccentric crankpin *b* to the valve stem *gh*. To reverse, it is only necessary to raise the radius rod *og* to *ge*. Then when the main pin *c* is turned in the reverse direction, the eccentric crankpin *b* moves the eccentric rod and the bottom of the link backwards, and as the link pivots at *f,* the upper end *e* of the link, the radius rod *ge* and the valve move forwards or to the right. When the valve moves to the right, the steam enters the cylinder through the front steam port and the engine will continue to run backwards.

9. By the use of a link *ed,* Fig. 1 *(b),* with a fulcrum at the middle, the same movement is imparted to the valve as if the eccentric were given one-half turn and moved from *b* to *b'*. In other words, the effect of moving the radius rod to the upper end of the link is precisely the same as if the rod were left at *go* and the eccentric crank were moved to *b'*, a distance of one-half turn. Therefore, when it is desired as in this case to develop a valve gear with but one eccentric crank for both motions, a link arranged as shown in view *(b)* forms a convenient means of reversing for the reason that it makes one eccentric perform the work of two.

10. Effect of Lap and Lead on Arrangement.—A steam engine cannot be operated economically unless provision is made for moving the piston by the expansive force of steam, and this can be done only by giving the valve steam lap. The steam lap makes the valve wider than the steam port and, therefore, introduces an interval between cut-off and release, instead of these valve events occurring simultaneously as in Fig. 1 *(a)* and *(b)*, and during this interval the steam expands in the cylinder and moves the piston. The effect on the arrangement in view *(b)* when the valve is given lap and lead, will be considered next. The changes in the valve gear when lap and lead is given will be seen by referring to Fig. 1 *(c)*.

In view *(c)* the same parts are used as in view *(b),* but the valve has been given steam lap and lead. This requires the eccentric to be moved away from its right-angle position at *ab* to *ai* in order to displace the valve from mid-position the amount of the lap and lead and so have the front steam port open when the piston is at the end of the forward stroke. The reason is if the position of the eccentric crank were not changed, the valve would be in mid-position with the piston at the end of either stroke. In this event the steam edges of both ports would be blanked the amount of the steam lap, and steam would not be admitted to the cylinder until the piston had moved some distance on its stroke. The change in the position of the eccentric pulls the link *de* out of the position shown in view *(b)* to the position shown in view *(c)*.

11. The arrangement in Fig. 1 *(c)* will cause the valve to operate properly so long as the engine is moving forwards, but will not permit the motion to be reversed, for the following reason: If the radius rod is moved from *go* to the top of the link, the effect as already explained is the same as if the eccentric crank *ai* were given one-half a turn which in this case, would bring it to *aj*. The valve, instead of being the amount of the lap and the lead to the right of the mid-position with the eccentric at *ai*, is now the amount of the lap and lead to the left of its mid-position, and hence will open the back steam port to steam. The piston cannot move backwards because it is blocked by the steam behind it.

12. In Fig. 1 *(b)* the eccentric crank *ab′* must be the same distance behind the main pin *c* when the engine is running backwards as the crank *ab* is behind the main pin with the engine running forwards. The effect of moving the radius rod from *go* to *ge* is the same as if the eccentric crank were given one-half turn from *ab* to *ab′*. It is only when the eccentric crank and the main crank are at right angles to each other that the eccentric crank, when given one-half turn, will still be the same distance from the main crank. For example, with the eccentric crank set at *ai*, view *(c)*, the effect when the engine is reversed is to give the eccentric crank *ai* one-half a turn to *aj*, and the crank at *aj* is farther from the main crank *ac* than at *ai*.

The foregoing shows that when one eccentric is used for both motions, it cannot be given angular advance or moved to displace the valve, the lap and the lead, because if it is, the engine cannot be reversed. Therefore, the right-angle setting of the eccentric crank must be adhered to, as shown in *(b)*, and some other means must be found to obtain the lap-and-lead displacement of the valve.

13. Arrangement for Displacing Valve Lap and Lead.—The right-angle setting of the eccentric crank can be adhered to and the lap-and-lead displacement of the valve with the engine on the dead center can be obtained theoretically by the use of another crank set at right angles to the eccentric crank. Thus, in Fig. 1 *(d)*, the lap-and-lead crank (with a throw equal to the lap and the lead) is set 90 degrees ahead of the eccentric crank with the locomotive moving forwards; and 180 degrees from the main crankpin, the valve being outside admission. Let it now be assumed that the two cranks are both connected to the valve by some mechanism that duplicates the action of the actual gear. The position of the lap-and-lead crank displaces the valve from its mid-position an amount equal to the lap and the lead; then the front steam port will have the required lead opening. When the main crankpin has made one-half turn, the lap-and-lead crank will have displaced the valve the required amount to obtain the lead opening at the back port.

14. With an inside admission valve, Fig. 2, the lap-and-lead crank must be set 90 degrees in advance of the eccentric crank with the locomotive moving forwards. However, as the eccentric crank in this case is assumed to follow the main crankpin, the lap-and-lead crank will be at no angle to the main crank instead of being at an angle of 180 degrees from it as with an outside admission valve.

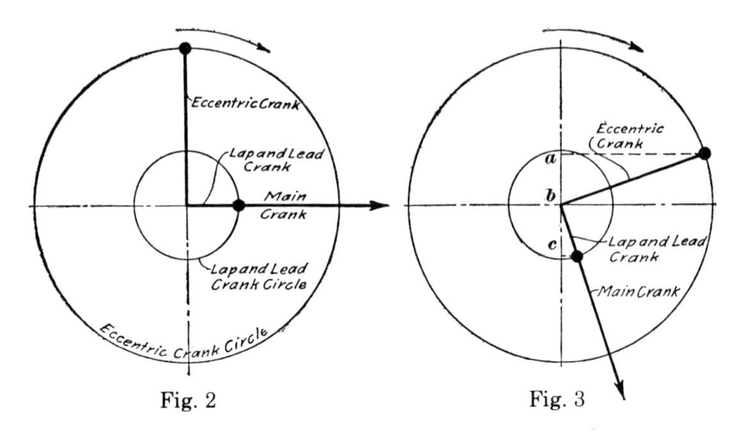

Fig. 2 Fig. 3

15. If the lap-and-lead crank was alone acting on the valve, its movement would be equal only to twice the lap and the lead for each one-half turn of the driving wheel. If the eccentric crank was alone acting on the valve, it would move its specified travel, but it would not be timed properly with respect to the piston, that is, the valve would be in mid-position with the piston at the end of its stroke. However, the combined action of both cranks will bring the valve to its proper position.

16. Action of Cranks.—The action of two cranks of unequal length on the movement of an inside admission valve is as follows: With the main crankpin passing the front dead center, Fig. 2, the eccentric crank acts to move the valve forwards, the lap-and-lead crank tends to pull the valve backwards, as this crank is now leaving its forward dead center. The action of the longer crank predominates but the forward movement of the valve is retarded by the lap-and-lead crank. This action continues until about full port opening is obtained or until the cranks arrive in the position shown in Fig. 3. At this point the valve stops because the action of the eccentric crank is now equal to a crank of a length *ab*, which in turn is equal to the length of the equivalent lap-and-lead crank *bc*. Hence, the eccentric crank is now moving the valve ahead at the same rate as the other crank is moving the valve back. Also, the valve is moving slowly when the cranks are approaching and leaving these points.

17. In Fig. 4 both cranks are acting on the valve to draw it backwards and close the front steam port and this action continues until the cranks assume the position shown in Fig. 5. The eccentric crank in the absence of the lap-and-lead crank would place the valve in mid-position, but with this crank acting with the eccentric crank the speed of the valve is increased, and this combined with the setting of the lap-and-lead crank places the valve one lap plus one lead from mid-position with the crankpin on the center.

The action of the cranks on the valve as the main crankpin passes the back center is similar to their action when the crankpin passes the forward center; that is, the eccentric crank is moving the valve back to open the back steam port farther and the lap-and-lead crank is delaying this movement. This action continues until the cranks arrive in the position shown in Fig. 6. The valve now stops momentarily for the same

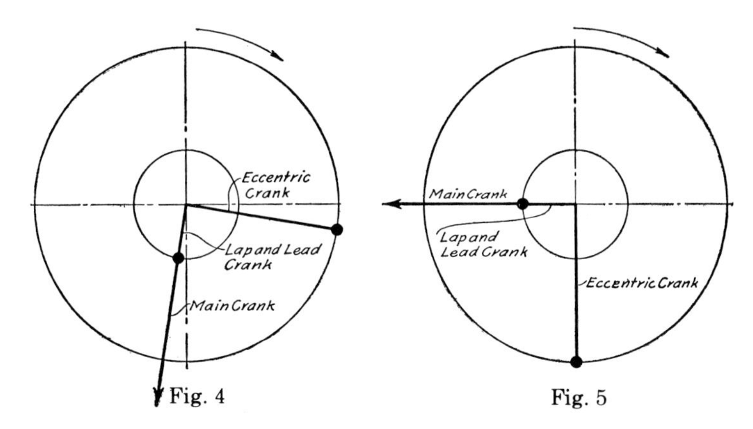

Fig. 4

Fig. 5

reason as given when considering Fig. 3; that is, the action of the eccentric crank on the valve is now equal to that of the lap-and-lead crank, owing to their effective crank lengths being the same in this position.

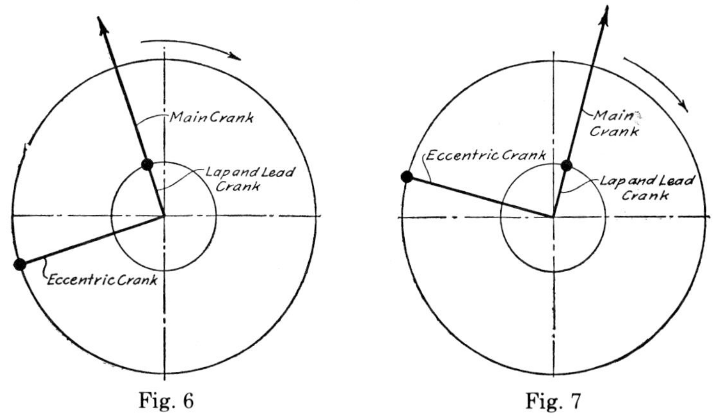

Fig. 6

Fig. 7

In Fig. 7, both cranks are acting on the valve, its velocity is thereby increased to such an extent that the valve is displaced an amount equal to its lap and lead, from mid-position with the crankpin on the forward dead center and the piston at the end of its stroke. It will be noted that the action of the cranks causes the port to open slowly for the admission of steam but to close rapidly. In order that the cranks may increase the speed of the valve when they are both operating together, it must be assumed that the two cranks are connected to the valve in such a manner as to cause them to act on it independently of each other. Then when so connected the valve will have a higher speed when the cranks are working together than when working in opposition.

18. If the eccentric crank were shortened, this giving in combination with the other crank an effect equivalent to hooking up the actual gear, the valve travel would be shortened, but the action of the lap-and-lead crank would be the same; that is, it would still act to vary the speed of the valve. With the eccentric crank shortened to less than the lap-and-lead crank, the latter crank would limit the valve travel to double the lap and the lead; the eccentric crank would then act to vary the valve speed. If the length of the eccentric crank were decreased to zero, the total valve travel would still be obtained from the lap-and-lead crank. It will be noted, after the eccentric crank has been shortened to less than the lap-and-lead crank, that any further shortening does not affect the valve travel; it remains equal to that imparted by the lap-and-lead crank.

19. It would be difficult or probably impossible to design the necessary mechanism to connect the eccentric crank and the lap-and-lead crank to the valve. However, it is possible to use the main crank for a lap-and-lead crank, provided the movement of the main crank or the crosshead (because they both move together) is reduced to the proper amount. This is accomplished by means of a lever known as the lap-and-lead lever, or the combination lever, with one end connected to the crosshead at *l*, Fig. 8, and its upper end connected to the radius rod at *m*. The reduction in the movement of the crosshead or the main crank to twice the lap-and-lead movement is brought about by connecting the valve rod to the lever at the point *g′*. This point is so located between the ends of the lever that the crosshead movement at the lower end of the lever is reduced to a movement equal to twice the throw of the lap-and-lead crank at the valve rod. With the point *m* held fixed, the movement of the crosshead and the lower end of the lever from *l* to *l′* moves the valve from *g′* to *g*, or twice the lap and the lead.

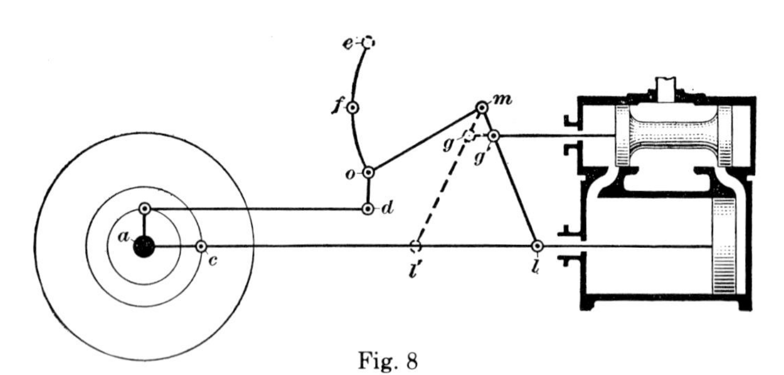

Fig. 8

20. The arrangement of the eccentric crank and the combination lever will produce practically the same action on the valve as the two cranks already described. That is, as the combination lever replaces the lap-and-lead crank, this lever, except as explained below, varies the speed of the valve and causes it to be displaced, its lap and lead from mid-position with the piston at the beginning of its stroke. The variable speed imparted to the valve by the combination lever results in the port opening slowly after the lead opening has been obtained, and closing rapidly. Therefore, in full gear as well as in the ordinary working cut-offs, the lap-and-lead lever neither adds to nor substracts from the valve travel to any extent.

The movement imparted to the valve by the eccentric crank decreases as the reverse lever is drawn up until finally a point is reached on the quadrant where this movement becomes less than twice the lap-and-lead travel; the eccentric crank is then cut out. The condition that now exists is similar to that referred to in Art. 18; that is, the total valve travel is then obtained from the combination lever and the eccentric crank acts only to vary the speed of the valve. The valve travel now remains the same as the lever continues to be drawn back, even to mid-gear, but in the latter position the action of the eccentric crank in varying the speed of the valve is eliminated.

Generally, the valve movement is derived from the combination lever at cut-offs less than 15 per cent. At these cut-offs the steam is cut off from the cylinder later in the stroke than in mid-gear, although the valve travel is practically the same. The reason is that the short arm of the eccentric crank that acts with the combination lever except in mid-gear counteracts the movement of the lever on the valve. This causes the valve to remain stationary at the end of its stroke through a longer arc of the main crankpin than with the combination lever alone acting as in midgear; hence the cut-off is later.

With the combination lever all the way forward as with the crosshead at its extreme forward movement, Fig. 8, the equivalent lap-and-lead crank would be on its front dead center. With the lever all the way back, the crank would be on the back dead center, and, with the crosshead halfway in the guides, the crank would be on the quarter.

21. In designing the gear, the eccentric crank is designed with a throw that will give the valve its specified travel, 7½, 8 inches, etc., or whatever it may be. As the movement of the eccentric crank is transmitted to the link some distance below the link block, the throw of the crank must necessarily be considerable in excess of the valve travel. Its throw generally exceeds the valve travel by at least three to one with the reverse lever in full gear.

GENERAL DESCRIPTION

22. Views of Valve Gear.—The ordinary type of steam locomotive has two valve gears, or one for each valve, so connected as to be moved by one reverse lever. Each valve gear consists of the same parts, and, therefore, when identifying the parts, both gears do not have to be shown in the same view. However, it is necessary to show the two valve gears in order to understand how the movement of the reverse lever is transmitted to both. In Fig. 9 (on page 404A) *(a), (b),* and *(c),* are shown three side views of a complete single Walschaert valve gear when an inside-admission valve is used, and Fig. 10 is a partial view which shows how the connection is made between the gears on each side so that the movement of the reverse lever and the reach rod can be transmitted to both. Similar parts in all of the views have the same reference figures.

23. Names of Parts.—The names of the parts of the gear shown in Fig. 9 are as follows: *1,* the eccentric crank; *2,* the eccentric rod, *3,* the link with the link foot *4; 5,* the link block; *6,* Figs. 9 and 10, the reverse shaft; *7,* the reverse-shaft crank with arms *7′* and *8′; 8,* Fig. 10, the reverse-shaft arm; *9,* Fig. 9, the radius rod or radius bar; *10,* the radius-rod hanger; *11,* the combination lever, combining lever, or lap-and-lead lever; *12,* the union link; *13,* the valve stem; *14,* the valve-stem crosshead guide; *15,* the valve-stem crosshead; *16,* the gear frame or the link support here shown cast with the guide yoke; *17,* the reach rod; *18,* the counterbalance spring and casing. The valve gear is shown in forward gear in Fig. 9 *(a),* in mid-gear in *(b),* and in backward gear in *(c).*

24. General Arrangement of Parts.—The general arrangement of the parts of the Walschaert valve gear will be explained by referring to Fig. 9. A detailed explanation of the arrangement and construction at the different points in the gear will be given further on.

The eccentric crank *1* is placed on the end of the main crankpin *a.* The bolt *b,* which passes through the eccentric crank, and a circular slot in the side of the crankpin, keeps the crank from coming off the pin, and the key *c* prevents the crank from turning on the pin. The rear end of the eccentric rod is placed on a pin in the eccentric crank which is usually made in one piece with the crank and stands at right angles to it. The front end of the eccentric rod is forked and is connected to the link foot *4* by the pin *d.* The gear frame *16* is here shown as being made in one piece with the guide yoke *e,* but it is usually made separate. The guide yoke is bolted to the end of a cross-tie *e′*, which extends across the frames to the other side and serves as a support for the other combined yoke and gear frame. A bracket or knee is bolted to each frame at the point where the cross-tie crosses it and the cross-tie is bolted to this knee. The link *3* is carried in the gear frame on the link trunnions *f* (one on each side). The radius rod *9* is forked where it passes through the link, and the link block *5* sets between the forks and is held in the rod by the link-block pin *g.* The link block *5,* which moves in a slot in the link, shown by dash lines, when the radius rod is raised and lowered, is used to transmit the backward-and-forward movement of the link to the radius rod. The rear end of the radius rod swings on the radius-rod hanger *10* which is connected at the upper end to the arm *8′* of the reverse-shaft crank, while the front end of the rod is connected by the pin *i* to the top of the combination lever *11.* The connection between the valve-stem cross-head *15* and the combination lever *11* is made by the pin *j* and the lower end of the lever is connected to the union link *12* by the pin *k.* The back end of the union link works freely on the outer end of the wristpin *l.* The valve-stem crosshead guide *14* is cast in one piece with the back-valve chamber head and the valve stem *13* is connected to the crosshead *15* by a key, not shown. The end of the reach rod *17* is forked and is connected to the arm of the reverse-shaft crank *7′* by a pin *n.*

25. The valve gear on each side of the locomotive is connected by the reverse shaft *6,* Fig. 10, which turns in boxes *d* on the gear frames when the reverse lever is moved. The reverse-shaft crank *7,* made in one piece, is keyed or bolted to one end of the reverse shaft, and the reverse-shaft arm *8* to the other end. The two reverse-shaft arms *8* and *8′* are connected by the radius-rod hangers *10* to the radius rods *9;* hence, when the reverse lever is moved, the reach rod *17* through its connection with the reverse-shaft crank *7,* turns the reverse shaft *6,* and, depending on the direction of movement, raises or lowers the arms *8* and *8′* and the radius rods *9.* The

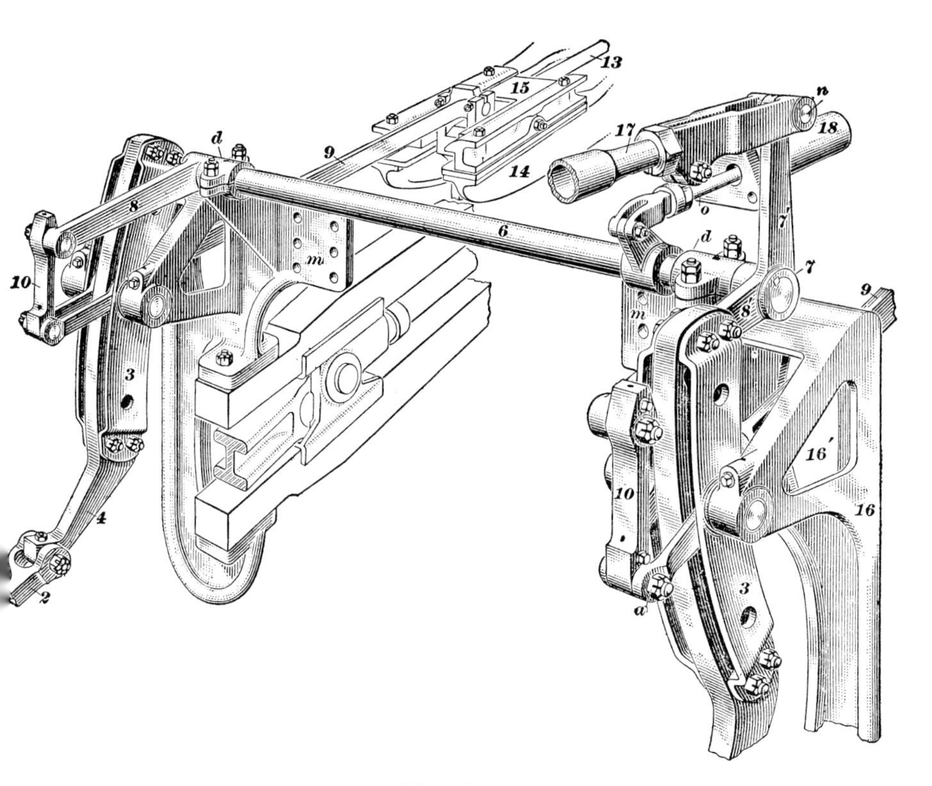

Fig. 10

combined guide yoke and gear frame *16* bolts at the points *m* to the rectangular plate or casting *e'*, Fig. 9, which extends across and is connected to the frames by brackets. Strictly speaking, the casting *e'* is the guide yoke, and the part *16* is the gear frame and guide-yoke end. The purpose of the counterbalance spring *18,* Fig. 9, which is connected to the reverse shaft *6*, Fig. 10, by an arm and a rod *o*, is to make it easier to move the reverse lever against the weight and friction of the valve gear. The spring *18,* Fig. 9 *(c),* is compressed when the radius rods are lowered to the bottom of the links, hence the expansion of the spring helps to lift the rods to the top of the links as shown in view *(a).*

26. A complete locomotive valve gear, by which is meant the two gears a part of which is shown in Fig. 10, may be considered as made up of two parts. One part is directly concerned with valve movement and is in operation when the locomotive is in motion, while the other part is used to transmit movement to the first part through the reverse lever. The eccentric cranks, the eccentric rods, the links, the link blocks, the radius rods, and the combination levers, there being two of each of the above, are the parts of the gear which have to do with moving the valves. The other part of the gear comprises the two radius-rod hangers, the reverse-shaft crank, the reverse-shaft arm, and the reverse shaft. These parts are used to transmit the movement of the reverse lever equally to the valve gear on each side of the locomotive.

27. Movement Imparted By Reverse Lever.—Any movement of the reverse lever changes the position of the link blocks and radius bars in the links. The action that occurs when the reverse lever is moved is as follows: In Fig. 9 *(b)* and 10, when the reverse lever and the reach rod *17* are moved forwards, the reverse-shaft crank arm *7′* moves in the same direction and turns the reverse shaft *6*. The reverse-shaft arms *8* and *8′*, Fig. 10, move upwards and the radius-rod hangers *10,* the radius rods *9* and the link blocks are raised to the position shown in Fig. 9 *(a)*. Therefore, in this case, the link blocks are in the upper half of the links, in forward gear. When the reverse lever, Figs. 9 *(b)* and 10, is moved from the front to the back corner of the quadrant, the reverse-shaft crank *7* turns the reverse shaft *6* and the downward movement of the reverse-shaft arms *8* and *8′* lowers the radius-rod hangers, the radius rods, and the link blocks to the lower half of the links. The extent of the valve movement when the reverse lever is moved depends on the position of the main pins. With the main crankpin *a* on the bottom quarter, moving the reverse lever from the front to the back corner of the quadrant causes the valve to close the front steam port and open the back steam port.

28. The cut-off is longest with the link blocks in the ends of the links and decreases as the blocks are brought nearer to the center of the links. With the link blocks in the center of the links, the port opening obtained with the engine on the dead centers is equal to the lead. The valve begins to close the port as soon as the main pin leaves the dead center. The movement of the reverse lever does not always place the link block and radius rod in the same end of the link. With the arrangement given in Fig. 11, and in the detail of the gear given in Fig. 12, the link block and radius rod are in the bottom of the link with the reverse lever in the forward corner of the quadrant and in the top of the link as shown with the lever in the back corner.

29. General Operation of Gear.—The general operation of the valve gear is as follows: When the locomotive is in motion, the rotation of the main crankpin *a*, Fig. 13, imparts a circular movement to the eccentric crank *1,* and the back end of the eccentric rod *2,* and a vibratory movement in an arc to the front end of the rod. This causes the link *3* to swing backwards and forwards in the gear frame on the link trunnions *f*, and thereby transmits a back-and-forth movement to the radius rod *9*, the back end of which swings in an arc on the bottom of the radius-rod hanger *10.* The movement of the radius rod merges at *i* with that of the combination lever *11* before it is transmitted to the valve stem *13* and to the valve.

The bottom end of the combining lever has a backward and forward movement, in an arc, and this, when transmitted to the top of the lever in combination with the action of the radius rod at this point, imparts the required movement to the valve to keep the locomotive in operation. The combination lever acts on the valve in opposition to the crank when passing dead centers.

30. Oiling Points in Valve Gear.—The eccentric-rod pin in the front end of the eccentric rod is oiled through the oil cup shown on the link foot, Fig. 9. In Fig. 10, black circular marks at the top and bottom of the radius-rod hangers, in the gear frames, in the reverse-shaft boxes and in the upper end of the arm 7′ indicate the oil holes. Other oiling points are as follows: In the top of the link block to oil the link-block pin, sometimes in the top of the link to oil the wearing surface of the link slot, in the front end of the radius rod to oil the radius-rod pin, in the top of the guide bars to oil the valve-stem crosshead, in the valve-stem crosshead to oil the

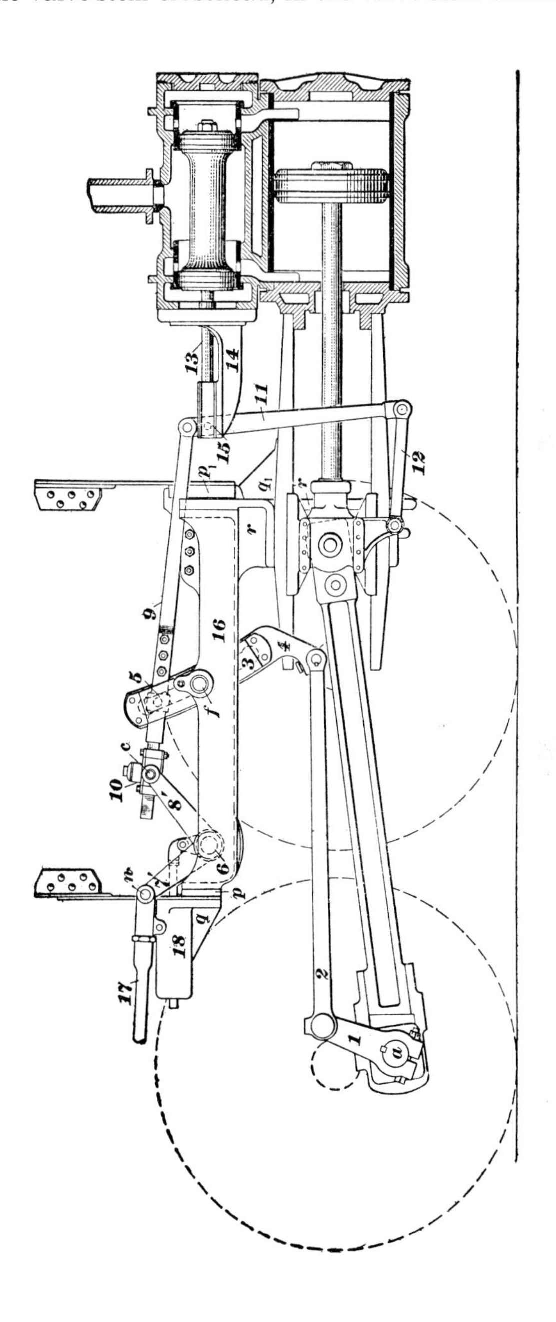

Fig. 11

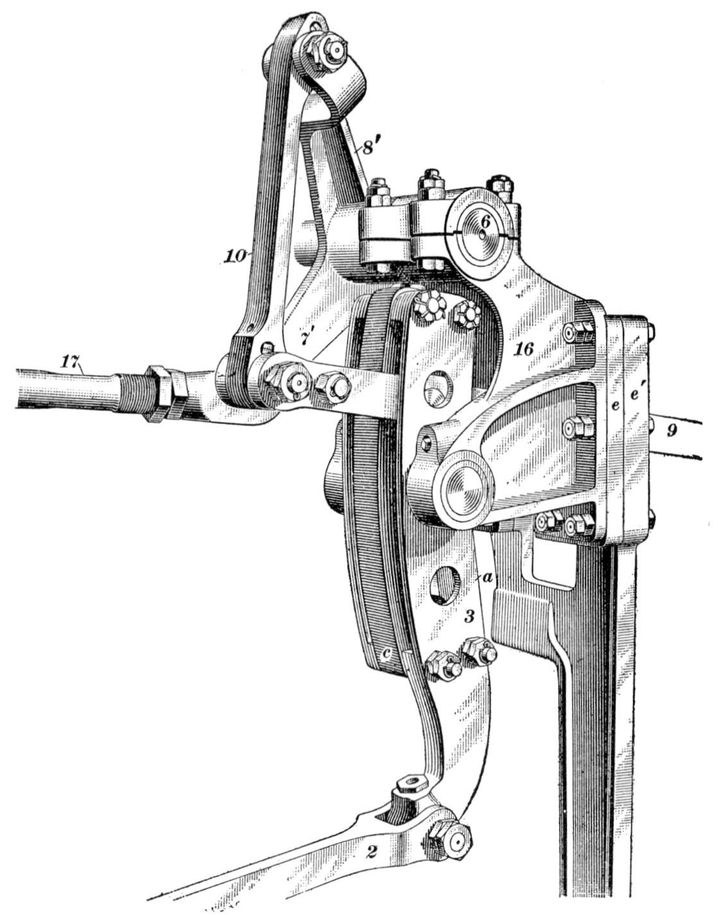

Fig. 12

combination-lever pin, in the lower end of the combination lever to oil the union-link pin, and at the rear of the union link to oil the bearing on the wristpin. Hard grease is generally used to lubricate the back end of the eccentric rod.

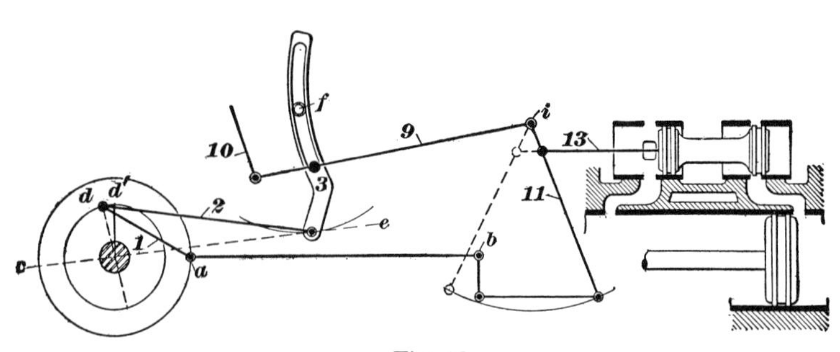

Fig. 13

31. Location of Radius-Rod Connection With Outside- and Inside-Admission Valves.—The radius rod is not connected to the combining lever at the same point with an outside-admission valve as with an inside-admission valve. In Fig. 14, with an outside-admission

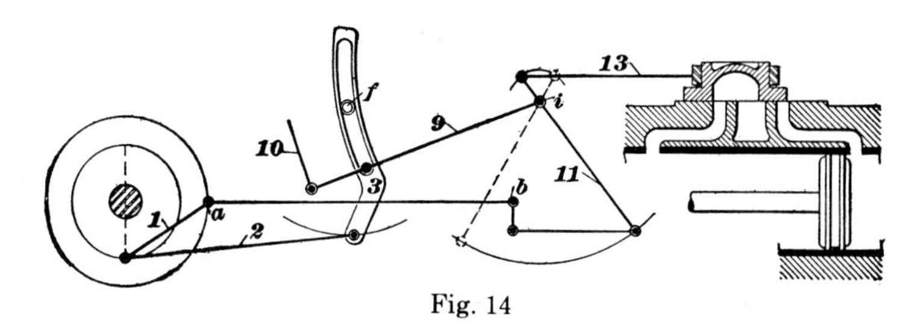

Fig. 14

valve, the radius rod *9* is connected to the combination lever *11* at *i* between the ends, and with an inside-admission valve, Fig. 15, the connection is made at the upper end. The reason is that an outside-admission valve during the lap-and-lead movement, moves in the direction opposite to the piston, while an inside-admission valve moves

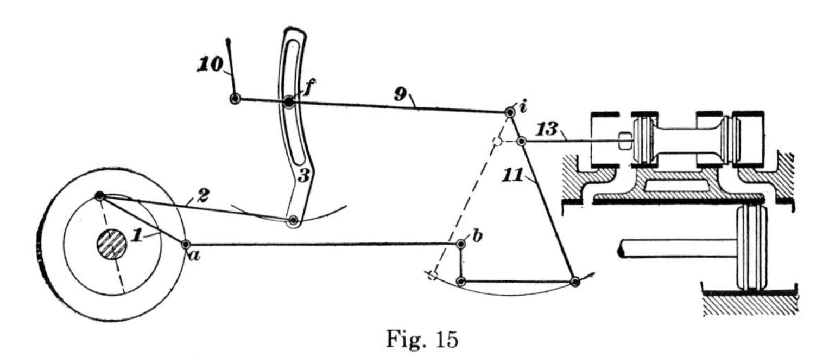

Fig. 15

in the same direction; therefore, the radius-rod connection, which acts as a fulcrum or fixed point, must be located accordingly. It will be noted that the lap-and-lead movement refers to the movement of the valve from the lead opening at one port to the lead opening at the other port.

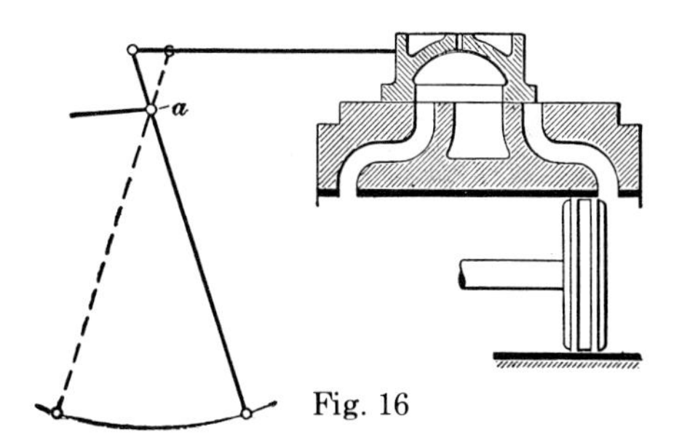

Fig. 16

In Fig. 16, when the piston moves from the front to the back end of the cylinder, the total movement of the valve has to be forwards or in a direction opposite to that of the piston in order to close the front steam port, and to open the back port the amount of the lead. Therefore, the fulcrum or fixed point of the combination lever must be between the ends, or at *a*, so as to cause the movement at the bottom of the lever to be converted into a movement in the reverse direction at the top.

32. In Fig. 17 with an inside-admission valve and the piston making the same movement, the whole movement of the valve, in order to close the front steam port and open the back port the amount of the lead, must be backwards or in the same direction as the piston. Therefore, the radius-rod connection or the fulcrum point of the combination lever must be located so as to cause the valve to move in the same direction as the piston, and this requires that the fulcrum point be located at the upper end of the lever at *a*.

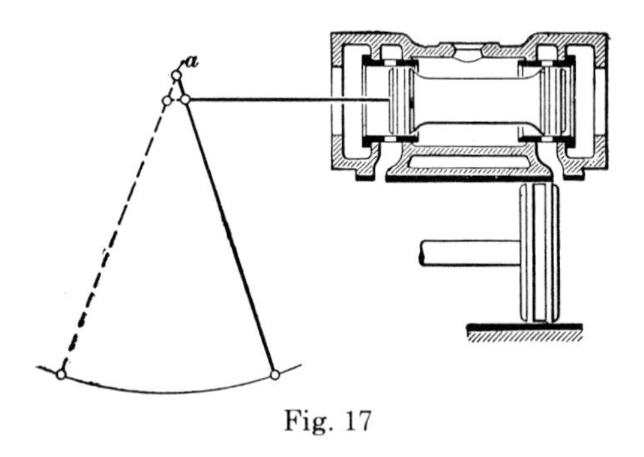

Fig. 17

The kind of valve used, if of the piston type, can then be identified by observing where the radius rod is connected to the combining lever. If between the ends, the valve is outside admission; if at the end, the valve is inside admission. A slide valve that is always outside admission can be readily identified by the shape of the steam chest.

DIFFERENT ARRANGEMENTS OF GEAR

33. There is very little difference in the arrangement of the Stephenson valve gear as applied to locomotives of different types. This does not apply to the Walschaert gear, as at certain points the design of the valve gear is influenced by the type and construction of the locomotive. The principal variations in the gear arrangement are found in the type of gear frame used to support the link, in the method of carrying the rear end of the radius rod, and in the arrangement of the arms of the reverse-shaft crank.

TYPES OF GEAR FRAMES

34. Reason for Different Types.—The type of gear frame used depends on the location of the guide yoke, for the reason that the complete gear frame, or at least one end of it, is always connected to the yoke. The location of the guide yoke varies because it has to be connected to the main frame of the engine, and hence must be placed at a point where the frame can be reached without interference from the driving wheel, and this fixes its position either in front or back of the first driving wheel.

When a locomotive has a two-wheel engine truck, the first driving wheel is so close to the cylinder that the guide yoke has to be placed at the extreme rear of the guide bars in order to clear the wheel and connect to the frame. In this case the type of gear frame shown in Fig. 9 is generally used because the guide yoke is far enough from the steam chest to permit a proper design of the parts of the gear forward of the yoke.

When a locomotive has a four-wheel truck, the rear-truck wheel is almost entirely behind the cylinder, and this brings the first driving wheel between the rear of the guide bars and the main frame. For that reason the guide yoke has to be placed ahead of the first driving wheel and quite close to the steam chest in order to connect to the frame. In this event the type of the gear frame shown in Fig. 9 cannot be used because the frame is too close to the steam chest to permit the valve gear to be properly designed. To bring the link about the same distance from the steam chest as before and thereby proportion properly the valve gear, the type of gear frame *16* shown in Fig. 11 is used. With this type of frame, the front end is bolted to the guide yoke p^1 to which the guide-yoke end r is also connected. The guide yoke is connected to the frame by the bracket q^1. The back end of the frame is connected to a cross-tie p laid across and connected to the main frames by a bracket q and extending far enough on each side to permit the gear frames to be connected to the outer ends. The link *3* is placed in the gear frame about midway between the ends. Therefore, the type of gear frame used depends largely on whether the engine has a two-wheel or a four-wheel engine truck.

TYPES OF RADIUS-ROD HANGERS

35. There are two ways of connecting the back end of the radius rod to the reverse-shaft crank, either by a hanger or by a radius rod lifter of the slip-block type. In Figs. 9 and 12 the radius rod *9* is connected to the arm *8′* of the reverse-shaft crank by a hanger *10*, and in Fig. 11 the rear end of the rod slides in a block *10* pivoted between the two arms *8′* of the reverse-shaft crank. One of these arms only can be seen, as the other lies directly behind it. The hanger arrangement is in more general use because the arrangement show in Fig. 11, being made of four parts, is difficult to maintain. The slip-block confines the end of the radius rod and the link block to a straight-line movement.

TYPES OF REVERSE-SHAFT CRANKS

36. Arrangement of Crank.—Unlike the Stephenson valve gear in which the link block is always in the upper part of the link in forward gear, the link block of the Walschaert gear, depending on the arrangement of the reverse-shaft crank may be in either the upper or the lower part of the link when the locomotive is running forwards. The link block and the radius rod may be in the upper part of the link in forward gear and in the bottom part in backward gear, as shown in Fig. 9; or, as shown in Figs. 11 and 12, these parts may be in the bottom of the link in forward gear and in the top of the link in backward gear. However, as with any type of locomotive valve gear, the locomotive runs forwards with the reverse lever in any position forward of the center notch of the quadrant, and backwards with the lever anywhere back of the center notch.

37. The position of the link block and the radius bar at the link depends on the arrangement of the reverse-shaft crank. In Fig. 9 the reverse-shaft crank arm *7′*, to which the reach rod connects, points upwards and a forward movement of the reverse lever and the reach rod places the link blocks and the back portion of the radius rods in the upper half of the links for forward gear as shown in Fig. 9 *(a)*, while a backward movement of the lever beyond the center notch of the quadrant places these parts in the lower half of the links for backward gear as shown in view *(c)*. In Fig. 12, the crank arm *7′* to which the reach rod is connected, points downwards, and when the reverse lever and the reach rod *17* are moved far enough ahead, the link block and the back end of the radius rod *9* move to the lower half of the link for forward gear, and to the upper half of the link as shown when the reverse lever is moved to the back part of the quadrant. With the reverse lever in the center of the quadrant, the link block and the radius bar are in the center of the link in both arrangements.

38. As will be fully explained later, the arrangement shown in Figs. 11 and 12 is to be preferred to that shown in Fig. 9. The reason for using the arrangement in Fig. 9 is that it is generally more convenient to connect the reach rod to an arm of the reverse-shaft crank which points upwards than to one which points downwards, as the connections at the power reverse gear and the crank arm are then more nearly in line.

The operation of the valve gear with the link block in the upper half of the link is confined generally to the type of gear frame shown in Fig. 9. With the type of gear frame shown in Fig. 11, the arm *7′* of the reverse-shaft crank to which the reach rod connects is more directly in line with the power-reverse gear when pointing upwards, and this places the link block in the lower part of the link in forward gear.

39. Disadvantages.—In road service where locomotives run mostly forwards, there are certain disadvantages connected with the operation of the gear when designed to work with the link block in the upper half of the link when the locomotive is running ahead. With switch engines, the position of the link block is immaterial, because these engines are worked as much in one gear as another. The disadvantages of having the link block in the top of the link when running forwards, are that there is a greater thrust and, therefore, more wear on the link trunnions and the bushings in the gear frame in which they work, than if the block was in the lower part of the link; and the slip of the link block, and hence the wear on the link, is greater. Besides, if the radius-rod hangers, hanger pins, or reverse-shaft arm should break when running at high speed, this radius rod would drop into backward gear, and if the reverse lever or reach rod should fail, both radius rods would fall into back gear. Under these conditions, the reversal of one or both valve gears would be certain to involve serious consequences.

40. Difference in Thrust on Link-Trunnion Bearings.—The purpose of Fig. 18 *(a)* and *(b)* is to show the difference in the thrust on the link fulcrum or trunnion bearings with the link block above and below the fulcrum point of the link in forward gear. It is assumed in *(a)* that the center of the link block is 9 inches below the center of the link fulcrum, and that it is 27 inches from the center of the link fulcrum to the center of the eccentric-rod pin hole. It is also assumed that it requires a force of 100

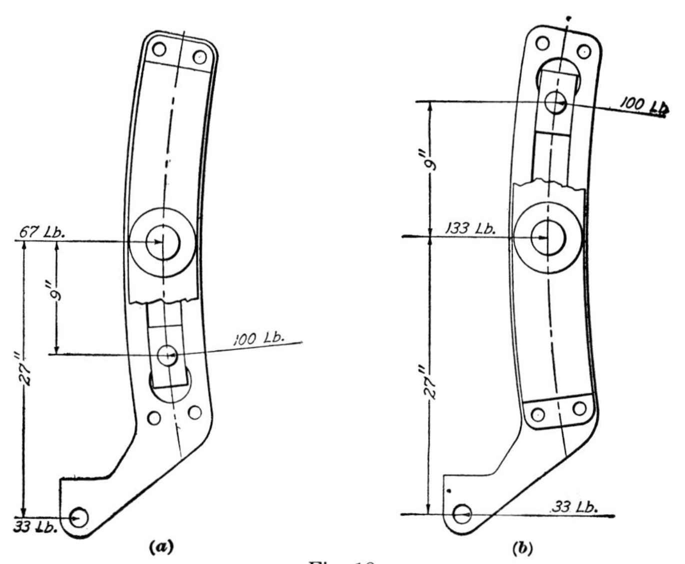

Fig. 18

pounds to move the valve. The thrust exerted by the eccentric rod will be 100x9/27 = 33 pounds, and the thrust on the link trunnion or fulcrum bearings will be 100—33 = 67 pounds.

In view *(b)* with the link block the same distance, or 9 inches, above the link fulcrum, the thrust exerted by the eccentric rod will be the same as before, or 100x9/27 = 33 pounds. The thrust on the link-trunnion

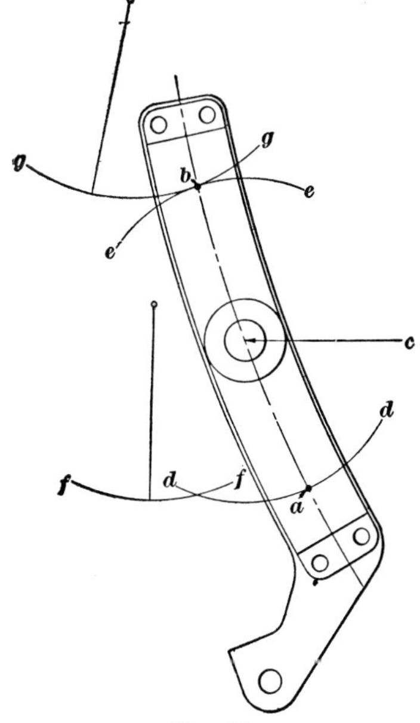

Fig. 19

bearings, however, will be the sum of the forces on the ends of the link, or 133 pounds. Hence, the wear on the link trunnions and the trunnion bushings is doubled when the gear is designed to operate with the link block in the top of the link in forward gear. This requires the link to be reground oftener than otherwise would be necessary.

41. Link-Block Slip.—The reason why the link block slips less in the link, with the block in the lower half of the link than in the upper half, is shown in Fig. 19. The center *a* of the link block follows the arc *d*, and the bottom radius-rod hanger pin follows the arc *f*, when the link swings on the fulcrum point *c*. The link block when in the top of the link follows the arc *e*, and the radius-rod pin the arc *g*. As the arc *d* conforma more nearly to the arc *f* than the arc *e* to the arc *g*, it is evident that the link block will slip less when in the lower part of the link and will wear less than when in the upper half.

ECCENTRIC CRANK

42. Development of Eccentric Crank.—With a locomotive, the crank action necessary for the rotation of the main driving wheel is obtained by setting the main crankpin in the wheel at the required distance from the center of the axle. It would, therefore, seem that the crank action required for the movement of the back end of the eccentric rod could be obtained in the same way by placing a pin in the driving wheel at the proper distance from the axle center. However, this cannot be done because two crankpins with a rod connecting to each cannot be placed on the same side of the wheel without one of the pins striking the other rod, and some other arrangement must be used.

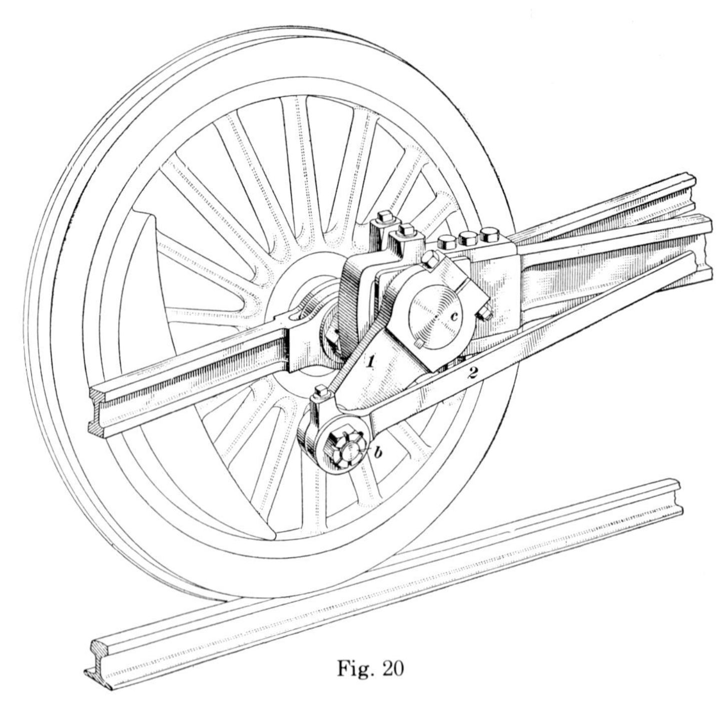

Fig. 20

For example, if the eccentric rod was connected to a crankpin shorter than the main crankpin, the main crankpin would strike the eccentric rod, when near the front dead center. The parts can be prevented from interfering by placing the eccentric crank *1* on the end of the main crankpin *c*, as in Fig. 20, and this method of driving the eccentric rod is used with outside valve gears like the Walschaert and the Baker, each of which requires an eccentric crank on the outside of the driving wheel.

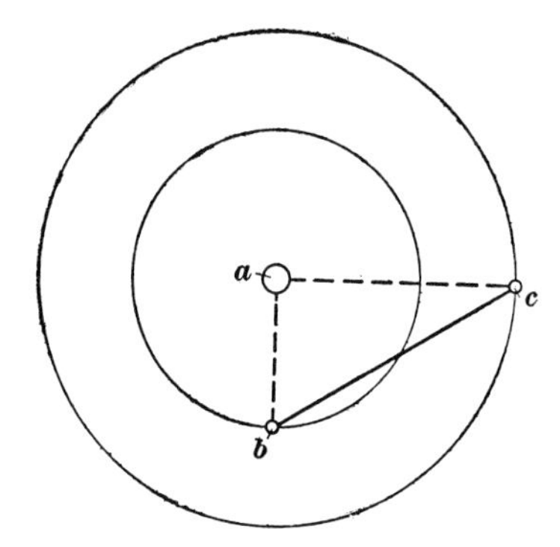

Fig. 21

43. Action of Eccentric Crank.—The action of the eccentric crank can be more readily understood from Fig. 21, in which *a* is the center of the axle, *c*, the center of the main crankpin, and *b*, the center of the eccentric crankpin. The eccentric crank is shown by the line *bc*, the main crank by the line *ac*, and the distance between the center of the axle and the center of the eccentric crankpin by the line *ab*.

The eccentric crank *bc* is merely a straight arm with a crankpin *b* on the outer end. The arm is of such a length that when properly set, the center of the crankpin *b* stands far enough from the center of the axle *a* to give the required movement to the eccentric rod when the driving wheel turns. The length of the crank formed by the eccentric crank *cb* is equal to *ab*, the distance between the center *a* of the axle and the center *b* of the eccentric crankpin. The throw of the eccentric crank or the maximum backward-and-forward movement imparted to the front end of the eccentric rod is equal to twice *ab*. If it were practicable to do so, the eccentric crank *bc* could be dispensed with, and the same effect obtained by inserting a crankpin in the main driver at the point *b*.

44. If it is desired to apply terms with the same meaning to the eccentric-rod connection at the rear end as to the main-rod connection, it will be found that the term *eccentric crank* as heretofore used is incorrect. Considering the main crankpin, the term *main crank* means the distance between the center of the axle and the center of the main pin or, in other words, the term *crank* means the amount the pin is out of center. Reasoning in the same way, the term *eccentric crank* would mean the distance *ab*, Fig. 21, between the center of the pin *b* on which the eccentric rod is placed and the center of the axle *a*, and the part connected to the main pin would be properly called the *eccentric-crank arm*. However, usage has definitely fixed the use of the term *eccentric crank* when identifying the part connected to the main crankpin, and to prevent confusion, the same practice will be followed in this Section.

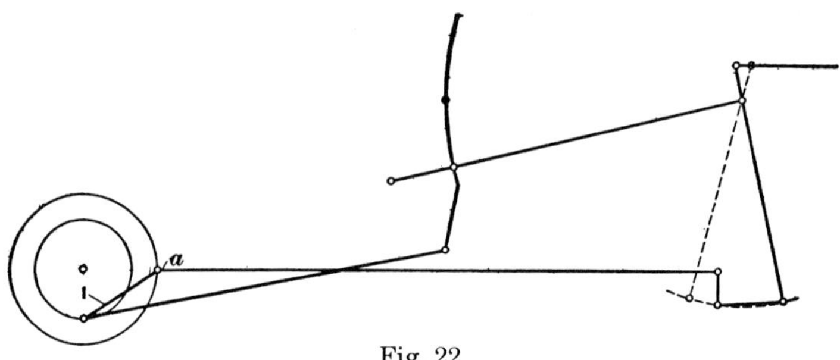

Fig. 22

It will be found more convenient to consider a line between the center of the axle and the center of the eccentric crankpin as the eccentric crank if the relative position of the main crank and eccentric crank is measured by an angle. The action of the eccentric crank in imparting movement to the eccentric rod and its position with respect to the main pin can also be more readily understood if the above dimension is considered.

POSITION OF ECCENTRIC CRANK

45. Direct and Indirect Motion.—The motion of a valve gear or the gear is said to be *direct* when the eccentric rod moves in the same general direction as the valve. The motion is indirect when these parts move in opposite directions. The Walschaert valve gear is direct with the engine moving with the link block in the lower half of the link because the movement of the eccentric rod can then be delivered directly to the radius rod and to the valve. The gear is indirect with the link block in the upper half of the link because, as the link is pivoted at the middle, the movement of the eccentric rod and the lower end of the link is necessarily in a direction opposite to that of the upper end of the link and the radius rod.

The Stephenson valve gear is either direct or indirect in both gears, but with a gear like the Walschaert with the link swinging on a pivot, the movement, if direct in one gear, must be indirect in the other. The Walschaert valve gear is usually direct in forward gear and hence indirect in backward gear, although the reverse applies to the arrangement in Fig. 9.

46. Center Line of Motion.—The center line of motion with the Walschaert valve gear is a line *ce,* Fig. 13, drawn through the center of the axle and the center of the eccentric rod pin at the link foot. The eccentric crank is always set with respect to the center line of motion and not with respect to the main crankpin.

If the link foot were extended down until the center of the eccentric rod pin was on the line drawn through the center of the main axle and the center of the cylinder, or the center line of the engine, the center line of motion, and the center line of the engine would coincide. In this event the eccentric crank would occupy the same position with respect to the main crankpin as it would to the center line of motion.

However, it would be impracticable to lengthen the link foot to such an extent, as it would require an eccentric crank of excessive throw for the link to swing far enough to give the required movement to the link block and the radius rod. With the link foot above the center line of the engine, the center of the eccentric crankpin *d* with the arrangement in Fig. 13 is slightly behind the top quarter position *d'*, when the main crankpin is on the front dead center.

The center of the eccentric crankpin comes very near the quarter positions with the main crankpin on the dead center, and hence is about one-quarter of a turn from the main pin. Therefore, the position of the eccentric crank will be identified by stating its location with respect to the main pin rather than to the center line of motion.

47. Conditions on Which Position Depends.—The position of the eccentric crank depends on whether the valve is an outside- or inside-admission valve, and whether the motion is direct or indirect.

48. Outside-Admission Valve and Direct Motion.—With an outside-admission valve and direct motion in forward gear, Fig. 22, the eccentric crank *1* is about one-quarter of a turn ahead of the main crankpin *a* when the locomotive is running forwards, and the same distance behind it when the locomotive is running backwards. With the main pin on the front dead center and the engine moving forwards, the eccentric crank will move the valve back or to the left, and this is the valve movement desired. In Fig. 14 is shown the eccentric crank in the same position as in Fig. 22.

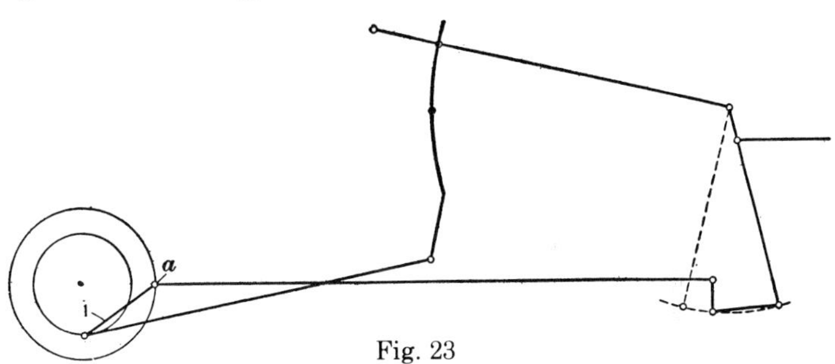

Fig. 23

49. Inside-Admission Valve and Indirect Motion.—With an inside-admission valve and indirect motion in forward gear, Fig. 23, the eccentric crank is set in exactly the same position as in Fig. 22.

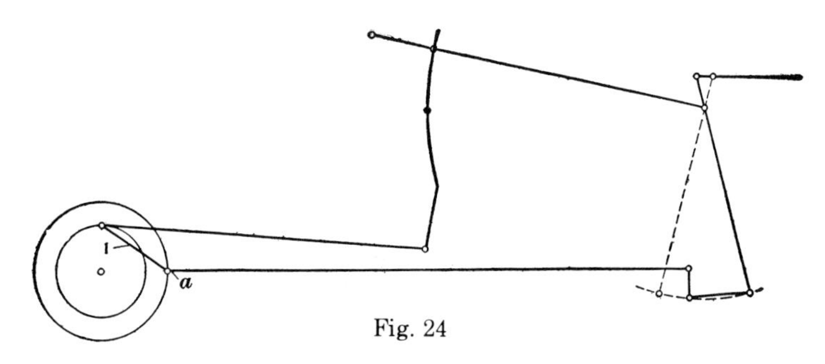

Fig. 24

50. Outside-Admission Valve and Indirect Motion.—With an outside-admission valve and indirect motion in forward gear, Fig. 24, the eccentric crank *1* is about one-quarter of a turn behind the main crankpin *a* when the locomotive is running forwards and the same amount ahead of the pin when running backwards. The eccentric crank when so set moves the eccentric rod ahead when the main pin turns forward from the front dead center, and this causes the top of the link and the radius rod to move to the left and pull the valve back or in the direction required.

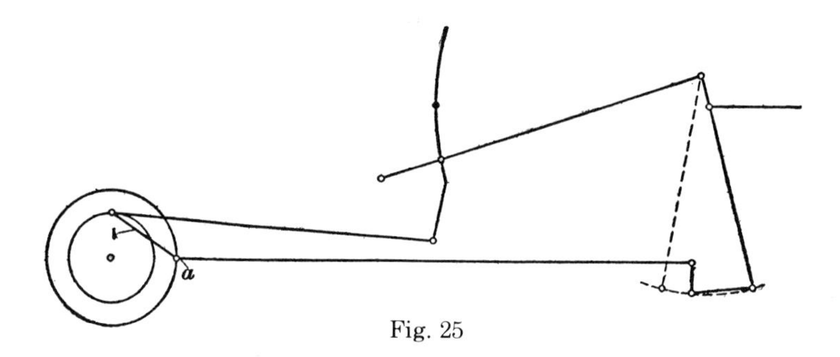

Fig. 25

51. Inside-Admission Valve and Direct Motion.—With an inside-admission valve and direct motion, Fig. 25, the eccentric crank is set in exactly the same position as with an outside-admission valve and indirect motion, Fig. 24. In Fig. 11 is shown a gear arranged for direct motion and inside admission. It will be noted that the center of the eccentric crankpin is either about one-quarter turn behind the main pin or about the same amount ahead of it, and that the difference between the positions is about one-half a turn of this crankpin.

LINK ARC AND LINK RADIUS

52. The link arc is a curved line drawn through the middle of the slot in the centerpiece of the link and forms a part of a circle. The radius of the link is the length of the line taken to scribe the link arc, or it is the radius of a circle of which the link arc is a part. The foregoing terms are illustrated in Fig. 26. Thus the link arc *ab* may be assumed to be a part of the circle *abc,* and *de* to be the radius of the link or the radius of the circle of which the arc *ab* is a part. If enough links were put together, the link arcs would form a circle of which the radius rod would be the radius or half the diameter. The point *e* is the center of the link arc, or it is the center of the circle from which the link arc is described. The center of the link arc of the Walschaert link is taken at the center of the pin in the front end of the radius rod, and the link radius or the length of the line taken to describe the link arc is the distance between the center of the radius-rod pin and the center of the link-block pin.

53. Constant Lead.—The link is said to be in central position when the center of the link arc falls on the center of the radius-rod connection to the combination lever, and the link occupies this position when the engine is on either dead center. Therefore, the reason that the lead remains unchanged or constant is that the link is in central position on the dead centers, and the center of the link-block pin when the radius rod is raised and lowered conforms with or follows exactly the curvature of the link arc.

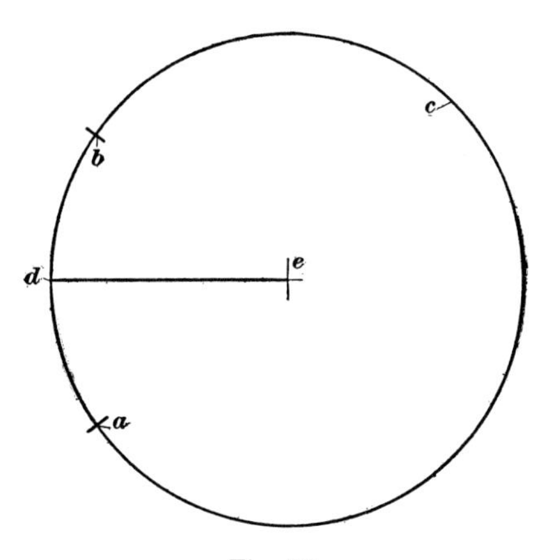

Fig. 26

The link arc and the arc described by the link-block pin do not coincide except on dead centers and the radius rod and valve will move with the reverse lever because the link block will be constrained to follow the slot in the link. Any movement of the valve on dead centers when the reverse lever is moved does not indicate that the link arc is struck with an improper radius, but rather that the link is not in central position or that it does not occupy the proper position for the link block to move without moving the front end of the radius rod. For example, if the eccentric rod is not the proper length the link will be slightly inclined on the dead centers, and the front end of the rod and, therefore, the valve will move when the reverse lever is shifted.

54. Curvature of Link With Walschaert and Stephenson Gears.—The curvature of the link with the Walschaert valve gear is in the direction opposite to that of the link of the Stephenson valve gear, owing to the position of the point from which the link arc is struck, or to the location of the center of the link arc.

With the Walschaert valve gear, the center of the link arc is in front of the link or at the center of the radius-rod pin, hence the link arc curves toward its center or forwards as in Fig. 26, where the arc *ab* curves toward the center *e*.

With the Stephenson valve gear, the center of the link arc is taken at the center of the eccentric and is behind the link, and the link radius is equal to the distance between the center of the eccentric and the center of the link. The link arc curves toward the link-arc center, therefore the link curves in the direction opposite to that of the Walschaert link.

55. Variable Preadmission With Constant Lead.—The term *preadmission* refers to the period or interval of time during which steam is admitted to one end of the cylinder while the piston is still moving toward that end of the cylinder. The term *point of preadmission* refers to the position of the valve when preadmission is about to begin, or at the instant the valve is about to open the port. The fact that the lead is constant with the Walschaert valve gear does not mean that the point of preadmission is constant or that it occurs with the piston in the same position in the cylinder for all positions of the reverse lever. The point of preadmission is a valve event and is influenced by the movement of the reverse lever in the same way as other valve events, such as cut-off and release, etc. The valve events all occur earlier in the stroke of the piston when the lever is drawn up, and later in the stroke as the lever is dropped down. Therefore while the lead is always the same with the piston at the end of the stroke, yet the position of the piston when the steam begins to enter the cylinder varies, or the point of preadmission varies, according to the position of the reverse lever.

CROSSED LEAD

56. Purpose of Lead.—The chief purpose of lead is to increase the port opening at running cut-offs. At maximum cut-off the port opening is ample and lead is then of no particular advantage; rather it is a handicap, because it hastens the cut-off and thus impairs the ability of the locomotive to start heavy trains. Passenger locomotives require good lifting power so as to start trains smoothly; also, as these locomotives are run at comparatively short cut-offs, the port opening should be ample to maintain speed. For these reasons it is sometimes the practice to design the Walschaert valve gear with crossed lead. In such a case the minimum lead is obtained in full gear, thus insuring a longer maximum cut-off and hence more starting power; a greater lead is secured at short cut-offs, thereby giving ample port opening. Crossed lead also delays the point of closure, thereby reducing the compression and hence increasing the mean effective pressure.

57. Classification of Lead.—Lead may be either constant, variable, or crossed. Constant lead is the same in all cut-offs, both forward and backward motions. Variable lead is the least in full forward and in full backward gears and increases to the maximum with the lever in the center. The Stephenson valve gear is a variable lead gear. Crossed lead is the least in full forward gear and the greatest in full backward gear. Crossed lead with the Walschaert and Baker valve gears is sometimes incorrectly termed variable lead.

58. Design for Crossed Lead.—The Walschaert valve gear is usually designed to give a constant lead; such a design gives a constant lead at all cut-offs in both motions. Crossed lead can only be obtained by varying from the ordinary design, or it is obtained by offsetting the eccentric crank from its right-angle position the amount necessary to change the lead. However, it must not be assumed that the valve travel will be the same as before, even though the crank circle is the same; the travel will be greater in one gear than in the other. The average of the valve travels in both gears will be very close to the specified valve travel.

59. In Fig. 27 the top line indicates the original position of the eccentric crank for direct motion in forward gear, and inside admission valves; the other lines indicate the position of the crank when making the change to crossed lead. Moving the center of the eccentric crank from *a* to *b* would decrease the lead in forward gear to the required amount, but the inner eccentric crank circle then scribed by the crank would not give the proper crank circle, as the crank would be too short. To preserve the original circle, the center of the crank must follow the large eccentric-crank circle. This requires the center of the crank to be located at *c*, which

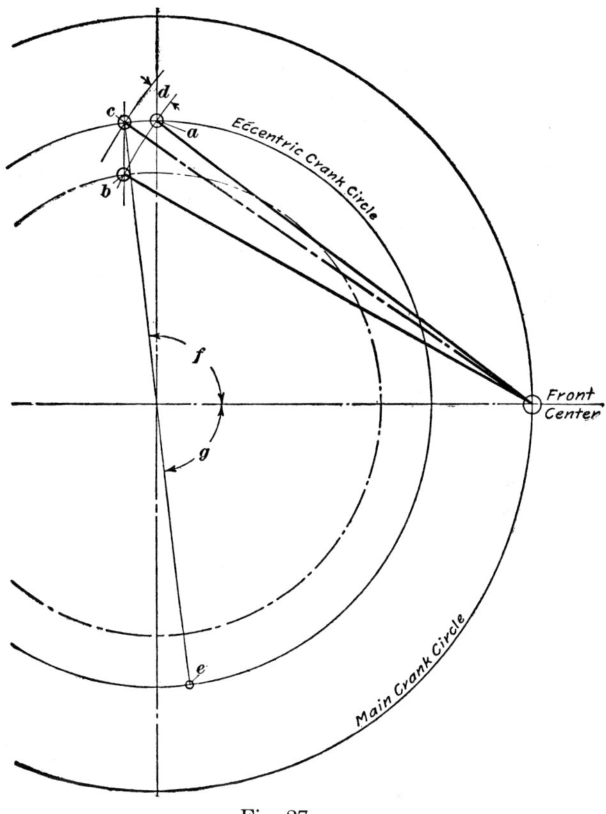

Fig. 27

in turn necessitates a crank of a greater length than the original one by an amount equal to *d*.

With an outside admission valve and direct motion in forward gear, the eccentric crank will have to be shortened because a decrease in the lead requires the crank to be moved away from the center of the axle as in Fig. 28. The eccentric crank, after its center has been moved from its original position at *a* to the point *b,* must be shortened an amount equal to *d* and the center located at the point *c* in order to bring it on its original circle.

The gear is never set for crossed lead until the correct length of the eccentric crank has been determined in the drafting office. Then, when setting the valves, all that is necessary is to locate the crank the proper distance from the center of the axle to give the specified lead in full gear or the correct eccentric-crank circle. A gauge especially adapted for this purpose is generally provided in railroad shops.

60. Increase in Lead.—If, as with crossed lead, the eccentric crank is moved to decrease the lead in full forward gear, the lead will continue to increase until the lever is in full back gear, where it reaches its maximum. The heavy lead in full back gear is comparatively unimportant with passenger locomotives because they are seldom run in back gear.

Let it be assumed that the valve gear was originally designed for a constant lead of ¼ inch; that is, that the lap-and-lead lever was proportioned to give this lead, and the required changes were made to obtain a crossed lead of 1/16 inch in full forward gear. The center of the link-block pin will then be drawn back 1/4 minus 1/16, or 3/16 inch. The lead will now increase from 1/16 inch in full forward gear to 1/4 inch in mid-gear; the mid-gear lead will always be the amount that the lap-and-lead lever was designed to give. As the link-block pin is 3/16 inch to the rear of its original position when in the bottom of the link, the pin will be the same amount forwards of this position when in the top of the link. Hence, the lead in full back gear will be equal to 1/4 inch, the mid-gear lead, plus 3/16 inch, or a total of 7/16 inch.

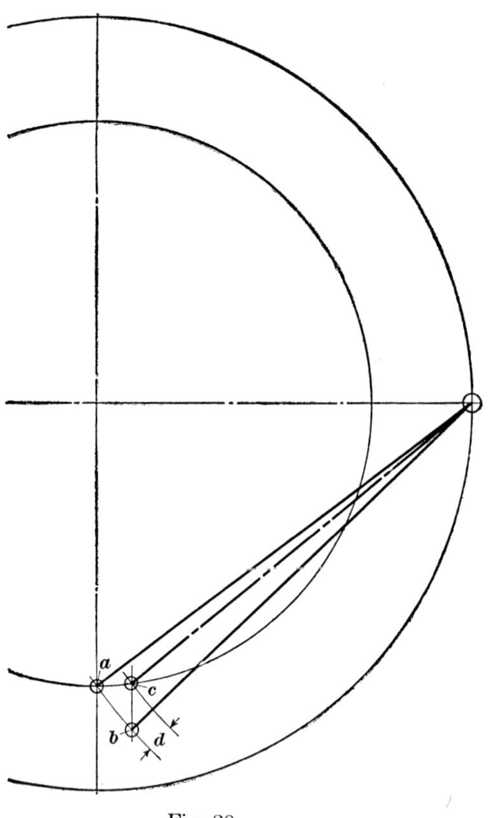

Fig. 28

61. Unequal Valve Travel.—Although with crossed lead the diameter of the eccentric-crank circle is unchanged, yet the valve travel will be less in forward gear than in back gear. This can be more readily understood by assuming a lap-and-lead crank as heretofore. Thus, in Fig. 27, the angle between the eccentric crank and the lap-and-lead crank in forward gear is equal to *f*; the angle between these cranks in back gear is less or is equal to *g*. Of course, reversing implies that the eccentric crank has been shifted 180 degrees, or from *c* to *e*. With a greater angle between the cranks, the action of the lap-and-lead crank in neutralizing the action of the eccentric crank on the valve occurs earlier in its travel than when the angle is less; this results in a lesser valve travel and a smaller steam-port opening in forward gear. The smaller angle between the cranks in back gear causes the lap-and-lead crank to begin to neutralize the action of the eccentric crank later, and hence the valve travel is increased. Therefore, the shorter travel in forward gear and a greater travel in back gear is due to the angle between the cranks being greater in the former case than in the latter. As already explained, the action of the lap-and-lead crank is equivalent to that of the combination lever. The foregoing applies to inside admission and direct motion and outside admission and indirect motion as in Fig. 27. With the conditions shown in Fig. 28, the reverse will hold true.

RADIUS-ROD MOVEMENT AND VALVE TRAVEL

62. Movement of Radius Rod.—The movement of the front end of the radius rod is never equal to the travel of the valve; this movement exceeds the valve travel for an inside admission valve and is less than the valve travel for an outside admission valve. In other words, the lap-and-lead lever with an inside admission valve reduces the movement that the radius rod transmits to the valve; this lever with an outside admission valve increases the movement. Therefore, with the same link swing, the

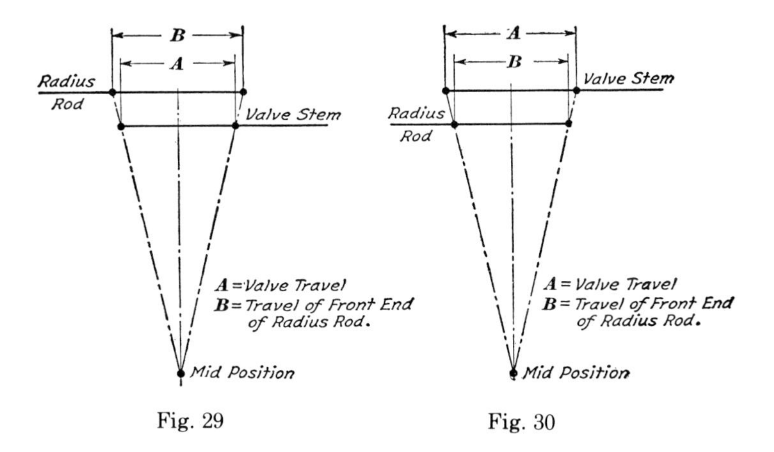

Fig. 29

Fig. 30

valve travel is longer with an outside admission valve than with an inside admission valve. The foregoing is illustrated in Figs. 29 and 30. The valve movement as well as the movement of the front end of the radius rod when the reverse lever is shifted from one corner of the quadrant to the other with the crosshead in central position, is as shown. The movement of the radius rod exceeds the movement of the valve for inside admission and is less than the valve movement for outside admission.

63. Changing the Lead.—To increase the lead it is necessary to make the short end of the combination lever longer; to decrease the lead, this end must be made shorter.

The following rule gives the length of the short end of the lever for either a decrease or an increase in lead.

Rule.—*First add 1/32 inch to the lap and to the lead desired to allow for lost motion; next multiply this lap and lead by the long end of the lever for outside admission and by the complete lever for inside admission, then divide by one-half the stroke.*

EXAMPLE.—Let it be assumed that it is required to decrease the lead of the valve gear shown in Fig. 31 from 3/8 to 1/4 inch. Find the length of the short end of the lever.

SOLUTION.—From the rule $1\frac{1}{2}+\frac{1}{4}+\frac{1}{32}=1\frac{25}{32}$. Then $\frac{1\frac{25}{32} \times 31}{16} = 3.45$. To convert .45 to thirty-seconds, multiply by 32, .45x32=14.40 or 14/32 or 7/16. The length of the short end is then 3 7/16 inch, or this end must be shortened 1/4 inch.

If it is found that changing the short end of the lever raises or lowers the radius rod too much, it may be better to apportion the change between the two ends. This is done by assuming how much the long end of the lever can be changed without affecting the action of the gear materially, and then use this new length of the long end in the above rule instead of the original length. It should be noted, when following the last plan, that the long end is to be shortened when calculating an increase in lead and lengthened when calculating a decrease.

Any alteration requires a new combination lever, and a change of this nature must, of course, be sanctioned by the proper authority.

DIMENSIONS OF WALSCHAERT VALVE GEAR

64. Copy of Print.—A copy of a print that shows the dimensions of all of the parts of the Walschaert valve gear designed to give a long valve travel is shown in Fig. 31. The size of the cylinders, the valve travel, the steam lap, and the lead are also indicated. The distance between the center of the main driving axle and the vertical center of the cylinders is 185 inches.

65. Advantages of Long Valve Travel.—The advantages of a long valve travel over a shorter valve travel may be enumerated as follows: *(a)* a longer maximum cut-off, thereby increasing the starting ability of the locomotive; *(b)* a wider steam-port opening at all cut-offs, hence insuring less wire drawing of steam and a high cylinder pressure with large cylinders; *(c)* a wider steam lap, hence greater expansion; *(d)* a wider exhaust port opening, thereby decreasing the back pressure; *(e)* increased lead, thereby providing more port opening at short cut-offs without increasing the preadmission.

66. Factors Limiting Valve Travel.—One of the factors that limit the valve travel with the Walschaert gear is the swing of the link as governed by the diameter of the eccentric-crank circle. The swing of the link can be increased by increasing the throw of the eccentric crank until the inclination of the link becomes excessive. With an excessive swing the link assumes more of a horizontal position and the slot comes more in line with the radius rod. At the maximum angle the action of the link in moving the link block is greatly reduced; in fact, the link will merely rotate the block on the pin. The link slot should be at right angles to the radius rod to be the most effective, the effectiveness of the link decreases as it varies from this angle. Therefore, the valve travel will not increase once the link swing reaches a certain amount. The greatest inclination of the link or its swing should not exceed 30 degrees from the vertical.

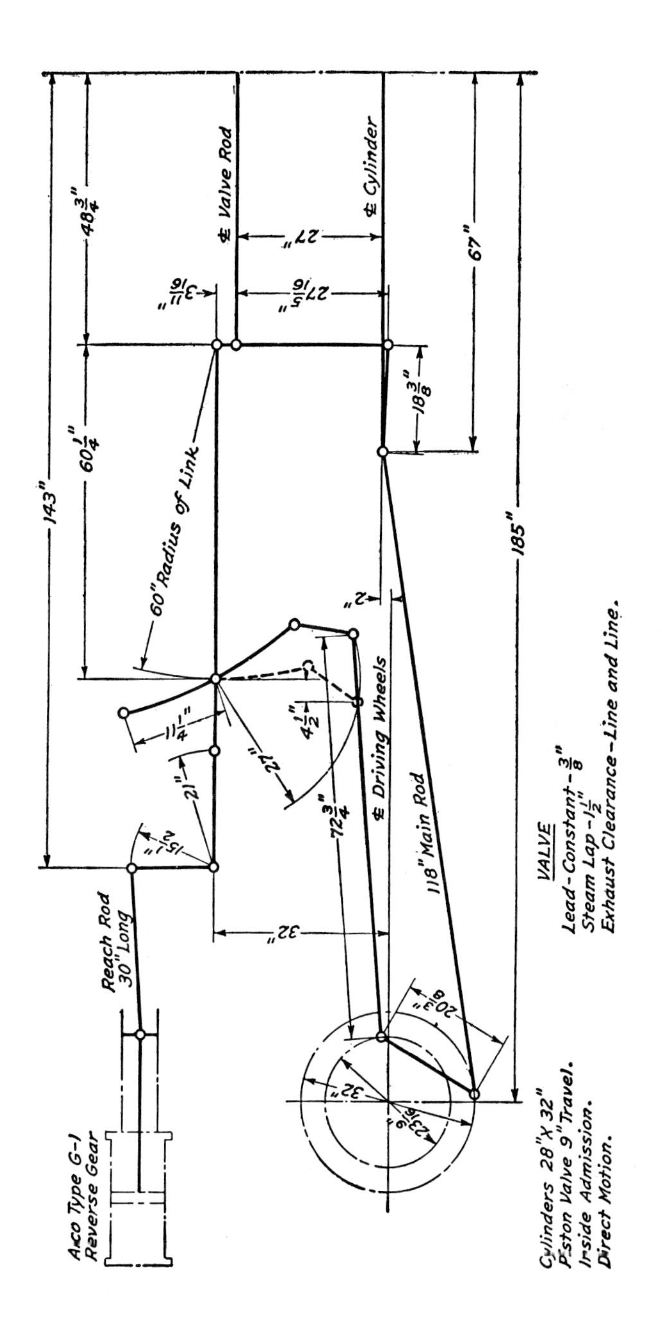

Fig. 31

WALSCHAERT VALVE GEAR
(PART 1)

EXAMINATION QUESTIONS

Notice to Students.—*Study the Instruction Paper thoroughly before you attempt to answer these questions.* **Read each question carefully and be sure you understand it;** *then write the best answer you can. When your answers are completed, examine them closely, correct all the errors you can find, and* **see that every question is answered.**

(1) Why is the curvature of the link in the reverse direction to that of the Stephenson link?

(2) Name the advantages of the Walschaert valve gear.

(3) (a) At what point is the radius rod connected to the combination lever with an outside-admission valve? (b) Where with an inside-admission valve?

(4) Define *direct* and *indirect motion.*

(5) What is the position of the eccentric crank with an inside-admission valve and direct motion?

(6) What is the link arc?

(7) Explain the general operation of the valve gear.

(8) What two general methods are used to connect the radius rod to the reverse-shaft crank?

(9) Why must the radius rod be connected to the combination lever at a different point with an outside than with an inside-admission valve?

(10) Why cannot the Stephenson valve gear be applied successfully to modern locomotives?

(11) On what does the type of gear frame used depend?

(12) Name the parts of the Walschaert valve gear.

(13) Where are the principal variations in the arrangement of the valve gear found?

(14) Generally speaking, why does a different type of gear frame have to be used with a locomotive with a two-wheel engine truck than with a four-wheel truck?

(15) What are the disadvantages of arranging the gear to work with the link block in the upper half of the link in forward gear?

(16) What movement is imparted to the valve by the combination lever for each stroke of the crosshead with the reverse lever in midgear?

(17) Name the oiling points in the valve gear.

(18) What change in the positions of the radius rods and link blocks is brought about by the movement of the reverse lever?

(19) In what position of the reverse lever is the eccentric crank prevented from imparting movement to the valve?

(20) What is the position of the eccentric crank in relation to the main crankpin with an outside-admission valve and direct motion?

Walschaert Valve Gear
Part 2

By
J. W. Harding

4003B Printed in U.S.A. Edition 3

1946 Edition

WALSCHAERT VALVE GEAR
(PART 2)

ARRANGEMENT AND OPERATION (Continued)

DETAILS OF GEAR

GEAR FRAME

1. Purpose.—The purpose of the gear frame is to carry or support the link and one end of the reverse shaft. The gear frame is sometimes called the *link support.*

2. Reason for Different Types.—Gear frames are of two general types. The gear frame *16,* Fig. 1, is generally used with freight locomotives. With these locomotives, on account of the wheel arrangement, the guide yoke *e* has to be located between the first and second driving wheels in order to connect to the frame, and when so placed it provides a support for the gear frame at a point that permits a proper design of gear.

In Fig. 2, with a passenger locomotive, the guide yoke *e* has to be placed about forward of the first driving wheel in order to connect to the frame, and this brings the type of gear frame shown in Fig. 1 too close to the valve to permit a valve gear of proper design. Therefore, the link has to be placed farther to the rear, and hence the type of gear frame shown in Fig. 2 is used.

3. Construction.—The gear frame *16,* Fig. 1, is a steel casting bolted to the guide yoke *e,* or it may be cast in one piece with the yoke. The upper end of the guide yoke is rectangular as shown and extends across the frames to the other side where the arrangement of guide bars and gear frame is the same. The guide yoke is connected to the frame by knees, or brackets, *e′*. The construction of the type of gear frame shown in Fig. 2 varies to such an extent that it is impossible to cover all the designs. Therefore, only two fairly representative designs will be considered.

In Fig. 2 the frame consists of two link supports *19* between which the link swings, each support being provided with a bearing in which the link trunnions turn. The front ends of the link supports are bolted to the part of the guide yoke *e* that extends across the frames, and the back ends are bolted to a cross-tie or cross-bearer *20.* The guide yoke and the cross-tie are connected to the frame by knees, or brackets, *e′* and *21.* The reverse shaft is carried in a bracket *16* which is bolted to the inside link support *19.*

4. In Fig. 3 is shown a gear frame of one-piece construction. With this design the two side pieces *16* are welded to end pieces, the whole frame forming a single rectangular part. The front end of the frame is bolted to the part p^1 of the guide yoke *r* that extends across the frames to the other side of the locomotive and supports the front end of the gear frame on the other side. The rear end of the frame is bolted to the cross-bearer *p,* which also extends across the engine and supports the rear end of the gear frame on the other side. The cross-bearer and the guide yoke are connected to each of the side frames by knees, or brackets, q and q^1.

A view of the rear part of the gear frame in Fig. 3 which shows the one-piece construction, is given in Fig. 4. The link trunnions *f,* one on each side of the link, are carried in bearings in the gear frame. Each trunnion

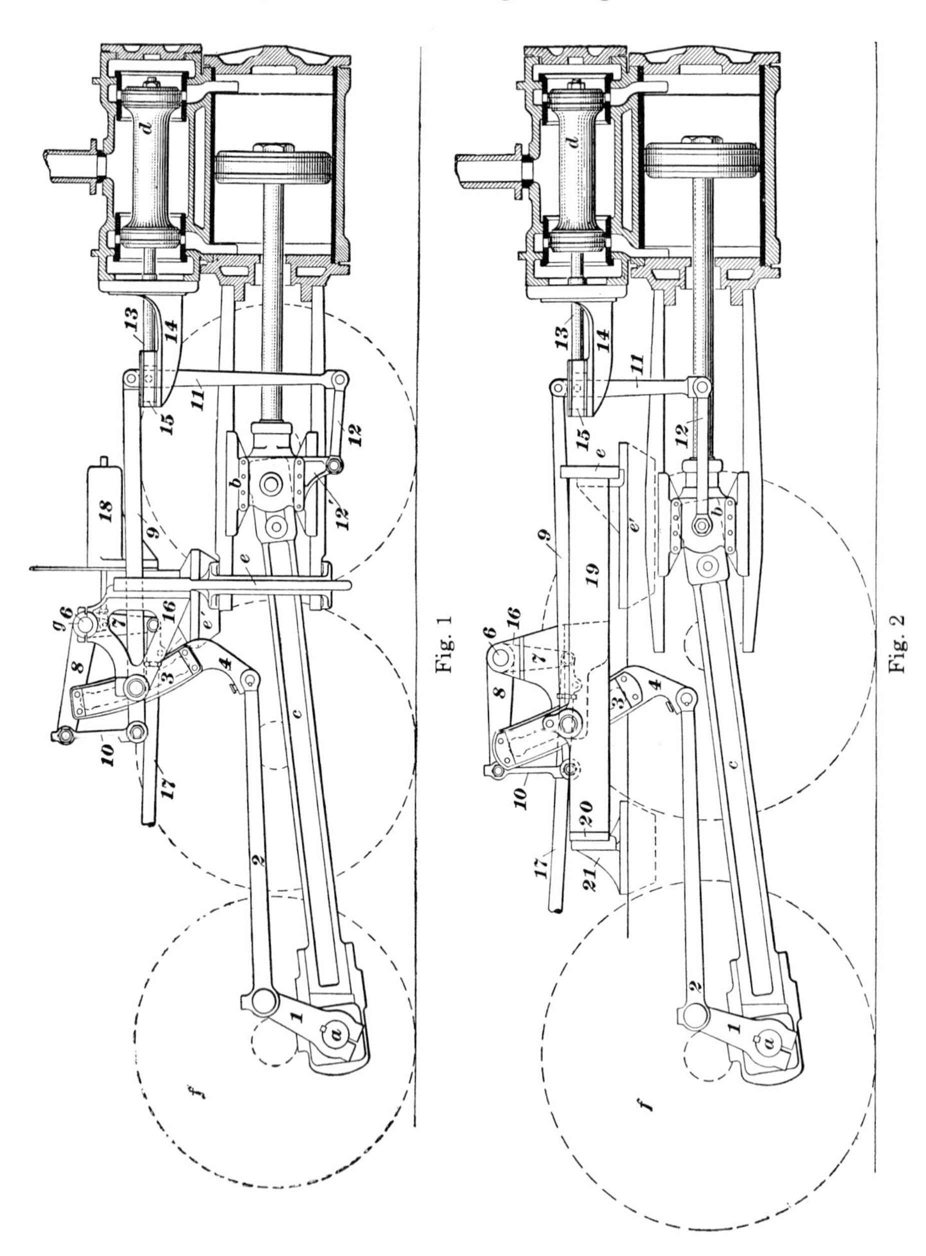

Fig. 1

Fig. 2

is lubricated from an oil cellar in the frame which is packed with waste by removing the plug shown. Oil is applied to the cellar through an oil hole in the top of the frame.

The reverse shaft *6* is carried in a box *e* bolted to the inner side of the frame. The reverse-shaft crank *7*, with the arms *7′* and *8′* is made in one piece and is keyed to the end of the reverse shaft *6*. The radius-rod lifter *10* is carried on the trunnions *c* between the arms *8′*.

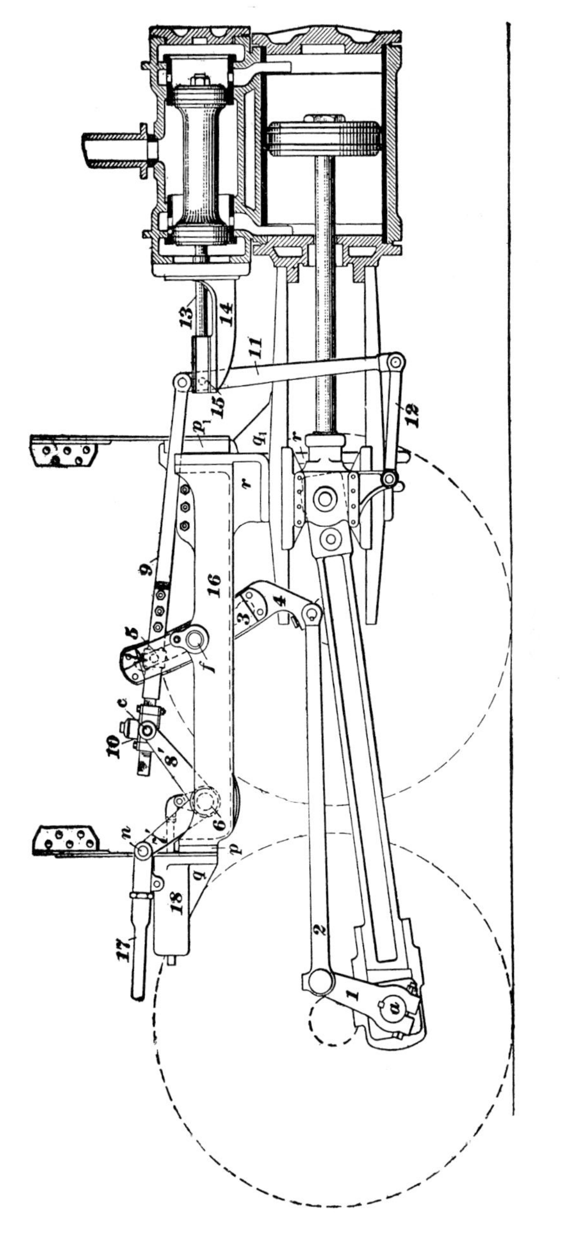

Fig. 3

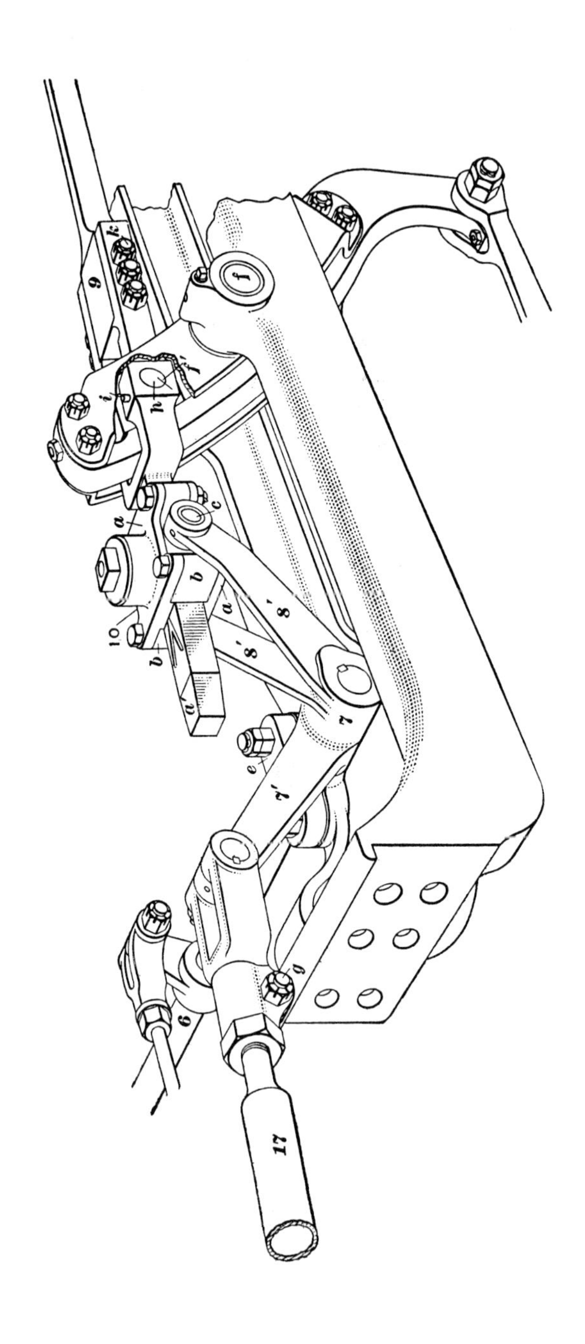

Fig. 4

5. Purpose.—The purpose of the cross-ties is to connect the gear frames to the main frames of the locomotive. They also serve as frame braces and as a means of connecting the boiler to the frames.

The primary purpose of the guide yoke is to provide a rigid part to which the rear end of the guide bars can be connected. In addition to forming a convenient support for the gear frame, the guide yoke also acts as a frame brace and as a means of connecting the boiler to the frames.

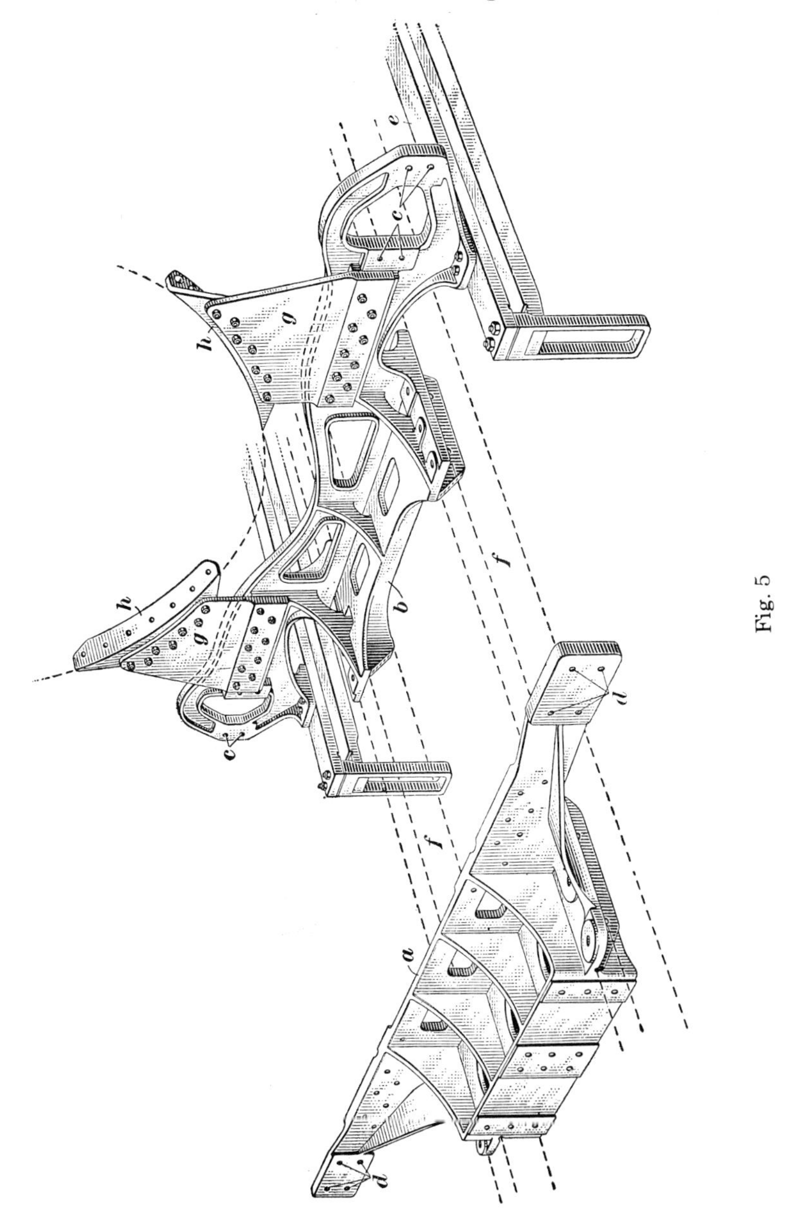

Fig. 5

6. Description.—The types of cross-ties and the guide yoke used vary so much that one representative design only will be shown. In Fig. 5 is shown a design of cross-tie *a* and guide yoke *b* used with the type of gear frame in Figs. 2 and 3 that differs in some respects from those given in these illustrations. The front end of the gear frame bolts to the guide yoke at *c* and the rear end to the cross-tie at *d*.

The guide yoke not only supports the rear end of the guide bars *e*, but it also acts as a frame brace because it is bolted to the frames *f*. The guide yoke also serves to connect the boiler to the frames, the connection between the two being made by the waist sheets *g*, which are bolted at their upper ends to the boiler angle pieces *h*. The cross-tie *a* in addition to supporting the gear frame also acts as a frame brace and as a means of connecting the boiler to the frames by waist sheets similar to those at the guide yokes.

ECCENTRIC CRANK

7. Purpose.—The purpose of the eccentric crank is to give the valve, with engine in full gear, a movement equal to the specified valve travel. It also serves as a collar to hold the main rod on the main crankpin. The eccentric crank consists of a short arm rigidly secured to the outer end of the main crankpin from which it obtains its movement. It is set in such a position that the end moves in a path having a diameter smaller than

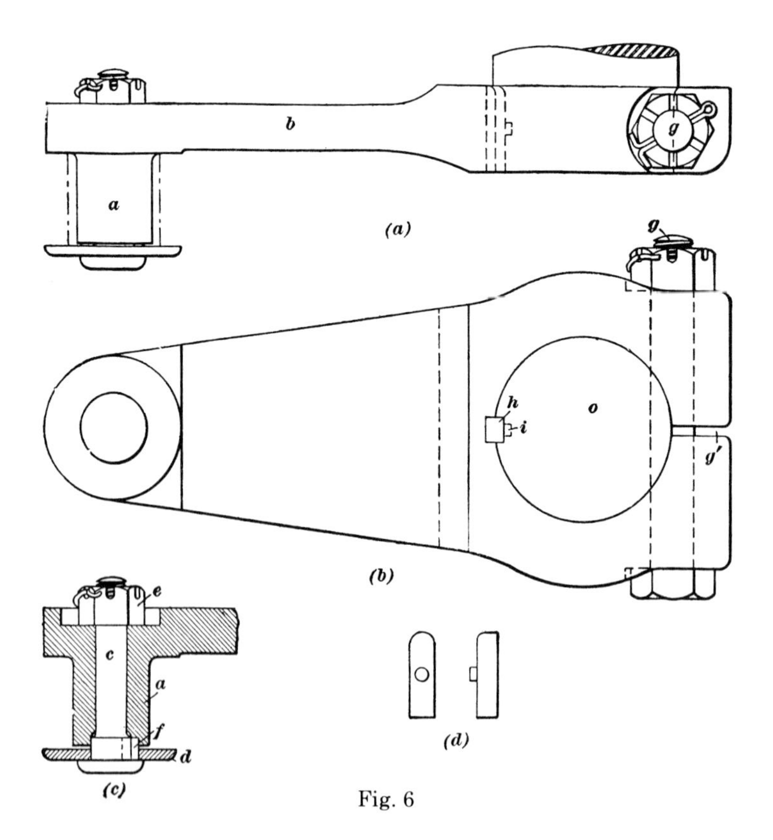

Fig. 6

that of the main pin. Due largely to the fact that the swing of the link is much less than the swing of the link foot, the eccentric crank has to be designed with a throw greater even than the total valve travel, so as to obtain, with the locomotive in full gear, the valve travel specified in the design.

8. Construction.—Eccentric cranks differ in construction, and the two types in general use are given in Figs. 6 and 7, which show the cranks removed from the main pins. In both illustrations, view *(a)* shows the cranks as viewed from the top, and view *(b)* as seen from the side. The eccentric crank in Fig. 6 *(a)* is of cast steel and the crankpin *a* is made in one piece with the arm *b*.

A bolt *c*, view *(c)*, passes through the center of the crankpin *a*, and a washer *d* on the outer end and a castle nut *e* with a cotter on the inner end serve to hold the eccentric rod on the pin. The inside face of the arm *b*, where the bolt passes through, is dished out so as to permit the nut when applied to seat deeper in the arm than it otherwise would. This is necessary in order to have the nut clear the main rod. The bolt *c* is prevented from turning when applying or removing the nut, by a square key *f*, which fits in a slot in the bolt *c*, the washer *d*, and the pin *a*.

9. The eccentric crank is held on the end of the main pin by the binding bolt *g* and a nut, views *(a)* and *(b)*, Figs. 6 and 7. A slot *g′* ¼ inch wide is cut from the end of the crank into the opening *o* for the main crankpin. When the nut on the bolt *g* is tightened up, the width of the slot

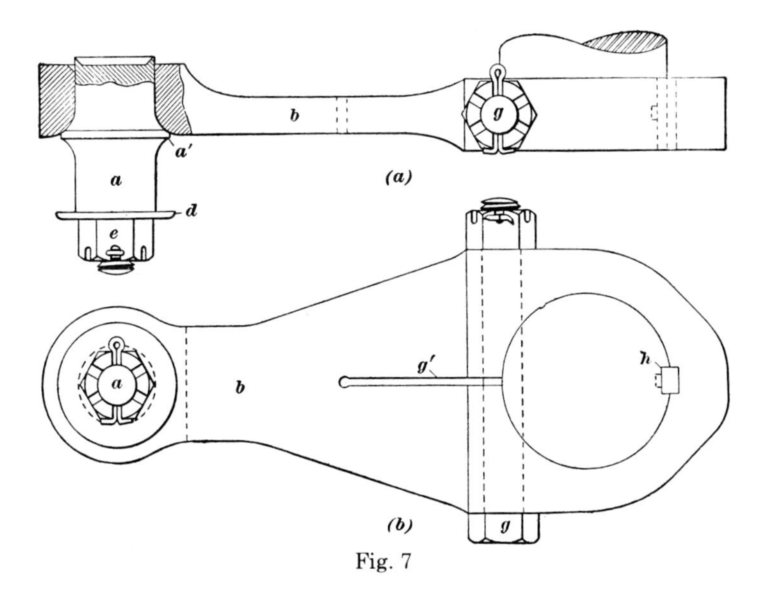

Fig. 7

is reduced and the eccentric crank is clamped on the pin. A part of the middle section of the bolt *g* extends into the circular opening *o* in the crank, and the bolt at this point fits into a semicircular slot in the main pin, and thereby prevents the eccentric crank from working off. The eccentric crank is prevented from turning on the main pin by the key *h*, Fig. 6, *(b)*, here shown in the crank. A part of the key fits in the slot in the crank, and the other part in the main pin. The key, which is placed in its keyway in the main pin before the eccentric crank is applied, is prevented from working out and striking the eccentric rod by the projection *i* which fits in a hole in the bottom of the keyway in the pin. Two views of the key are given in view *(d)*. Sometimes the key is held in position by a dowel pin in the bottom of the keyway in the main pin, in which case the outer end of the pin fits into a hole in the key.

10. The general practice is to cut the slot for the key in the eccentric crank on the center as shown so that the crank on one side can be used on the other side and to avoid the necessity of carrying right and left cranks in stock. A crank must be turned over when taken from one side of the engine to the other and the slot when cut in the center comes in the same place on the other pin.

In Fig. 7 the crankpin *a* is made separate from the arm *b*, and is pressed into it until a shoulder *a′* on the pin strikes the side of the arm. The inner end of the crankpin now projects about ¼ inch beyond the other side of the arm as shown, this allowance being provided for riveting over. The side of the arm is countersunk as shown and the end of the pin when riveted over is flush with the side of the arm. The outer end of the pin *a* is turned to a smaller diameter than the pin proper and is threaded for the castle nut *e*. The eccentric rod is held on the pin by the nut which is prevented from coming off by a cotter key, and by the washer *d*.

The main crankpin end of the crank differs from that shown in Fig. 6, in that the positions of the slot *g′* and the keyway have been reversed. This arrangement brings the binding bolt *g* in front of the main crankpin. The eccentric crank is prevented from turning on the axle by the same type of key *h* as shown in Fig. 7 *(b)*. With the slot between the ends of the eccentric crank, it is more difficult to draw the crank back into position again by the binding bolt after it has been spread to apply it to the main pin, than when the slot is on the outside, as in Fig. 6. On this account, the pin sometimes has to be built up to the required size with an electric weld.

With the eccentric crank made in one piece as in Fig. 6, the nut *e*, on account of being countersunk, is difficult to tighten thoroughly, and the nut may work off and strike the main rod. On account of the hollow crankpin, and the side of the arm being countersunk, this type of eccentric crank is not as strong as the one shown in Fig. 7.

11. Applying and Removing Eccentric Crank.—The eccentric crank is applied in the following manner: The key *h*, Figs. 6 and 7, is first placed in the keyway in the main crankpin. A wedge is then driven into the slot *g′* and the crank spread enough so that it can be slipped on over the end of the main pin. Next, the wedge is removed, the bolt *g* is driven through and the nut and cotter key are applied.

The eccentric crank is removed by driving out the bolt *g*, then driving a wedge in the slot *g′*, and prying the crank off. The end of the main pin on which the eccentric crank is place, may be either tapered or straight; a tapered pin makes the crank easier to remove.

ECCENTRIC ROD

12. Purpose.—The purpose of the eccentric rod is to transmit the movement of the eccentric crank to the link.

13. Construction.—A top view of an eccentric rod is shown in Fig. 8 *(a)*. View *(b)* shows the rod as viewed from the side and views *(c)*, *(d)*, and *(e)* show the eccentric-rod pin. A solid brass bush *a* about ½ inch thick is pressed into the back end of the rod where it fits on the eccentric crankpin. The bushing is pressed out and renewed when it becomes worn. This end of the rod has a grease cellar *b*, the upper end of which is normally closed by a grease plug *b′*. The bottom of the grease cellar is provided with a hollow steel plug *c* which is screwed into the bottom of the cellar and extends nearly to the wearing surface of the brass bushing. The plug provides a means by which the grease from the cellar is applied to the eccentric crankpin when the grease plug is screwed down, and it also prevents the bushing from turning. The front end of the eccentric rod, view *(a)* is forked and when the rod is connected up, the link foot is placed between the two jaws *d* and *d′*. The holes *e* and *e′* in each jaw for the eccentric-rod pin are tapered toward the outside of the rod and are shown by dash lines as they cannot be seen when viewed from the top. The position of the pin between the jaws is shown by broken lines. The jaws are usually made about 1⁄16 inch wider than the link foot.

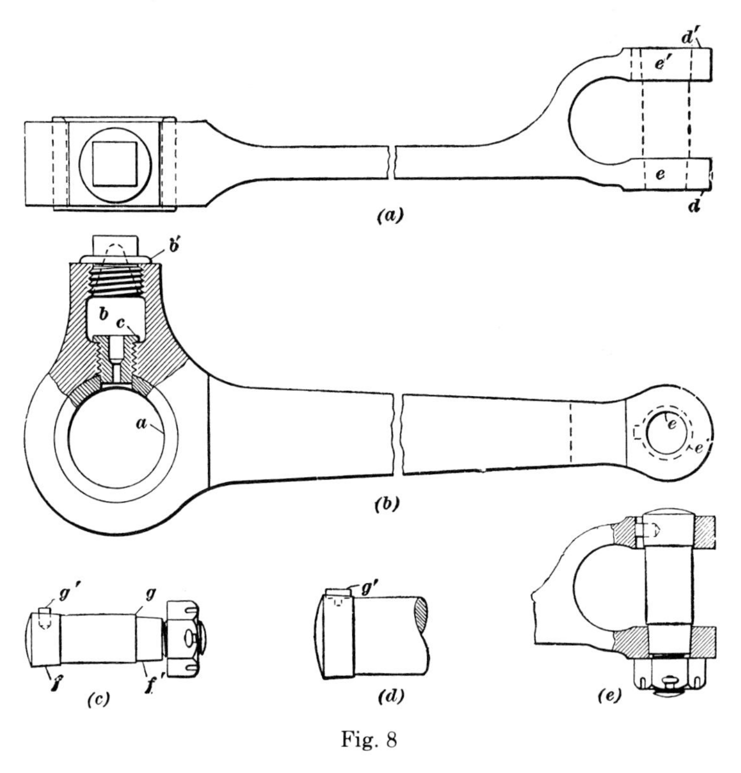

Fig. 8

14. The fact that the holes in the rod are tapered is also shown in Fig. 8 *(b)*, in which the larger hole *e′* is dotted in, thereby indicating that this hole is in the inside jaw and cannot be seen when the rod is viewed from the outside. The eccentric-rod pin, view *(c)*, which serves to connect the eccentric rod to the link foot, is case-hardened, and as the pin holes are tapered toward the outside the pin has to be applied from the inside of the rod, and this fact makes it more convenient to drive out, as well as to apply, the nuts. The ends *f* and *f′* of the pin, which come in the jaws of the rod, are tapered to fit the taper of the holes, and the portion between the jaws is straight. The tapered ends permit the pin to be easily driven out, as the slightest movement loosens it. The pin has a shoulder at *g*, which prevents the jaws of the rod from being cramped on the link foot when the castle nut on the pin is tightened up. A dowel pin *g′* about 3/8 inch square is driven into the side of the pin and fits into a slot in the side of the rod when the pin is applied.

15. The dowel pin serves two purposes; it prevents the pin from turning when the nut is being applied or removed, and it keeps the pin from turning when the engine is moving, and thereby confines the wearing surface to the portion between the jaws and not the whole pin, as would be the case if it could turn. As the pin does not turn, no bushings are required in the holes in the rod. Sometimes the pin is prevented from turning by a key and a dowel pin made in one piece, as shown in detail in view *(d)*, Fig. 8. This arrangement offers more resistance to shearing off than the plain dowel pin, as the key extends into the side of the head of the pin. The pin is lubricated from an oil cellar in the link foot packed with waste or hair. This cellar, as well as others in the gear, is usually packed with hair in the winter and with waste in the summer. Waste hardens in cold weather, and prevents oil from reaching the pin. while hair permits the oil to feed too rapidly in the summer.

The eccentric-rod pin is shown applied to the rod in view *(e)* in which a nut and a cotter key are used to prevent the pin from coming out. The eccentric rod is made as long as possible because the irregularities due to the angularity of the rod are less when the rod is long than when it is short.

PURPOSE AND CONSTRUCTION OF LINK

16. Purpose.—the purpose of the link is to transmit the movement of the eccentric crank and the eccentric rod to the radius rod. It also permits the locomotive to be reversed and the cut-off changed.

17. Construction.—The construction of the link used with the Walschaert valve gear will be explained by referring to Fig. 9 in which *(a)* is a side view of the link; *(b)* a rear view, and *(c)* a side view of the center piece or slotted part of the link. The link is usually made in three parts, the two cheek pieces or bracket pieces *a*, one on each side, view *(b)*, including the fulcrum pins or link trunnions *b* by means of which the link is carried in the gear frame and a center piece *c* with a slot *c′*, view *(c)*, in which the link block operates. The entire link is case-hardened.

18. The purpose of the bracket pieces *a*, view *(b)*, Fig. 9, is to connect the fulcrum pins or link trunnions to the center piece *c* in such a way as to permit the radius rod to move the link block up and down in the slot in the center piece. Therefore, the bracket pieces, view *(b)*, must be bolted to the ends of the center piece because, if bolted to the middle, the radius rod could not move the link block past the center of the link. Then, too, the bracket pieces must be made so that they stand away from the center piece, except at the ends, and thereby leave a space *d* about 1 9/16 inches in width on each side for the forked end of the radius rod. The slotted center piece merely forms a guide for the link block when the radius rod is moved by the reverse lever. The center piece is extended down much farther than the bracket pieces so as to bring the eccentric-rod connection at the lower end as near as possible to the horizontal line through the center of the main driver and the center of the cylinder. The lower portion *4* of the center piece is referred to as the *link foot,* which is sometimes made separate and bolted to the center piece instead of being made in one piece with it.

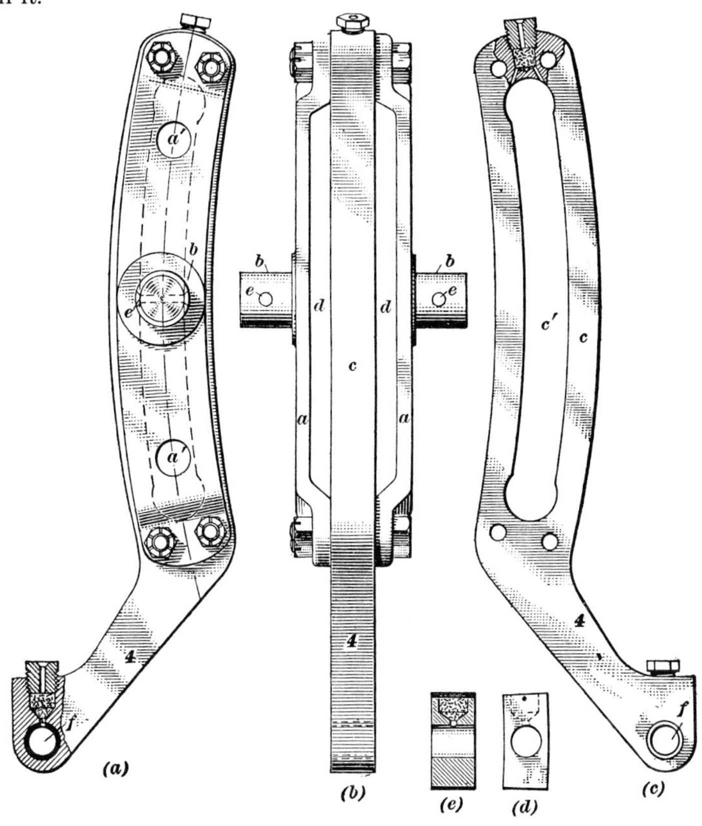

Fig. 9

19. Each of the bracket pieces, Fig. 9 *(a),* has a circular opening *a′* at the top and bottom. These openings are used to remove the link-block pin when it is required to take down or put up the radius rod without disassembling the link. They also provide a means for the direct oiling of the bearing surfaces of the link block and the link. The three parts of the link are held together by four tapered bolts with nuts and cotter keys, two at the top and two at the bottom of the bracket pieces.

The link trunnions *b* are provided with case-hardened bushings which are pressed on and held in place by 3/8 inch pins *e*. These pins, view *(a),* are driven all the way through and are riveted over at the ends as shown. The inner part of the bushings are made large enough to cover a part of the bracket pieces so as to prevent these pieces from coming in contact with the gear frame. The bushings when worn can be removed, thus making it unnecessary to discard the bracket pieces for new ones. A case-hardened bushing *f* is pressed into the link foot and provides a bearing for the eccentric-rod pin.

The top of the link is supplied with an oil cellar, view *(c)*, with oil holes to the slot in the center piece. The cellar is packed with hair or waste and is closed by a plug with a small hole through it as shown. The oil is applied to the oil cellar through this hole and lubricates the wearing face of the link slot and the link block. A similar arrangement, view *(a)*, is provided at the link foot to oil the eccentric-rod pin.

A side view of the link block which works in the slot *c′* in the center piece is given in view *(d)*, and view *(e)* shows a section through the center of the block. This part will be described more fully later.

20. One-Piece Link.—A view of a one-piece link used on the Pennsylvania Railroad is shown in Fig. 10. The link is carried in the gear frame on the trunnions *a* which are made in one piece with the two bracket pieces *b*, one on each side of the link and connected to the back of it by three bolts. As shown, the distance between each inside face of the

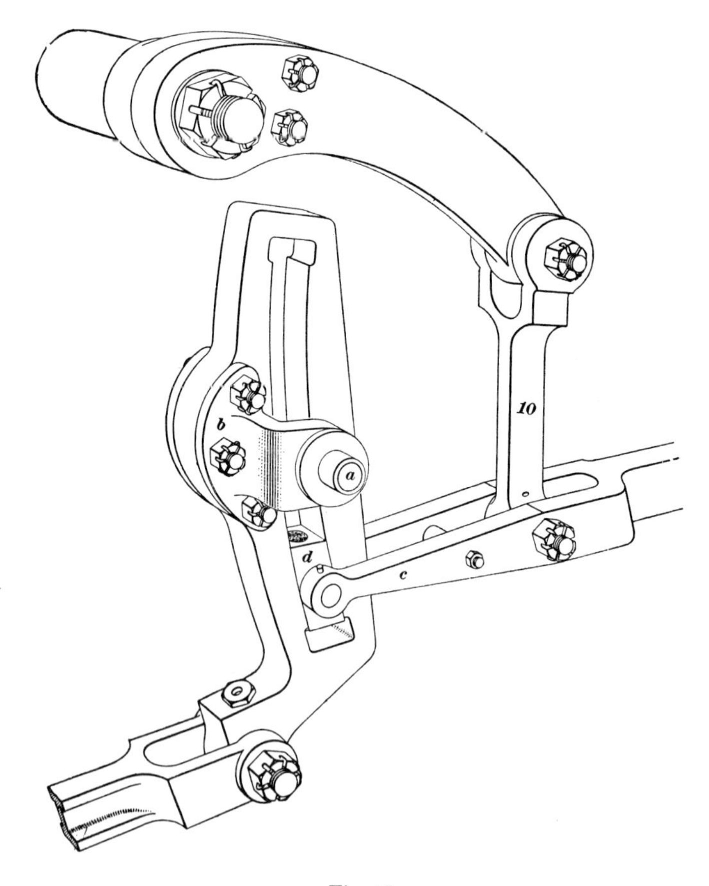

Fig. 10

jaw of the bracket pieces and the nearest sides of the link is sufficient to permit the radius rod *c* to move the link block *d* from one end of the link to the other. The radius-rod hanger *10* is connected to the radius rod in front of the link because the bracket piece *b* would interfere with the movement of the rod were it extended to the rear of the link.

21. Dimensions of Link.—The total length of the link is about 3½ feet, the width of a bracket piece, measured on the side is about 5 inches, and when measured at the rear about 1½ inches. The center piece is about 2½ inches wide and the total width of the link is about 6½ inches. The link trunnions are usually about 3 inches in diameter, and the length of the link slot is about 22 inches.

RADIUS ROD

22. Purpose and Construction.—The purpose of the radius rod, or *radius bar,* as it is sometimes called, is to transmit the movement of the link to a point on the combination lever. The radius rod may be made in one or two pieces, depending how it is carried at the rear end, but the one-piece construction is the more generally used.

Two views of a radius rod of one-piece construction are given in Fig. 11, *(a)* and *(b)*. View *(a)* shows the rod as viewed from the top, and *(b)* is a side view. Fig. 12 *(a)* is a top view of a radius rod of two-piece construction, and *(b)* is a side view.

The radius rod is made in one piece when the rear end is carried on a hanger as shown in Figs. 1 and 2, because with the hanger pin *a,* Fig. 11 *(c),* removed, the forked end of the rod can be inserted during assembly through the link from the front. The radius rod has to be made in two pieces in order to assemble it in the link when a radius-rod lifter is used to carry the rear end of the rod as shown in Fig. 3, because this end of the rod then has to be made in a straight piece. Accordingly, the rod is made in two pieces as shown in Fig. 12, and the construction is such as to leave the front of the rear portion forked so that it can be applied to the link when the other part is removed.

23. The forked portion of the radius rod, Fig. 11, carries two tapered case-hardened pins, the link-block pin *h* indicated by dash lines in view *(a)* and used to connect the link block to the rod, and the radius rod-hanger pin *a,* view *(c),* which connects the rod to the hanger. The pin holes in the jaws are indicated by dash lines, view *(a),* and they are tapered to conform to the shape of the pins toward the outside face of the rod, hence the holes *e* and *e′* in the inside jaw are larger than the holes *f* and *f′* respectively, in the outside jaw, as shown by the full and the dash circles in view *(b)*. The hanger pin is shown applied to the radius rod in the detail given in view *(c)*. The pin similar to the link-block pin is tapered so that it can be easily driven out and is applied from the inside. The shoulder *a′* prevents the pin from being drawn through and clamping the jaws on the hanger when the nut is tightened up. A dowel key *g* similar to the one used with the eccentric-rod pin, Fig. 8 *(d)* and *(e)* and applied in the same way, prevents the pin from turning and confines the wear to one part of the pin or to the straight section between the jaws. The keyway in the rod is shown at *g′*, Fig. 11 *(b)*. The hanger pin is held in the rod by a castle, or turret, nut which is prevented from coming off by a cotter key. The link-block pin is prevented from turning in the rod and wearing on the ends, by the pins *i* view *(a)* which are driven down all the way through the jaws and engage with semicircular slots in the side of the link-block pin. The ends of the link-block pin are flush with the sides of the jaws.

24. The case-hardened bushing *j*, Fig. 11 *(b)*, pressed into the front end of the radius rod provides a bearing for the pin which connects the rod to the combination lever. This bushing as well as bushings at the other points in the valve gear may be prevented from turning and closing

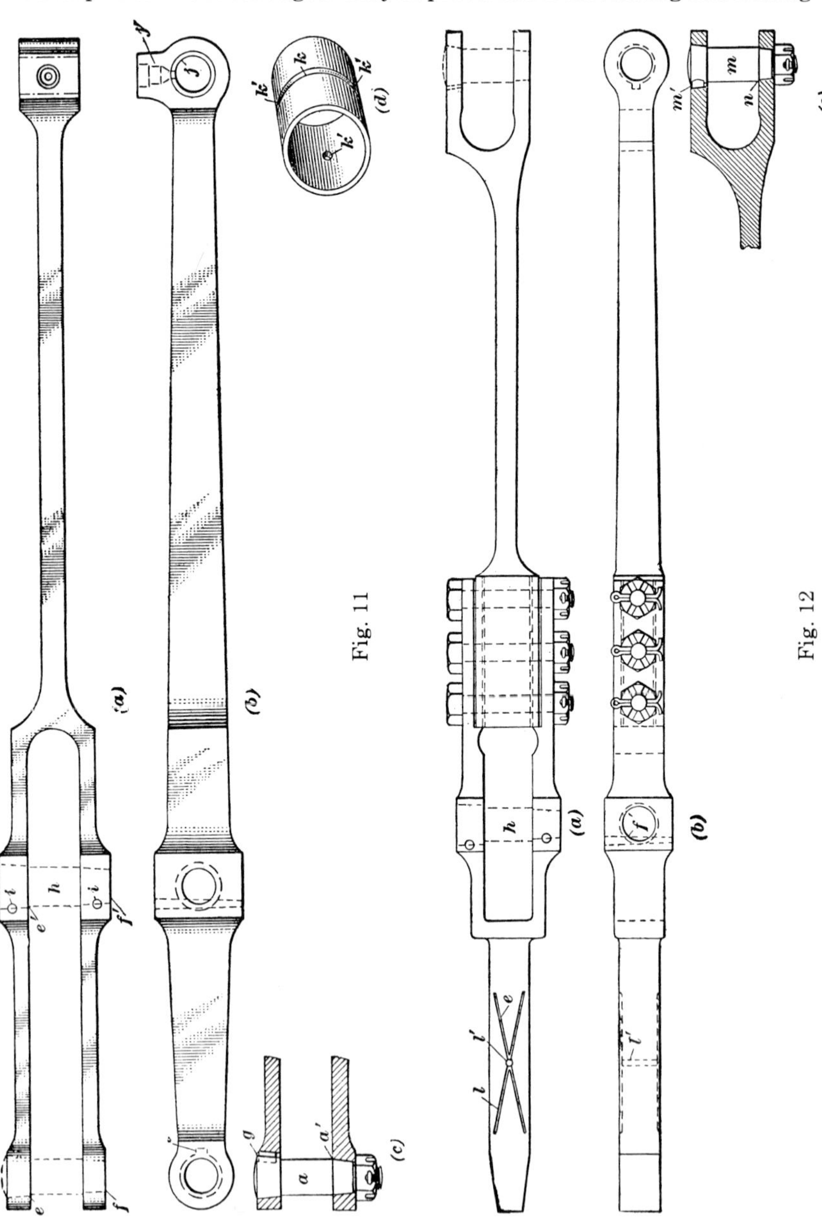

Fig. 11

Fig. 12

the oil hole by a round dowel pin, driven in, half of which comes in the bushing and the other half in the rod. If a dowel pin is not used, an oil groove *k* view *(d)*, is cut entirely around the outer face of the bushing so that the oil hole in the rod cannot be closed if the bushing turns. Usually three equally spaced oil holes *k'* are drilled from the oil groove to the bearing face so that, should the bushing turn, an oil hole will always be uppermost. The pin and the bushing are lubricated from a cellar *j'*, view *(b)*, shown by dotted lines, which is packed with waste or hair. The oil plug through which the cellar is packed is not shown, but it has a countersunk hole for applying oil, as in Fig. 9 *(a)*. The front end of the radius rod in Fig. 11 is straight and therefore the end of the combination lever to which it is connected is forked.

25. The two parts of the radius rod are connected by the three 1-inch bolts, shown in Fig. 12. The rear end of the rod is case-hardened and works in a slip block or radius-rod lifter on the reverse-shaft crank shown in Fig. 3. The top and bottom surfaces of this part of the rod have oil grooves *l* and *e* connected by an oil hole *l'*. The arrangement of the link-block pin *h* is the same as in Fig. 11 *(a)*. The front end of the rod is forked, hence the upper end of the combination lever is straight. The arrangement of the tapered radius-rod pin *m*, view *(c)*, with the key *m'* to prevent turning and wear on the ends, is exactly the same as the link-hanger pin *a* in Fig. 11. The pin is applied from the inside of the rod and has a shoulder at *n* to prevent the jaws from being clamped on the combination lever when the nut is tightened up. The radius-rod pin is held in the rod by a castle nut and a cotter key.

26. Assembly of Link, Link Block, and Radius Rod. The details of the assembly of the link, the link block, and the radius rod are given in Fig. 13 *(a)*, which shows the lower portion of the link *3* with a part of one

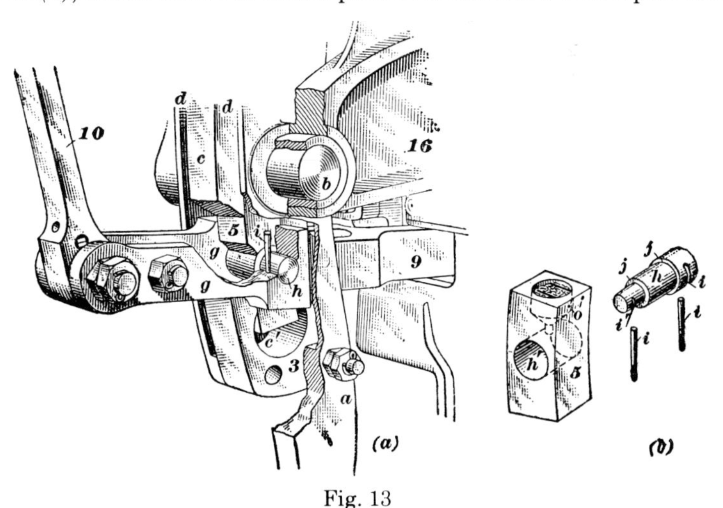

Fig. 13

of the outer bracket pieces *a* and the gear frame *16* broken away. The link block is marked *5*, the link-block pin is seen at *h*, and the jaws of the forked end of the radius rod *9* are indicated by *g* and pass through the link on each side of the center piece *c*; the jaws move up and down in the space *d* between each bracket piece and the center piece when the radius-rod hanger *10* is raised and lowered. The link block *5* is placed in the slot *c′* in the center piece *c* of the link and hence sets between the jaws *g* of the radius rod. The link-block pin *h* on which the link block turns when the link swings passes entirely through the jaws *g* and the link block, and pins the link block to the radius rod. The purpose of the link block and the link-block pin is to transmit the backward-and-forward movement of the link to the radius rod. The link block also guides the movement of the rear end of the radius rod when it is raised and lowered.

27. The link-block pin is prevented from turning in the jaws of the radius rod and thereby wearing on the ends, as well as from working out, by a 5/8 inch pin *i*, Fig. 13, one in each jaw. These pins are usually behind the link-block pin, as shown, so as to make their application easier. They are driven down from the top, and engage semicircular slots *i′*, view *(b)*, in the link-block pin where they pass it. With the link-block pin fixed, the wear is confined to the portion on which the link block turns. The pins *i*, view *(a)*, are split slightly and then spread at their lower ends after they are applied, to prevent them from working up and out. The link block *5*, the link-block pin *h* and the two pins *i*, are shown removed in view *(b)*. The link block is case-hardened and its wearing surfaces are slightly curved so as to conform to the curvature of the slot *c′*, view *(a)*, in the center piece when the block is being raised and lowered. The top of the link block has an oil well packed with waste. The waste retains the oil and permits a slow feed through the small opening shown in the bottom of the well to the link-block pin which passes through the circular opening *h′* in the link block. The oil from the oil well can also pass through the small holes *o* to the front face of the link block. The ends of the link-block pin *h* where they fit in the jaws of the radius rod, are tapered in the same direction or toward the outer face of the rod so that the pin can be easily driven out. The shoulders *j* on the tapered ends limit the distance the pin can be driven into the radius rod. As the link-block pin hole in the inside jaw of the radius rod is larger than the hole in the outer jaw, the pin is driven in from the side of the link next to the boiler and is driven out in the reverse direction.

COMBINATION LEVER

28. Purpose.—The combination lever serves the same purpose as a crank set 90 degrees from the eccentric crank and with a throw equal to the lap and the lead.

29. Construction.—In Figs. 14 and 15 are shown two views of combination levers used with inside-admission valves, which differ in construction at their upper ends, as one is straight and the other is forked. Views *(a)* show the levers as viewed from the side and views *(b)* as seen from the front.

The radius rod is connected by the radius-rod pin to the top of the lever at *a*, the valve-stem crosshead is connected to the lever by the valve-stem crosshead pin *b*, and the union link is connected to the bottom of the lever at *c*. The only difference between the levers in Figs. 14 and 15 is in the method used in connecting the radius rod to them. In Fig. 14 the top of the rod is straight and this requires the front end of the radius rod to be forked as shown in Fig. 12. In Fig. 15 the upper end of the rod is forked

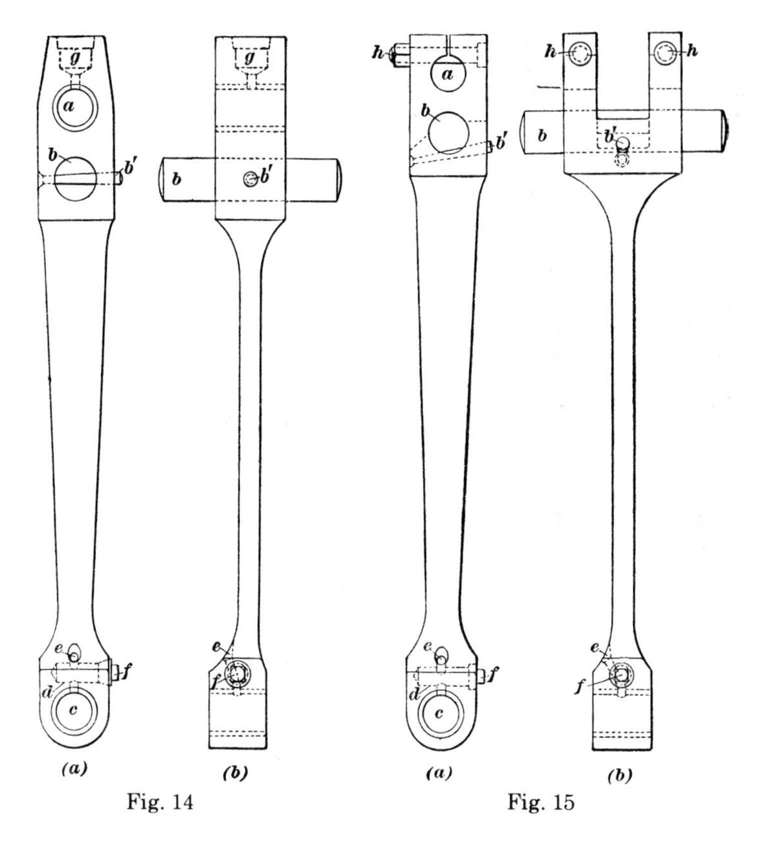

Fig. 14

Fig. 15

and this requires the front end of the rod to be straight as in Fig. 11. A case-hardened bushing is pressed into the lever at *a* and *c*, Fig. 14, and at *c*, Fig. 15, as the rod turns on pins at these points. The valve-stem crosshead pin *b* is about 1¾ inches in diameter and is held in the lever by a ½ inch tapered pin *b′* which is driven in from the front and is riveted over on the small end. The pin *b′* may pass through the center of the pin *b*, Fig. 14 *(a)*, or through the side, Fig. 15 *(a)*.

30. The pin connecting the union link to the combination lever at *c*, Figs. 14 and 15, is lubricated from a cavity *d* in the rod, which is provided with an oil hole *e*. The cavity is filled with hair or waste by removing the pipe plug *f*, which is a round plug with a square head. The radius-rod pin at *a*, Fig. 14, is lubricated in the same way except that the oil hole is in the plug, not shown, which closes the upper end of the waste cavity *g*. The upper end of the lever in Fig. 15 is forked, and this requires the front end of the radius rod to be made straight and supplied with a case-hardened bushing *j*, Fig. 11 *(a)*, for the radius-rod pin. The pin which is inserted at *a*, Fig. 15, is prevented from coming out and also from turning by the two binding bolts and nuts *h*. The bolts pass through slots in the side of the pin and draw the slotted upper ends of the lever together and thereby clamp them on the pin. The bushings in the lever may be prevented from turning by dowel pins, or an oil groove may be cut in the bushing so that should they turn, the oil holes in the lever will not be closed.

31. Proportions of Combination Lever.—The length of the piston stroke, the lap and the lead and the total length of the combination lever being given, the proportions of the lever or the distance between the center of the pin holes can be found from the following rule:

Rule.—*To find the short end of the lever or the distance between the center of the radius rod and the valve-stem crosshead pins with an outside-admission valve, multiply the whole length of the lever by twice the lap and the lead and divide by the sum of twice the lap and the lead and the stroke. In the case of an inside-admission valve, multiply the whole length of the lever by twice the lap and the lead, and divide by the stroke. To find the long end of the lever, subtract the length of the short end from the total length.*

EXAMPLE.—The stroke of the piston is 30 inches, twice the lap and lead is 2½ inches, and the length of the combination lever is 36 inches, the valve is of the inside-admission type. What is the length of the short and long arms of the lever?

SOLUTION.—From the rule, the distance between the center of the radius-rod pin and the valve-stem crosshead pin, or the length of the short arm is 36x2½/30=3 in. The length of the long arm is 36—3=33 in.

Ans.

When working examples of this kind, 1/16 inch is usually added to twice the lap and the lead to allow for lost motion in the connections.

COMBINATION OR UNION LINK

32. Purpose.—The purpose of the combination, or union, link is to transmit the movement of the crosshead to the lower end of the combination lever. The link is always forked at the front end and may or may not be forked at the back end, depending on whether it is connected to an arm on the crosshead or to the wristpin.

33. Construction.—Two views of the type of union link used when its rear end is connected to the crosshead pin or wristpin, is shown in Fig. 16, in which *(a)* is a top view, and *(b)* a side view. In view *(b)* a case-hardened bushing *a* is pressed into the back end of the rod at the wristpin connection, and is prevented from turning by a round 3/8 inch pin *b*, driven in. A cellar, which is packed with waste or hair by removing a ½ inch pipe plug *c* in combination with an oil hole *d*, with a countersunk opening, provides for lubrication.

A detail which shows how the back end of the rod is connected to the end of the wristpin is shown in view *(c)*. A collar *e* is placed over the end of the wristpin *f*, the inner side of which bears against the side of the crosshead *g*. The rear end of the union link *12* with the bushing *a* turns freely on the collar, and is held between the shoulder *h* on the collar and the nut *i*. The front end of the union link, view *(a)*, is forked to receive the lower end of the combination lever.

34. The pin holes *l* and *l'*, Fig. 16, in the jaws are tapered toward the inside jaw, hence the larger hole *l* is in the outside jaw as shown in view *(b)*. The ends of the pin are tapered to conform to the pin holes, but as shown by the broken lines, view *(a)*, the portion between the jaws is straight. The pin has to be applied from the outside because the piston rod interferes with its application from the inside. The slot in the outside jaw for a dowel key is shown at *k*. The key is similar to the one shown in

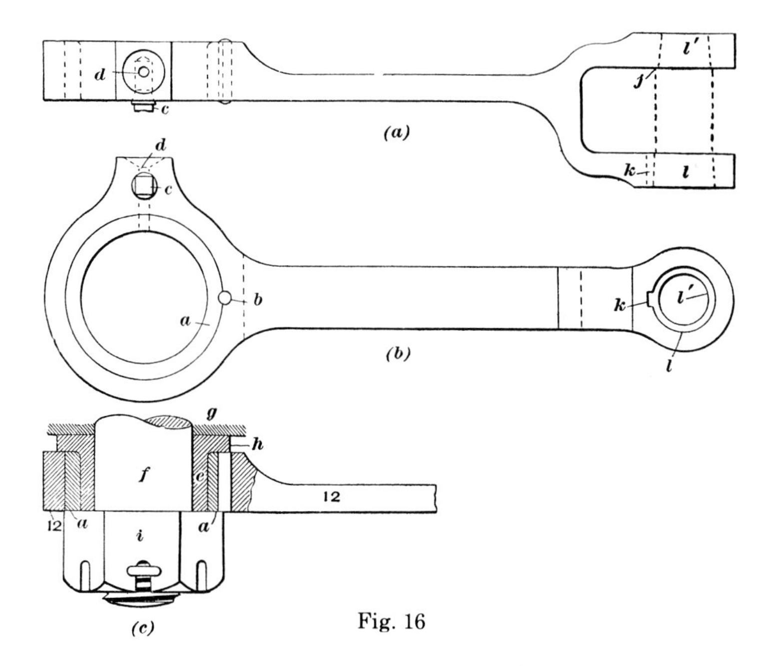

Fig. 16

Fig. 8 *(c)* or *(d)*, and is placed on the pin before it is placed in the link. Its purpose is to prevent the pin from turning when the nut is being applied or removed, and, in addition, it holds the pin fixed and thereby confines the wear to the portion between the jaws. The pin is held in position by a castle nut and a cotter key.

When the union link is connected to an arm on the crosshead, the construction at the rear end of the link is the same as at the front end.

RADIUS-ROD HANGER

35. Purpose.—The purpose of the radius-rod hanger *10*, Fig. 2, is to raise or lower the radius rod and link block. The lower end of the hanger is flexibly connected to the radius rod by the radius-rod hanger pin *a*, Fig. 11 *(c)*, and the upper end to the reverse-shaft crank or to the reverse-shaft arm, depending on the side of the engine considered.

36. Construction.—A side view of a radius-rod hanger is given in Fig. 17 *(a)*, and *(b)* is a front view. As both ends of the hanger turn on pins, a case-hardened bushing *a* is pressed into the lower end of the hanger and a similar one into the upper end. Oil is applied to the wearing surface of the bushings and the pins from oil cellars *c*, in the top and bottom of the hanger, shown by dash lines in view *(b)*, and oil holes *d*. The cellars are packed with hair or waste by removing the pipe plugs *e*, and are supplied with oil through the holes *f*.

37. Radius-Rod Hanger-Pin.—The pin used to connect a straight type of radius-rod hanger to the arm of the reverse-shaft crank is shown in Fig. 17 *(c)*. The tapered end of the pin is held in the arm of the crank between the collar *a* and the nut at the right. The hanger swings on the straight section of the pin between the collar *a* and the nut and washer *b*.

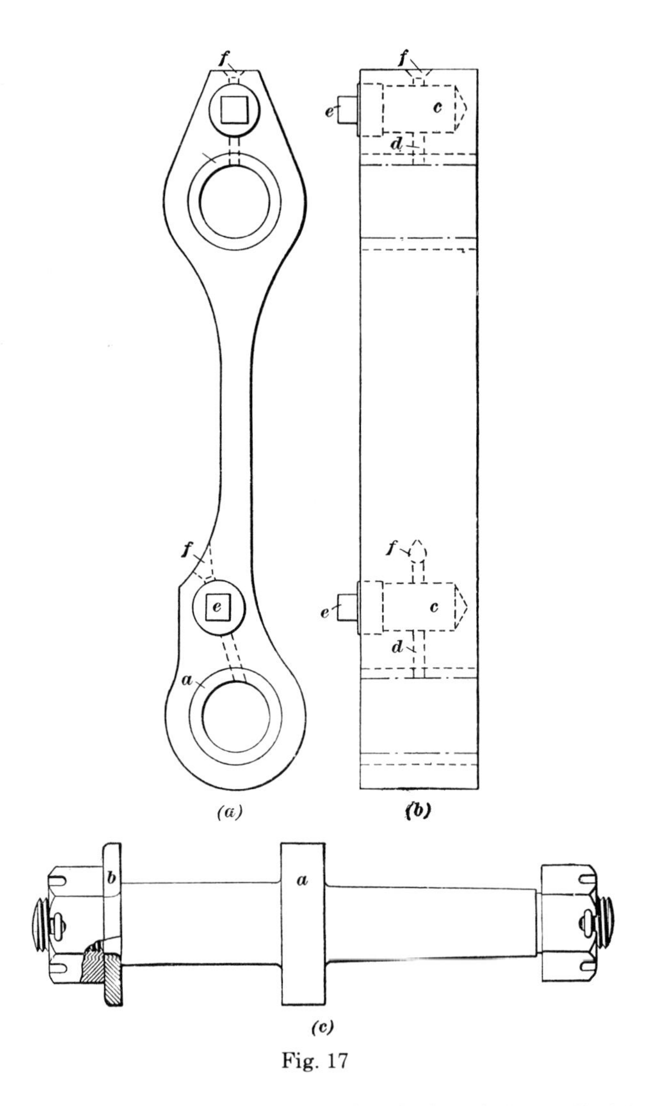

Fig. 17

The collar, made with the pin, prevents the pin from being pulled through the crank arm, and it also brings the hanger in line with the center piece in the link.

REVERSE-SHAFT CRANK AND REVERSE-SHAFT ARM

38. Purpose.—The purpose of the reverse-shaft crank *7*, Fig. 18, is to transmit the movement of the reach rod *17* to the reverse shaft *6* and to the radius-rod hanger on the right side of the engine. The purpose of the reverse-shaft arm *8* is to transmit the movement of the reverse shaft to the radius-rod hanger on the left side of the engine.

39. Construction.—The type of reverse-shaft crank used depends largely on the method used to connect the crank to the radius rod. The

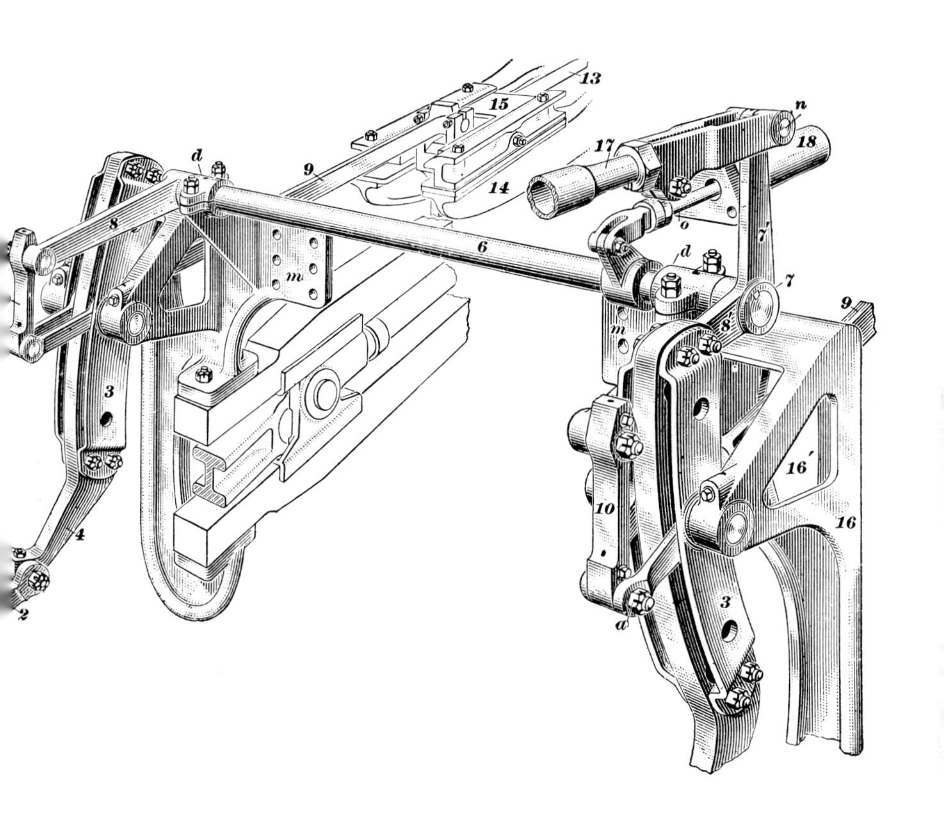

Fig. 18

type of crank shown in Fig. 18 is used when the radius-rod hanger is a straight bar as shown. The reverse-shaft crank in this case is held on the reverse shaft by a ⅝ inch square key, the end of which is shown. The key is the same length as the width of the crank at the reverse shaft. The reach rod *17* is connected to the vertical arm *7′* of the reverse-shaft crank, by a pin *n*, which has to be applied from the outside because the reach rod lies too close to the boiler to permit it to be applied from the inside. A case-hardened bushing is pressed into the reverse-shaft crank arm *7′* where the reach-rod pin passes through. The assembly of the reverse-shaft arm *8* on the reverse shaft *6* is similar to that of the reverse-shaft crank. In some cases a bolt which passes entirely through the reverse-shaft crank and the reverse shaft, is used to hold the crank on the shaft, and a similar arrangement is used with the reverse-shaft arm. The arrangement must be such as to prevent the crank from turning on the shaft.

40. Another type of reverse-shaft crank used when the rear end of the radius bar slides in a block, is shown in Fig. 4. The part of the crank next to the radius rod is forked and consists of an outer and inner arm *8′*, one on each side of the radius-bar lifter *10*. The arm *7′* is a single shaft, connected to the reach rod *17* by the pin shown. The crank is keyed to the end of the reverse shaft as in Fig. 18, and case-hardened bushings are pressed into the arms *8′* for the radius-rod lifter trunnions *c*. The reverse-shaft arm on the other end of the reverse shaft *6* is similar to the reverse-shaft crank except that the arm *7′* is not used. The arm is keyed to the reverse shaft similar to the reverse-shaft crank. In some cases the outer arm *8′* of the reverse-shaft crank is made separate from the inner arm, and the two are connected at the reverse shaft by studs. This arrangement makes it easier to apply the radius-bar lifter, as it can be assembled in the arms in one piece.

The reverse-shaft cranks in Figs. 1 and 2 are similar to the one shown in Fig. 18, except that the arm *7* to which the reach rod connects, points downwards.

REVERSE SHAFT

41. Purpose.—The purpose of the reverse shaft 6, Fig. 18, is to transmit the movement of the reverse lever to the valve gear on the other side of the engine. The shaft is carried in boxes *d* on the top of the gear frame *16*.

ASSEMBLY OF PARTS

42. Assembling Link.—The manner in which the link is placed in the gear frame depends on the construction of the frame. When the gear frame is made in one piece, as in Fig. 18, the link has to be assembled in it one part at a time, but with a built-up type of frame, Fig. 2, the complete link, if desired, can be placed in the frame before it is bolted together; if not, the link has to be applied in the same way as with a one-piece gear frame. The application of the link to a one-piece gear frame is as follows:

The trunnion *b*, Fig. 9, of one of the cheek pieces of the link is first placed in its bearing in the gear frame. The other cheek piece is next placed beside the first one and moved sideways until its trunnion enters the bearing in the other side of the frame. The center piece *c* with the link block assembled in it is then placed between the two cheek pieces, and the cheek pieces and the center piece are bolted together. There is sufficient room to place one cheek piece in position after the other one has been applied, because the length of a link trunnion is less than the width of the center piece.

Bearing boxes of the type shown at each end of the reverse shaft, Fig. 18, are sometimes used to carry the link trunnions, and the link can then be applied to the gear frame, assembled. Solid bearings, however, are preferable whenever they can be used, because there are no studs or nuts to work loose.

43. Assembling Radius Rod in Link.—The manner in which the radius rod is assembled in the link depends on whether the rod is made in one or two pieces. In Fig. 4 the rear end of the radius rod *9* slides in the radius-rod lifter *10*, and with this construction the rod must be made in two parts in order to assemble it in the link. To place the rod in the link, the two parts are separated by removing the three bolts *k*. The rear end *a′*

of the rod is placed in the lifter and pushed up as far as possible and the link is swung far enough forwards to permit the forked end of the rod to be inserted through the link. The rod is raised or lowered in the link, whichever is more convenient, until the link-block pinhole in the rod comes opposite the holes *a′*, Fig. 9, in the cheek pieces. The link block is next raised until the hole in the link block also lines up with the holes in the cheek pieces and in the rod. The link-block pin *h*, Fig. 13, is then driven through from the inside of the link until it is flush with the side of the rod and the pins *i*, one in each jaw, are driven down to keep the link-block pin from turning. The pins *i* are split and spread slightly at their lower ends to prevent them from working up. The assembly is completed by bolting the two parts of the radius rod together.

44. In order to insure that the pins *i* strike the slots *i′*, Fig. 13 *(b)*, when the link-block pins are driven down, the proper position of the pin in the rod to bring the slots and the holes in the rod in line is indicated by a mark *f′*, Fig. 4, on the end of the pin and the side of the rod. With the two marks in line, the pins can be driven down without trouble. When the rear end of the radius rod *9* is carried on a hanger *10* as in Fig. 18, the rod does not have to be made in two pieces in order to assemble it in the link. The rod is placed in the link by removing the pin *a* and passing the forked end of the rod backwards through the opening *16′* in the gear frame. The link block is connected with the radius rod, as already explained, that is, the link-block pin is inserted in the link through one of the circular holes in the inside-bracket piece and the tapered pins that hold the link-block pin from turning are driven down. The back end of the rod is then connected to the radius-rod hanger *10*.

It is not always necessary to assemble the link in the gear frame before assembling the radius rod in it. If desired, the cheek pieces can be placed in the gear frame, as already described. The center piece and the link block are then placed in the forked end of the rod and the link-block pin driven in and pinned, to prevent turning. Then the center piece with the radius rod and link block is raised to position and bolted to the cheek pieces.

45. Assembling Radius-Bar Lifter.—The radius-bar lifter *10*, Fig. 4, is made in four pieces, two side pieces *b*, each with a trunnion *c*, and a top and bottom plate *a* of brass, the parts being held together by four bolts. The trunnions *c* are carried in case-hardened bearings pressed into the arms *8′* of the reverse-shaft crank. When the link swings, the back end of the radius bar moves back and forth in the lifter and because the end of the rod swings in an arc, the lifter turns slightly on the trunnions *c*. Liners are provided between the parts *a* and *b* which are removed and replaced by thinner ones when the top and bottom pieces *a* wear. The oil cup with a screw-cap on the upper plate *a* is packed with hair, and oil is applied through the hole in the top of the cap. The oil grooves in the top of the radius rod serve to retain the oil. The wearing faces of the plates *a* are also supplied with oil grooves.

The radius-bar lifter is assembled in the arms of the reverse-shaft crank one part at a time. With one of the side pieces *b* in position, the other is placed beside it and moved sideways, until its trunnion enters fully the bearing in the arm. The top and bottom plates *a* are then bolted on. This type of lifter, on account of the number of wearing parts, is troublesome to maintain. A better design of radius-bar lifter when a cradle type of gear frame is used, is shown in Fig. 2.

46. Arrangement and Assembly of Combination Lever and Valve-Stem Crosshead.—In Fig. 19 *(a)* is shown the valve-stem crosshead *15* connected to the combination lever *11,* assembled in the crosshead guide *14.* View *(b)* shows the crosshead *15* removed from the guide with parts of the valve stem *13* and combination lever connected to it. The purpose of the valve-stem crosshead and guide bars *b* and *e* is to guide the movement of the valve stem and keep it central with the valve chamber. They also carry or support the combination lever and the parts connected with it.

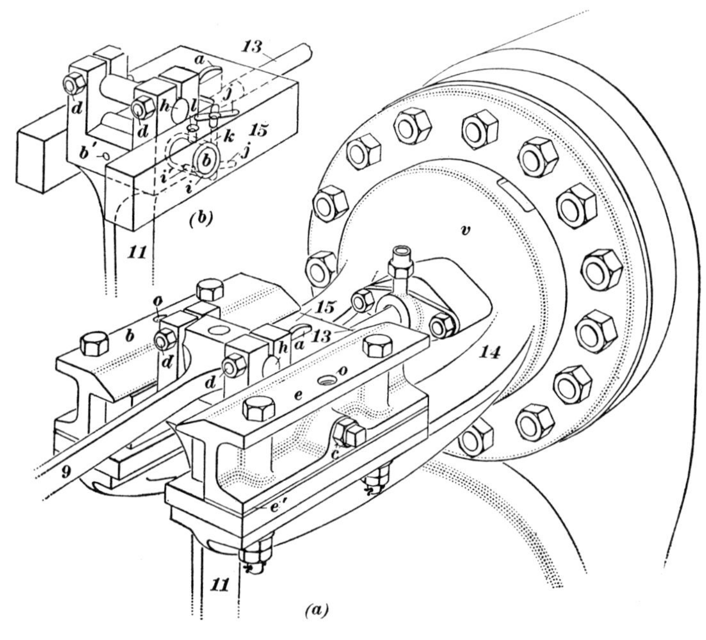

Fig. 19

In Fig. 19 *(a)* the valve-stem crosshead guide *14* is cast in one piece with the steam-chest head *v* of the valve chamber and the guide bars *b* and *e* are secured to the guide by four bolts. Brass liners *e'* are placed under each guide-bar and permit the guides to be closed when they wear. The valve-stem crosshead *15* which is connected to the valve stem *13* by the key *a* is oiled through oil cups screwed into the oil holes *o,* view *(a).* The setscrew *c* in the side of the outer guide bar, when screwed in, forces the valve-stem crosshead against the opposite guide bar and keeps it from moving. This screw, which has a locknut next to the guide bar, is only used when a breakdown in the valve gear or engine requires that the valve be held in a fixed position.

47. The combination lever *11,* Fig. 19 *(b),* is connected to the valve-stem crosshead *15* by the pin *b,* the ends of this pin turn in case-hardened bearings *i* in the guide. The bearings are pressed into the guide and are prevented from turning by round dowel pins *i'* driven below the surface, and the hole burred over. The pin *b* is prevented from turning in the combination lever and the wear on the pin, therefore, is confined to the ends by a ½ inch tapered pin *b'* which passes entirely through the lever as shown in Figs. *14* and *15*, and either engages a slot in the side of the pin, or passes through its center. The pin is applied from the front through the valve-stem hole in the crosshead so that it cannot work out when the valve stem is connected, and it is also riveted over on the outer end.

48. The oil grooves *j*, Fig. 19 *(b)*, in the top of the crosshead serve to retain the oil, and they are connected through the oil hole *k* to similar grooves in the bottom face of the crosshead. The oil feeds to the oil grooves from the oil cups on the oil holes *o*, view *(a)*. The ends of the pin *b* are oiled through the oil holes *l* direct, or indirectly through the oil holes *o* in the guides, in Fig. 19 *(a)*. The combination lever is connected to the valve-stem crosshead as shown in *(b)* before the crosshead is placed in the guides and connected to the valve stem. The combination lever, view *(b)*, is placed in the crosshead through the slot in the rear, and the pin *b* is driven through until the ends are flush with the sides of the crosshead. The pin *b'* is next driven in from the front through the valve-stem hole and the outer end is riveted over. Then the crosshead is placed in the guides *e*, view *(a)*, and is moved forwards on to the end of the valve stem, and the key *a* driven down. The radius rod *9* is connected to the combination lever by placing it between the jaws of the lever, then driving in the pin *h*, and tightening the nuts on the binding bolts *d*. In some cases the crosshead is lightened by drilling four holes in its lengthwise, two at the front and two at the rear.

49. Other Arrangements of Valve-Stem Crosshead and Guides.—There are two arrangements of valve-stem crossheads and guides, one as shown in Fig. 19, in which the guide is cast in one piece with the back-stem chest head, and the other as shown in Fig. 20, in which the crosshead *a* slides on a guide bar *b*. The type shown in Fig. 19 is in more general use because the guide does not have to be lined up with the center of the valve chamber when assembling as with the type shown in Fig. 20. The upper end of the combination lever is forked to span over the valve-stem crosshead to which it is flexibly connected by the pin *e*. The valve stem *d* is held in the crosshead by a nut *d'* on the rear end, and by the key *k*.

The crosshead is made in two pieces, the part *a* which slides on the top and sides of the guide bar and the part *a'* which wears on the under side of the bar. The two parts are connected by three bolts *c* with the nuts on the inside. The wearing surfaces of the part *a* are lined with brass or some other suitable bearing material.

50. The combination-lever pin is oiled through the oil cup *f*, Fig. 20, and the guide bar through the oil cup *f'* which has a hinged cover. These cups are packed with waste or hair so that the oil feed will be gradual. The front end of the guide bar is bolted to a lug on the back steam-chest head and the back end to a lug on the main guide yoke. The guide bar *b* is supplied with oil grooves *h*, and oil holes *h'* are drilled all the way through and connect with oil grooves on the under side, similar to the grooves *h*. The purpose of the oil holes and the oil grooves on the under side of the bar is to supply oil to the part *a'* of the crosshead, as the oil cannot be applied regularly to this part in any other way. The crosshead is assembled by placing the part *a* on the guide bar and bolting the part *a'* to it. Sometimes it is more convenient to assemble the crosshead and

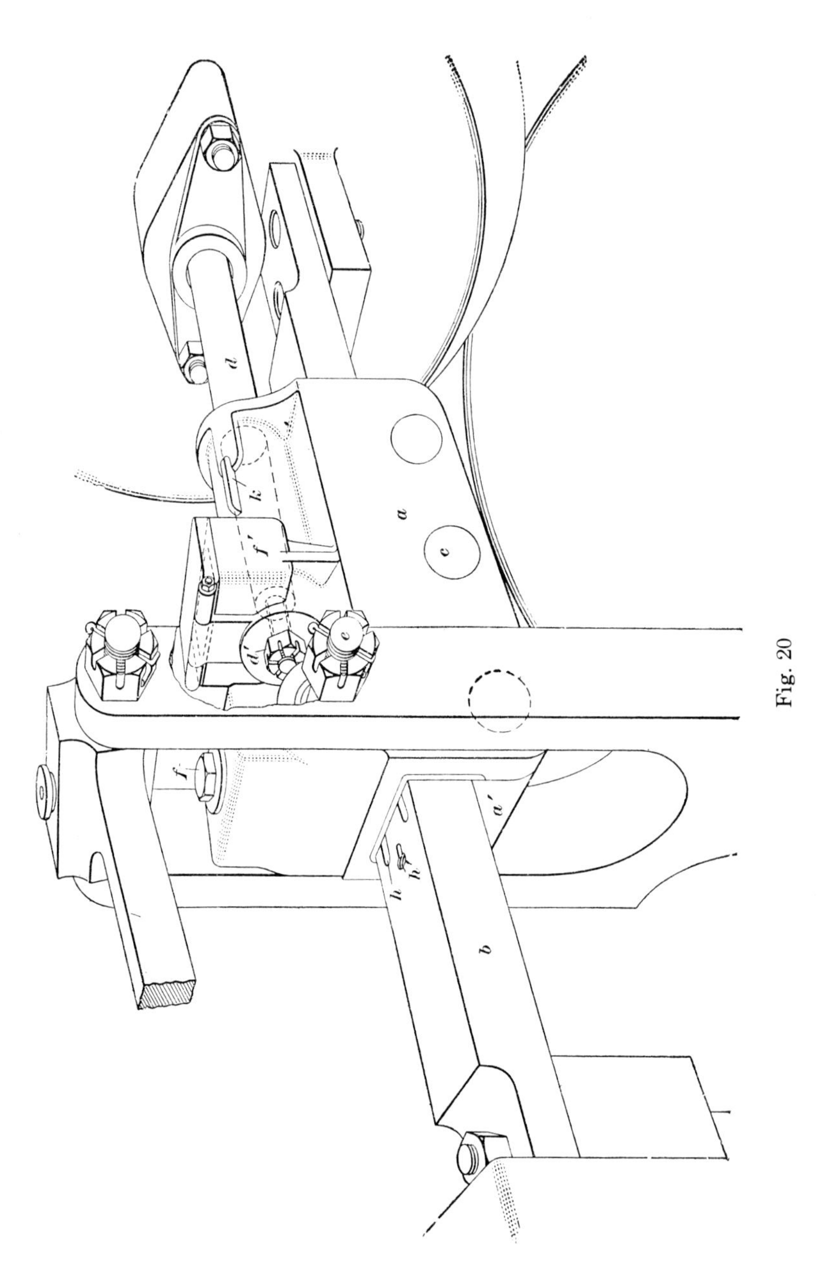

Fig. 20

place it on the guide bar before the latter is bolted in position. The valve stem and combination lever is then connected to the crosshead.

51. Arrangement of Valve-Stem Crosshead and Guide With Outside-Admission Valve.—The arrangement of the valve-stem crosshead when an outside-admission valve is used, is shown in the detail of the gear given in Fig. 21. The crosshead *a* slides on a guide bar *b* as in Fig. 20, but the crosshead has to be constructed somewhat differently because of the fact that the end of the combination lever and the valve stem have to be connected to the bottom of the crosshead instead of to the top, as shown in Fig. 20. The parts of the crosshead are bolted or riveted together as shown; the crosshead can be assembled on the guide bar, or assembled before the guide bar is set up, whichever is the more convenient. The valve stem is connected by the key *c* to a part *c′* which is forked at the rear and is flexibly connected to the crosshead by the pin *d*. This construction permits the valve stem to accommodate itself to slight differences in the height of the valve yoke or the valve, and will allow the valve to lift freely in the event of excessive compression in the cylinder, because the valve stem can turn freely on the pin *d*.

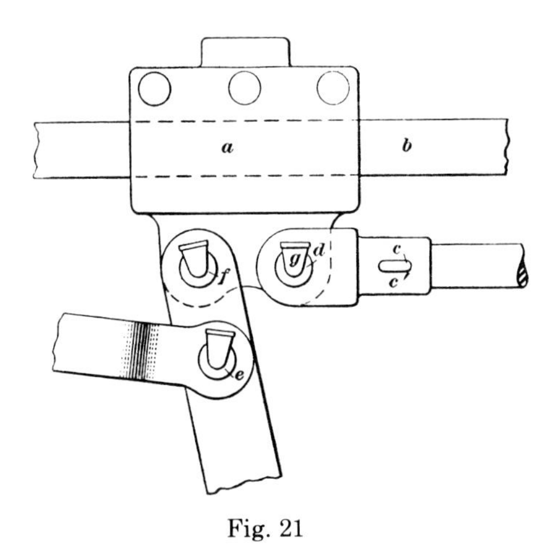

Fig. 21

52. As is the usual practice, the pin *d*, Fig. 21, the radius-rod pin *e*, and the combination-lever pin *f* are fixed in their respective parts and the wearing surfaces are confined to the portions of the pins between the jaws of the parts. The arrangement shown will not permit the pins to be oiled from the outside. For example, to oil the radius-rod pin and bushing would require the application of oil through the combination lever, and there is not room for an oil cup between the top of the lever and the radius rod. The same conditions exists at the other bearings. The wearing surfaces of the pins and bushings are, therefore, oiled from the center of the pins. A hole is drilled into the center of each pin from the outer end and passes through the greater part of the pin, and oil holes lead out from the central hole to the bearing surface of the pin. Elbows *g* are screwed into the central holes, and are filled with hair to permit a constant and gradual feed of oil.

REVERSING

53. General Explanation.—A locomotive with the Walschaert valve gear is reversed by moving the reverse lever to the back corner of the quadrant and thereby placing the link blocks either in the top or bottom of the links, depending upon the arrangement of the gear. A locomotive valve gear will not positively reverse one of the engines of a locomotive in all positions. Thus, it is well known that a locomotive operating on one side cannot be reversed if it is near or on the dead centers, although it can be reversed in all other positions. In order for a locomotive to be reversed in any position, the main pins must be one-quarter of a turn, or 90°, apart. With the pins more or less than this amount apart, there would be positions in which the locomotive could not be reversed. The reverse lever must be moved all the way in the quadrant to insure reversal, and even then the starting impulse in the opposite direction is frequently imparted to one side only, the starting impulse varying with the position of the main pins. In certain positions the slack has to be taken so as to obtain amore advantageous position from which to start the train.

54. Principle Involved.—the principal involved in the reversing of an engine will be explained by referring to Fig. 22. With the main pin *c* on the bottom quarter and the eccentric crankpin *b* on the back dead center, the engine can be made to run backwards by moving the pin *b* to the front dead center at *b′*. The reason is that this condition places the eccentric crank the same amount ahead of the main pin when running backwards as forwards, and opens the back steam port the same amount as the front port was opened before reversal.

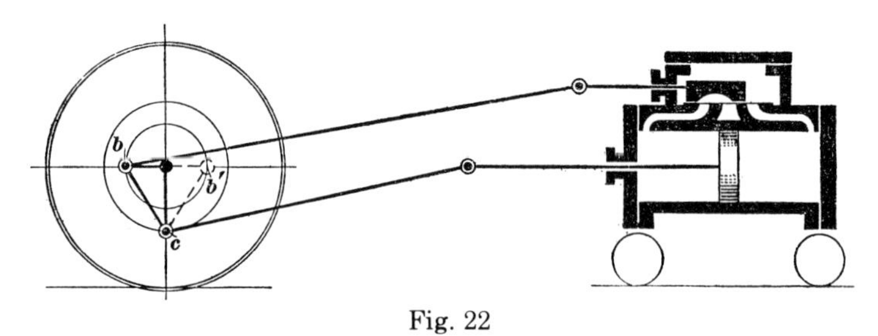

Fig. 22

55. The following fundamental principle can be derived from the foregoing: Reversal of motion requires eccentrics, or an effect equivalent to eccentrics at the same distance from the main pins, when moving backwards as when moving forwards. All locomotive valve gears are designed to conform to this principle when the reverse lever is moved from either corner of the quadrant to the other. With the Stephenson valve gear the back-up eccentric meets the requirement of an eccentric the same distance from the main pin when running backwards as when running forwards. With the Walschaert gear and but one eccentric crank, an effect is produced equivalent to changing the position of the crank one-half of a turn when the radius rod is moved from one end of the link to the other. Thus, although the eccentric crank is fixed, it can be considered as occupying the same position with respect to the main pin when the locomotive is running backwards as when it is running forwards. The effect of eccentric cranks at the same distance from the main pins when moving back as when moving ahead, places the valves in the proper position and also gives them the proper movement for the reverse direction.

56. Movement of Valves When Reversing.—The action equivalent to giving the eccentric cranks one-half of a turn when reversing with the Walschaert valve gear causes a certain definite movement of the valves. The valves will so move that steam will be admitted to the opposite end of one or both cylinders, depending on the position of the main pins, and the movement of the valves will be such when the engine starts as to keep it moving in the reverse direction. As already stated, the movement of the radius rods from the bottom to the top of the links has the same effect as giving the eccentric cranks one-half turn on the main pins. It will be found more convenient at first when studying the effect on the valves when a locomotive is reversed, to consider the movement of the eccentric cranks through one-half turn instead of shifting the reverse lever and radius rods.

57. In Fig. 23 *(a)* is shown the position of the valve gear when the main pin on the left side is on the top quarter, and view *(b)* shows its position on the right side with the main pin on the forward dead center. The valve gear on the right side will be considered first. If the eccentric crankpin is moved from *a* on the top quarter to the bottom quarter *a'* or one-half of a turn, the valve will be in the same position with the front steam port open the amount of the lead as shown. When the engine starts backwards, the valve moves ahead, or the movement is such as to keep the engine moving in the reverse direction. The same movement will be imparted to the valve if the radius rod *9* is moved to the position shown by dotted lines to the upper part of the link. However, as the steam is admitted to the same end of the cylinder, it is evident that the starting impulse backwards must be obtained from the left side.

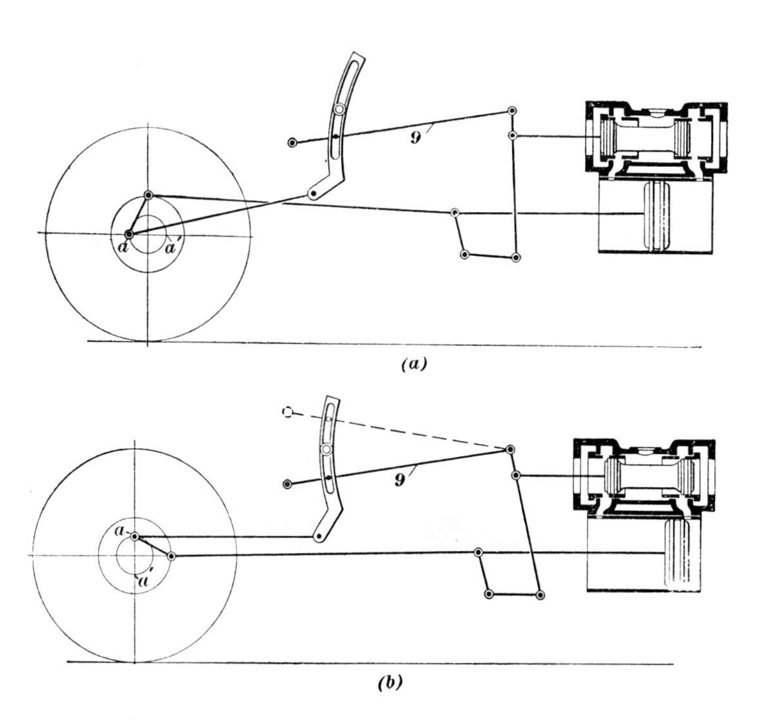

Fig. 23

In view *(a)* the back steam port is opened fully, as the eccentric crankpin *a* is on the back dead center. If the eccentric crank is given a one-half turn and moved to a *a′* on the front dead center, the valve will be moved to close the back steam port and open the front steam port an equal amount. Therefore, steam is admitted to the opposite end of the cylinder and the driving wheel will turn backwards. The valve when the engine starts will also move in the proper direction to keep the engine running backwards, that is, the valve will move back preparatory to opening the back steam port the amount of the lead when the piston reaches the back end of the cylinder.

The same effect is produced by moving the reverse lever and placing the radius bar in the top of the link. With the eccentric crankpin at *a* the link is inclined as shown, and the lower end is moved to its extreme position backwards and the upper end to its extreme position forwards. The radius rod when raised, will move ahead and the valve will be moved the same amount ahead of the center of the link trunnions as it was behind them before reversal. The valve then shifts the steam from the back end to the front end of the cylinder.

58. The more the links are inclined, the greater will be the horizontal movement of the radius rods and hence of the valves when the reverse lever is changed; and the less their inclination, the less will be the valve movement. The maximum and the minimum movements of the valve when a locomotive is reversed occur when the main pin on one side is on either dead center and the pin on the other side is, therefore, on the top or the bottom quarter.

In Fig. 23 *(a)* the valve on the left side opens the opposite steam port fully because the link is at its greatest inclination, on account of the eccentric crankpin being on the dead center. In view *(b)* the valve does not move because the eccentric crankpin is on the quarter and the link is in central position.

The foregoing can be summarized as follows: To reverse a locomotive it is necessary merely to produce an effect equivalent to placing the eccentric cranks in the same position relative to the main pins when running in one direction as in the other, for the reason that this action not only causes the valves to admit steam to the opposite end of one or both cylinders, but also gives the valves the proper movement to keep the locomotive moving. The valve movement depends on the position of the main crankpins, and is greatest with the pins on the quarters and the least on the centers.

59. As already stated, the amount that the valves will move when the engine is reversed will vary with each position of the eccentric crankpins and of the main crankpins. On account of practically the infinite number of positions in which the main crankpins can be placed, it is impossible to study the valve movement in all positions. However, in Figs. 24, 24, 26, and 27 are shown in the change in the position of the valves when the engine is reversed with the main pins in certain representative positions.

In Figs. 24, 25, 26, and 27 is shown the movement of inside-admission valves when a locomotive is reversed, with the main crankpins in different positions. The positions of the main crankpins are shown in view *(a)*, the position of the valve before reversal in views *(b)* and *(c)*, and their position after reversal in views *(d)* and *(e)*. The crankpins and the cylinders on the left side are marked *l*, and on the right side they are

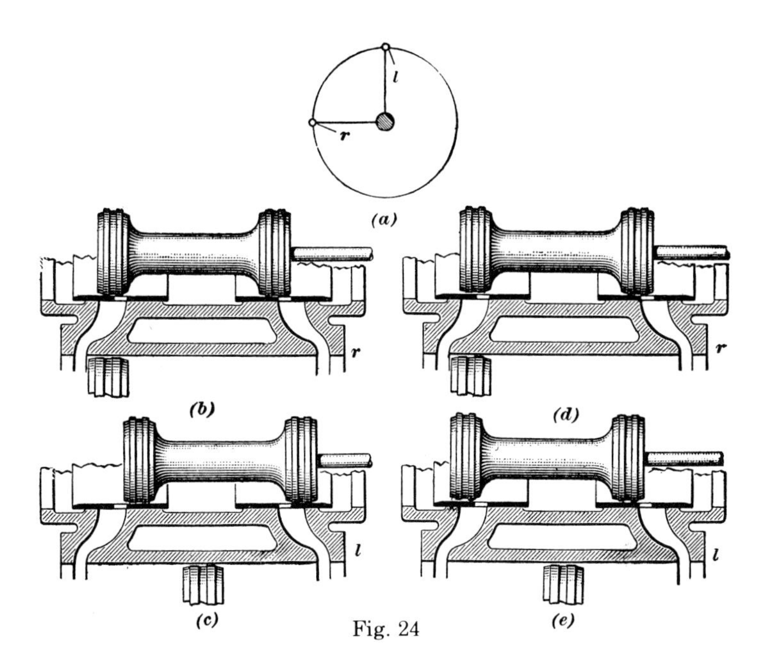

Fig. 24

marked *r*. As shown by the valve stems, the front of the engine points to the left.

It is assumed that the reverse lever is in full forward gear before reversal, and that it is in full backward gear after reversal. The lever is usually moved all the way back when reversing so that the maximum valve movement and the greatest port opening possible will be obtained.

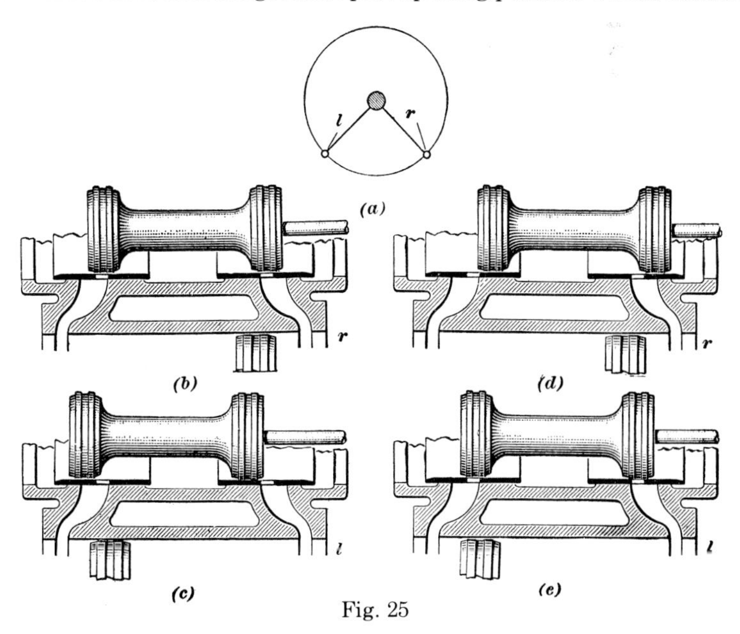

Fig. 25

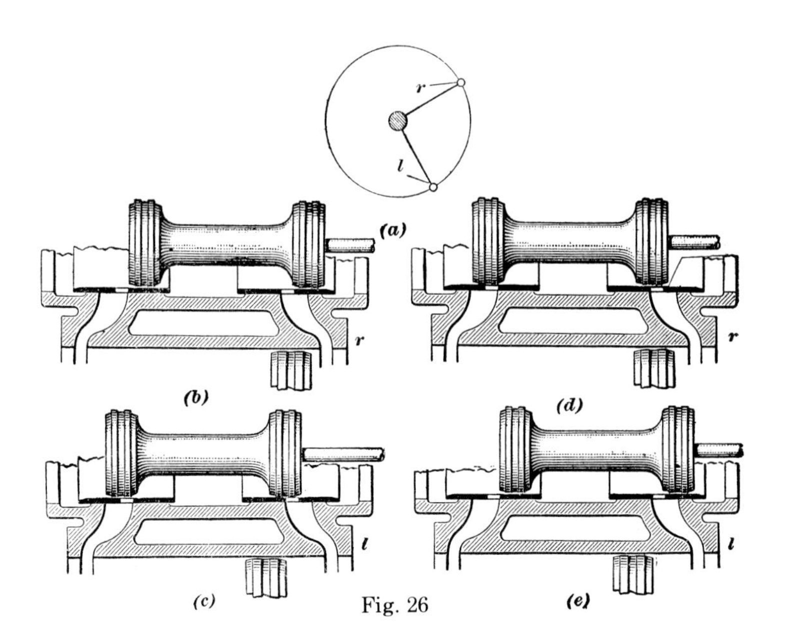

Fig. 26

With the main crankpins in the position shown in Fig. 24 *(a)*, the valve on the right side, view *(b)*, has the front steam port open the amount of the lead, and the valve on the left side, view *(c)*, has the back steam port wide open. The position of the valves is, therefore, the same as shown in Fig. 23. When the locomotive is reversed, the valve on the right side, view *(d)*, does not change its position. The valve on the left side, view *(e)*, opens the

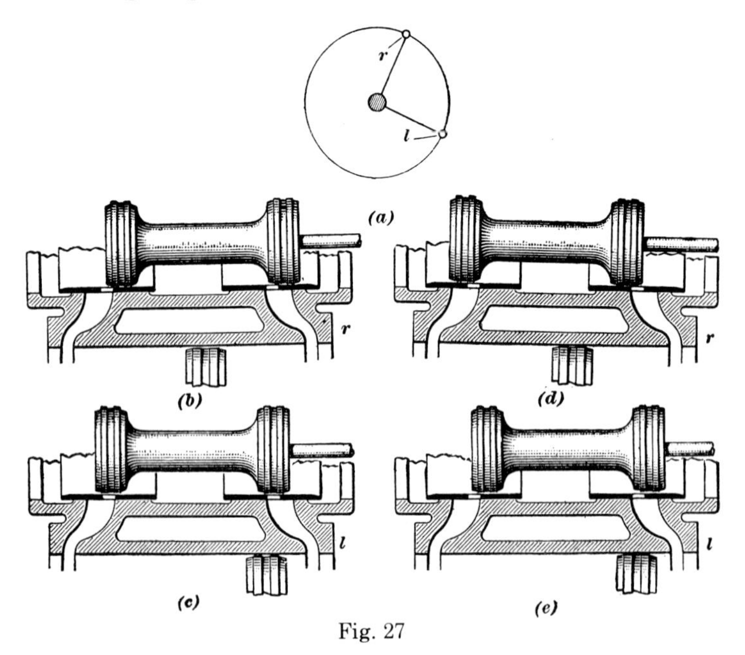

Fig. 27

front steam port the same amount as the back port before reversal. In this case the locomotive for the instant starts backwards from the left side.

60. The valve movement during reversal with the main pins as shown in Fig. 25, view *(a)*, is as follows: The valve on the right side, view *(b)*, is at cut-off at the front steam port, and the front steam port on the left side, view *(c)*, is fully open. When the locomotive is reversed the valve on the right side, view *(d)*, opens the back steam port fully and the valve on the left side, view *(e)*, moves to cut-off at the back steam port. In this instances the locomotive starts backwards from the right side.

With the crankpins in the position shown in Fig. 26 *(a)*, the valve on the right side has the back steam port opened fully and the valve on the left side, view *(c)*, is about at cut-off at the front port. When the locomotive is reversed, the valve on the right side moves to cut-off at the front port. The valve on the left side opens the back port fully and the locomotive starts back from this side.

61. With the crankpins in the positions shown in Fig. 27 *(a)*, the valve on the right side, view *(b)*, has the back steam port opened wide, and the valve on the left side is about to open the front steam port to exhaust. When the locomotive is reversed, the valve on the right side, view *(d)*, moves so as to open partly the front steam port, and the valve on the left side, view *(e)*, opens the back steam port fully. Therefore, the locomotive starts backwards from both sides. The foregoing shows that the locomotive when reversed usually receives a starting impulse backwards from one side. Generally a locomotive starts backwards from both sides providing that prior to reversal the locomotive had stopped with one main crankpin near to and approaching a dead center. Under other conditions the movement in the reverse direction is generally received on one side.

CHANGE IN CUT-OFF

62. General Explanation.—The term *cut-off* refers to the position of the valve when its steam edge is in line with the steam edge of the port, with the valve moving in a direction to close the port. A characteristic of all locomotive valve gears is that the cut-off can be varied or changed, that is, the valves can be made to cut off the steam to the cylinders at different points in the stroke of the pistons. A variable cut-off permits the power of the locomotive to be changed in accordance with the work being performed. The shortest or earliest working cut-off obtainable with any valve gear is secured with the reverse lever in the notch next to the center notch of the quadrant, and the longest cut-off is secured with the lever in either corner of the quadrant.

63. Reason for Shorter Cut-Off.—It is a comparatively simple matter to understand how an earlier cut-off is obtained with the Walschaert and Baker valve gears provided these gears are considered in their simplest form, namely, as being made up of a long eccentric crank and a shorter lap-and-lead crank spaced 90 degrees apart. Now a shorter cut-off implies that the valve has reversed the direction of its movement earlier in the stroke of the piston, and if it can be proved with a simplified form of gear that such an action occurs, the reason for a shorter cut-off with the actual gear as the reverse lever is drawn back will be apparent.

The proof of an earlier cut-off will then be based on the fact that such a cut-off implies an earlier reversal of the valve.

64. In Fig. 28 the theoretical movement of the valve in full gear for a rotation of the eccentric crankpin and the lap-and-lead crankpin through the spaces j and j' is found by drawing vertical lines through equally spaced points on the circles and then obtaining the difference between the widths of the spaces j and j'. As the space j exceeds the space j', the forward movement of the valve by the eccentric crank has not yet been

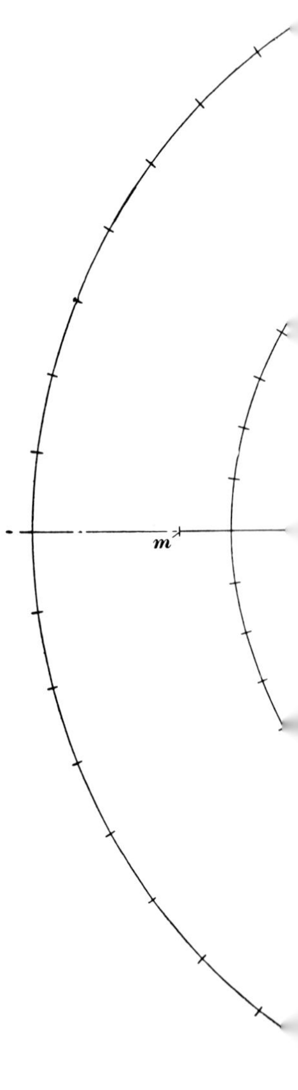

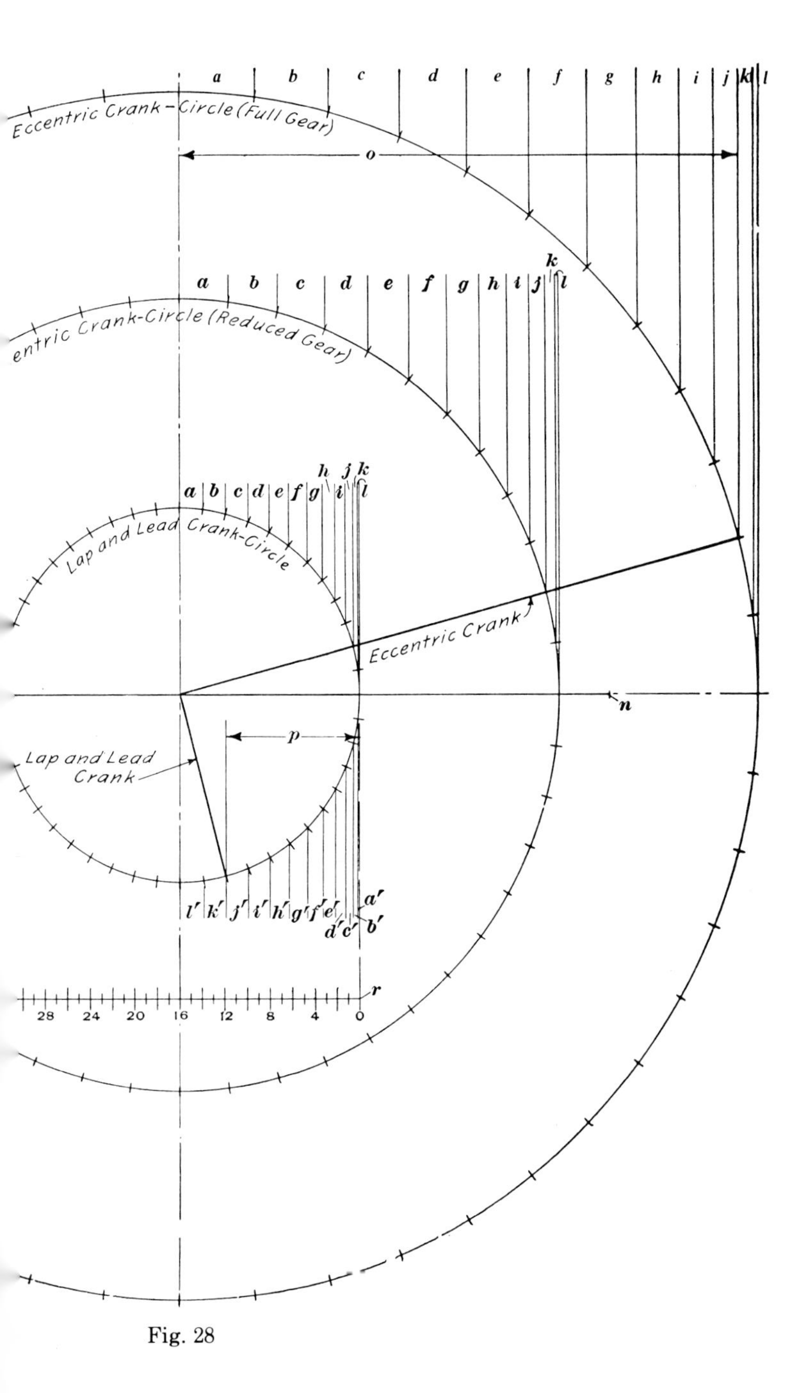

Eccentric Crank-Circle (Full Gear)
entric Crank-Circle (Reduced Gear)
Lap and Lead Crank-Circle
Eccentric Crank
Lap and Lead Crank
o
p
n
r
a b c d e f g h i j k l
l′ k′ j′ i′ h′ g′ f′ e′ d′ c′ b′ a′
28 24 20 16 12 8 4 0

Fig. 28

overcome by the backward movement imparted by the lap-and-lead crank, the net forward movement being equal to the difference between the spaces. For a rotation of the crankpins through the spaces *k* and *k'*, the space *k'* will be found to exceed the space *k;* this indicates that the influence of the eccentric crank on the valve has been neutralized and overcome by the lap-and-lead crank, hence the valve has reversed.

Reversal of the valve then occurs when its movement by the eccentric crank in one direction is overcome by the movement of the lap-and-lead crank in the opposite direction.

65. Next it will be shown what happens when the eccentric-crank circle has been reduced in length to that of a smaller circle as shown, this action being similar to raising the link block higher up in the link.

Reasoning the same as before it will be noted that space *i* on the second largest circle is greater than *i'* and that *j'* exceeds *j* so that the valve reverses at about where the line divides the spaces *i* and *j* or approximately 7½ degrees farther from the front dead center with the shorter crank than when it was longer. The reason is that the spacing of the vertical lines from the lap-and-lead crank remains unchanged, whereas the spacing of the lines from the eccentric crank comes nearer together as the length reduces, hence one must look farther back to find spaces of the same width.

Therefore, with two cranks of unequal lengths spaced 90 degrees apart, a progressive shortening of one crank with the length of the other left unchanged causes a correspondingly earlier reversal of valve movement and an earlier cut-off.

The reason, then, for the steam being cut off earlier in the stroke at all ordinary cut-offs with the Walschaert and Baker valve gears is that drawing up the lever diminishes the action of the eccentric crank on the valve without affecting the action of the lap-and-lead lever, the result being an earlier reversal of the valve and hence a cut-off earlier in the stroke of the piston.

66. With the eccentric crank reduced in length until it becomes the same length as the lap-and-lead crank and so falls on the lap-and-lead crankpin circle, the valve is reversed by the lap-and-lead crank when this crank is at the same angle in advance of its dead center position as the other crank is behind it or with both in their forward eighth positions. When the eccentric crank becomes of a lesser length than the lap-and-lead crank, the functions of the cranks reverse; the valve travel is then controlled by the lap-and-lead crank and valve reversal is accomplished by the eccentric crank. With the eccentric crank reduced in length to zero, as when the block is in the middle of the link, the valve reverses at the same instant as the piston, the reason being that no angle exists between the main crank and the lap-and-lead crank.

The reversal of the valve except in mid-gear, of course, only occurs with one crank above and the other below the horizontal center line of the circle. In Fig. 28 no account is taken of the angularity of the rods.

Although an earlier reversal of the valve brings about an earlier cut-off, this does not imply that cut-off will take place 7½ degrees earlier if reversal takes place this much earlier. Actually, shifting the influence from one crank to another has a greater effect on the cut-off than on the reversal. For example, if reversal occurs 5 degrees earlier, cut-off may be hastened by as much as 10 degrees, depending on the position of the reverse lever. A shorter interval of time takes place between reversal and cut-off in short cut-offs than in full gear.

67. Construction of Diagram.—In Fig. 28 the valve travel is assumed to be 9 inches, and the lap plus the lead 1⅞ inches. The length of the eccentric crank, which is the radius of the eccentric crank circle in full gear, has to be determined by trial for the reason that the influence of the short lap-and-lead crank subtracts from the movement of the long eccentric crank. In this instance, a circle with a radius of 6 1/16 inches was found necessary to obtain a travel of *m n*, or *9* inches. If the influence of the lap-and-lead crank had not taken away from the action of the eccentric crank, an eccentric crank circle of a radius of 4½ inches would be all that would be required.

The proper radius of the full-gear eccentric-crank circle to give the correct valve travel is found by first drawing a circle approximately 30 per cent larger in diameter than the valve travel, dividing it equally into degrees and locating the point at which the valve reverses. The horizontal distance *o* from this point to the vertical line through the center of the axle is then measured and from this is subtracted the distance *p*. If the result is 4½ inches, the circle is the correct radius; if less, a greater radius is required; if more, a smaller one.

By drawing various circles, the point of reversal for different crank lengths may be found, first, however, dividing the circles into arcs of the same number od degrees. In this case, points at intervals 7½ degrees apart were taken. More accurate results will be obtained with a large circle divided into two- or three-degree positions.

68. To find the position of the piston at the point of reversal, draw the line *q r* and divide it into thirty-two equal parts if the stroke is 32 inches; each part then represents 1 inch of stroke. A perpendicular drawn from a reversing point on the lap-and-lead circle will then show the theoretical position of the piston when the valve reverses; the angularity of the main rod will affect this slightly. For example, in the full-gear position a perpendicular from the line separating spaces *j′* and *k′* shows that the valve reverses after the piston has moved 12 inches.

In Fig. 28 the spaces between the division points on the eccentric-crank circles are identified by letters from *a* to *l* and the corresponding lap-and-lead spaces are lettered from *a′* to *l′*. The reason for doing this is to emphasize the right-angle position of the two cranks. When the eccentric crank is in the space *g* on any of the circles, the lap-and-lead crank must be in the space *g′*.

BACK-SET OF LINK FOOT

69. Angularity of Eccentric Rod.—The angularity of the eccentric rod affects the movement of the link foot and, therefore, of the valve in the same way that the angularity of the main rod influences the movement of the piston. This means that the link foot will not be midway between its extreme forward and backward positions with the eccentric crankpin on the quarters, halfway between dead centers. As a result, the valve will not move an equal distance on each side of its mid-position and the cut-off will not occur at the same point in the stroke for one stroke of the piston as the other unless a correction is introduced.

Thus, in Fig. 29 *(a)* when the eccentric crankpin moves from the forward dead center *a* to the bottom quarter at *b*, the bottom of the link foot, unless arranged differently than shown, will move from *c* to *d* which exceeds the movement from *d* to *e* through which the foot passes while the crankpin moves from the bottom quarter to the back dead center at *f*. In order to move the valve an equal amount on each side of mid-position and thereby make the cut-off equal at each end of the cylinder, it is necessary for the link-foot pin to move through the same space while moving from its extreme position forward at *c* to its central position *d*, as

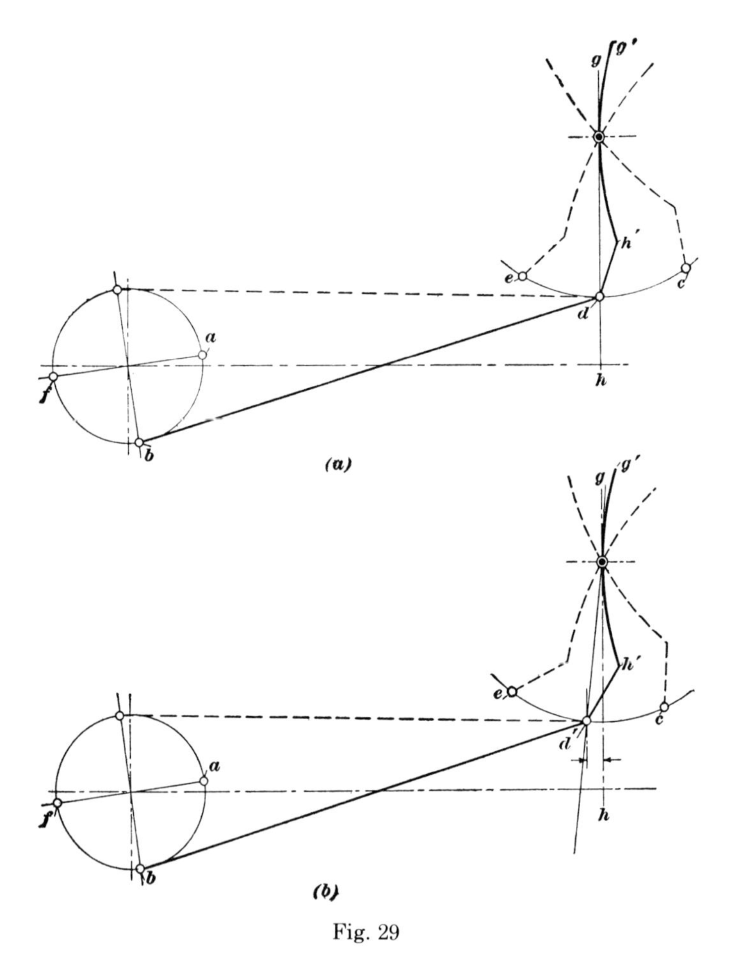

Fig. 29

when moving from *d* to its extreme position backwards at *e*. The link-foot pin is made to swing through an equal arc on each side of its central position and, thereby, equalizes the movement of the valve, by moving the link foot to the rear, or back-setting it.

70. Effect of Back-Setting Link Foot.—In Fig. 29 *(a)* the link foot is not backset because the line *g h* tangent to or touching the link arc *g′ h′* with the link in central position passes through the center of the link-foot pin *d*. Fig. 29 *(b)* shows the link *g′ h′* in central position as before, but with the link foot moved behind the line *g h* to *d′*. It might seem that setting the link foot back would merely decrease the forward swing by the amount of the back-set and increase the backward swing by the same amount, and thereby leave the condition the same as before. However, this is not the case, for while the forward swing of the link foot is restricted to an extend practically equal to the amount of the back-set, yet the backward movement is increased in excess of the back-set. The reason is that the back-set decreases the acuteness of the angle between the eccentric rod and the link foot as it is approaching the end of its rearward swing, thereby causing for any certain movement of the eccentric rod a greater swing of the link to the rear than would otherwise be the case. The effect in reality is a knuckle-joint action and is similar to a progressive shortening of the rod during the latter part of the rearward swing of the link. The forward swing of the link is not materially affected owing to the angle at this time remaining more nearly normal.

The result with the proper amount of back-set is that the arcs *c d′* and *d′ e* become equal and the link foot and the valve will move the same distance on each side of their center positions. The exact amount of the back-set is obtained by trial when designing the gear, that is the poind *d′* is moved backwards until it swings as far to the rear of its central position as it does in front of it.

An error is sometimes made in stating that the link foot is back-set to correct the angularity of the main rod. This is not the case, as the main-rod error is small, amounting with a rod 10 feet in length to about ¾ inch. This error is taken care of by a slight adjustment of the eccentric rod in setting the valves. This equalizes the cut-offs but only at the expense of port openings, the inequality amounting to about ⅛ inch in the running cut-offs, with the back port opening the wider. In actual practice, the main-rod error appears only while the valves are being set, when it is corrected without its significance being generally recognized.

BREAKDOWNS

71. Parts Most Easily Disconnected.—To expedite repairs so that the main track may be cleared as quickly as possible, one should be familiar with the parts of the Walschaert gear that are the most easily disconnected. Parts of the gear that fail and that are difficult to take off should be left alone and the required repairs made by removing more easily disconnected parts, that may be intact. The eccentric rod, the union link, and the pin in the top of the radius-rod hanger are the easiest parts of the valve gear to disconnect, and they can be very easily removed even with the limited tool equipment found on locomotives.

Practically all breakdowns of the Walschaert valve gear can be handled with safety to the engineer by the removal of these parts. On the other hand, it is practically impossible for the engineer to disconnect the radius rod from the top of the combination lever. First, the bolts *d*, Fig. 19, are very difficult to remove, then before the pin *h* can be driven out, one guide has to be taken off and this requires the removal of its two bolts. Also, work on this part of the valve gear places the engineer in a dangerous position astride of the radius rod with his feet resting on the crosshead; the performance of work in this manner is not in accord with safety-first principles.

72. Broken Eccentric Crank, Eccentric Rod, Link Foot, Radius Rod in Front of Link, Combination Lever, Union Link or Valve Stem.—Should any of the parts mentioned break, remove the eccentric rod, disconnect the radius-rod hanger from the reverse-shaft arm, disconnect the union link from the crosshead, tie the combination lever forwards out of the way, and secure the valve in mid-position. The best way to loosen the hanger pin without damage is to take out the cotter pin, partly unscrew the nut, and then strike it with a hammer.

In the event of a valve not stopping in a position to cover the ports when a valve gear fails, the valve can be moved to the desired position by moving the bottom of the link. To prevent the valve from moving out of this position, when drifting, loosen up the nuts on the crosshead-guide bolts, drive a thin wedge between the top of the crosshead and the guide, and tighten the nuts. If there is enough play, the wedge may be driven between the side of the crosshead and the guide.

Another method, should it be desired to retain the lap-and-lead movement of the valve with the radius rod, the combination lever, and the union link intact, is to place the reverse lever in mid-gear, thus bringing the link block in the center of the link, and drive pieces of wood through the link above and below the radius rod, Fig. 30. Then disconnect the radius-rod-hanger pin from its arm.

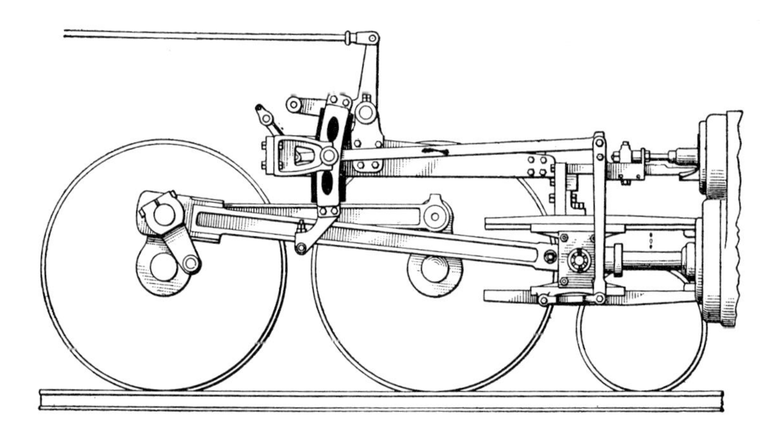

Fig. 30

73. The amount of steam admitted to the cylinder does not help the movement of the piston materially because the maximum port opening or the lead, which is about ¼ inch, occurs with the piston at the beginning of the stroke, and the valve begins to close the port as soon as the piston begins to move. However, the steam lubricates the piston and the cylinder. The locomotive should be stopped on the working side with the pin on the quarter, because, if it is stopped on the center, the valve on the partly disabled side will be in mid-position and the locomotive cannot be started unless the combination lever is disconnected at the lower end and the valve moved in the proper direction. If the back end of the radius rod is carried in a slip block, as shown in Fig. 4, the block will have to be removed by taking out the four bolts shown, before the radius rod can be blocked in the link.

The eccentric rod is the part of the valve gear that usually breaks and the failure usually occurs where the forked front end is welded to the rod. Another cause of failure of the eccentric rod is that the nuts work loose on the eccentric-rod pin, and the pin works out and strikes the main rod. A failure of the combination lever or union link is liable to lead to serious consequences because the valve stops moving; and should it stop in a position to trap steam in the cylinder, the compression may become high enough to blow out a cylinder head, bend the main rod, or even demolish the cylinder saddle. The nuts on the pins in the ends of the union link should be inspected frequently to see that they are not working off.

74. Broken Radius Rod Back of Link, Broken Radius-Rod Hanger, Reverse-Shaft Crank, or Reach Rod.—Should the radius rod break back of the link or should the radius-rod hanger break, block the link block at the position required to start the train, by placing blocking in the link above and below the radius rod. A little slack should be left in the blocking to provide for the slip of the block. However, if blocked tight, the movement of the block will soon make the required slack. The locomotive must not be reversed unless the blocking is first changed. The purpose of blocking over the radius rod is to prevent the link block from moving up to back gear. Should the reach rod or the arm of the reverse-shaft crank to which the rod is connected, break, block both radius rods at the required cut-off, and disconnect both radius-rod hangers.

75. Broken Front Cylinder Head.—When the front cylinder head breaks, see first that the good side is on the quarter so that the train can be started. Then remove the eccentric rod, disconnect the radius-rod hanger from the reverse shaft arm, center and clamp the valve, remove the union link, and tie the combination lever forwards. Steam from the other side of the engine will appear at the front cylinder head when the locomotive is in motion; this, however, should not be taken as an indication that the valve is not secured accurately. If thought necessary, the steam lap of the valve may be used to close the front steam port.

WALSCHAERT VALVE GEAR

(PART 2)

EXAMINATION QUESTIONS

Notice to Students.—*Study the Instruction Paper thoroughly before you attempt to answer these questions.* **Read each question carefully and be sure you understand it;** *then write the best answer you can. When your answers are completed, examine them closely, correct all the errors you can find, and* **see that every question is answered.**

(1) How is the pin that connects the combination lever to the valve-stem crosshead prevented from turning in the lever?

(2) What should be done if the radius-rod hanger breaks?

(3) In what positions of the main pins does the locomotive generally start backwards from both sides?

(4) What is the purpose of the gear frame?

(5) What should be done if the radius rod breaks?

(6) What is the purpose of the link?

(7) Why is a failure of the union link liable to lead to serious consequences?

(8) What is the purpose of the eccentric crank?

(9) Explain the valve movement that occurs when the locomotive is reversed with the right side on the forward dead center?

(10) How is the link-block pin prevented from turning in the radius rod?

(11) What should be done if the eccentric crank, eccentric rod, or link foot breaks?

(12) Explain the reason for using different types of gear frames.

(13) What is the purpose of the radius rod?

(14) What should be done if the combination lever or union link breaks?

(15) How is the eccentric crank held on the main pin?

(16) What is the purpose of the reverse shaft?

(17) What is the purpose of the radius-rod hanger?

(18) With the main pins in the positions given in Fig. 24, from what side of the locomotive does the starting impulse backwards come?

(19) What is the purpose of the eccentric rod?

(20) What is the purpose of the combination lever?

Baker Locomotive Valve Gear

By
J. W. Harding

5370 Printed in U.S.A. Edition 1

1945 Edition

BAKER LOCOMOTIVE VALVE GEAR

ARRANGEMENT AND OPERATION

INTRODUCTION

1. A locomotive valve gear is the mechanism used to operate the valves of a locomotive. The movement imparted to the valves by the valve gear is such as to cause the locomotive to be run either forward or backward; also, the action of the valve gear causes the valves to cut off the steam from the cylinders at any desired point in the stroke of the piston. A locomotive valve gear for a two-cylinder locomotive comprises two sets of identical mechanism, one for each engine, so connected by a shaft that a reversal of motion or a variation in cut-off can be accomplished by the movement of a single lever in the cab known as the reverse lever.

It is much easier to understand the operation of a locomotive valve gear by beginning with an elementary design and then tracing its development to the gear that is under consideration. Hence, the operation of the Baker locomotive valve gear with particular reference to the manner in which reversal of motion is obtained, will be explained by considering first a simple form of valve gear, as in Fig. 1, applied to an engine with a valve without steam lap. A valve gear without the reversing feature would be a simple device; it is the introduction of this feature, combined with a variable cut-off, that makes the operation of a valve gear more difficult to understand. These features require the introduction of additional parts and make the mechanism as a whole less easy to understand than otherwise.

UNDERLYING PRINCIPLES

2. Elementary Valve Gear.—Included in the valve gear shown applied to the elementary type of engine in Fig. 1 is an eccentric crank *ab* set on the axle *a* at right angles to the main crank *ac;* also, an eccentric rod *bf* connected to the lower end of a rod *fg,* which will be called a gear connecting-rod, the upper end of which is coupled to a bell-crank *h,* pivoted at a fixed point *i*. The gear connecting-rod is flexibly connected at *j* to a bar *jk* known as a radius bar, pivoted at *k* to a fixed point. A valve rod *lm,* one end connected to an arm of the bell-crank and the other end to the valve stem, completes the assembly. All parts are free to move in their paths of motion as shown by the arcs, with the exception of the fixed points *i* and *k*. With the engine on the dead centers, the ends of the main rod are shown curved to avoid confusing the rod with other lines and points.

3. Position of Valve With Respect to Piston.—To start the piston moving when the wheel is given a slight turn in the direction of rotation,

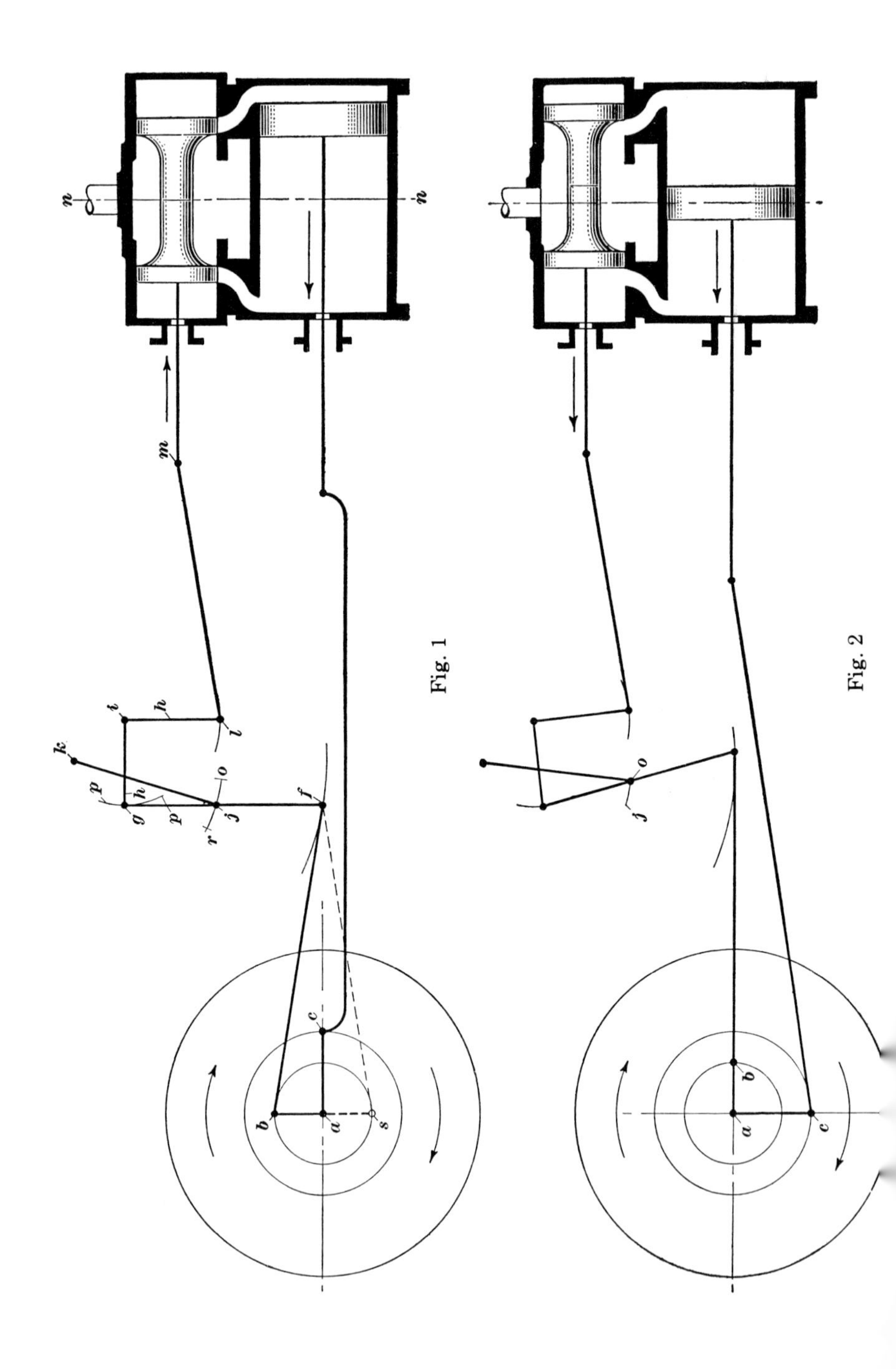

Fig. 1

Fig. 2

which, in this case, is forward, the valve, with the piston at the beginning of the stroke, must be at the point of admitting steam to the cylinder in front of the piston, as shown in Fig. 1. In this position, the valve is at mid-stroke or in mid-position, because the line *nn* drawn through the center of the valve (halfway between the outer ends) comes midway between the steam ports in the valve seat. The valve when in mid-position is one-half stroke behind the piston, and, if this difference in position is maintained, the steam will be admitted to and exhausted from the cylinder in such a manner as to keep the piston, and therefore the driving wheel, in motion as long as steam is supplied. To keep the valve one-half stroke behind the piston, on the assumption that the driving wheel is turning forward, the eccentric crank must be placed one-quarter turn behind the main crank, because the eccentric crank, in moving the valve, is similar to the main crank in moving the piston, where one-quarter turn of the crank moves the piston one-half stroke.

4. Operation.—The operation of the valve gear in keeping the engine in motion is as follows: When the main crankpin *c*, Fig. 1, turns in the direction of the arrow, the rotation of the axle *a* will impart a circular movement to the eccentric crankpin *b*, which in turn will move the eccentric rod and the lower end of the gear connecting-rod to the right. But as the gear connecting-rod is pivoted to the radius bar at *j*, this point, which will be referred to as the pivot pin, will be constrained to follow the arc *jo* and move downward. The end of the bell-crank will be pulled down and will follow the arc *gp*, thus causing the end *l* of the crank to move in an arc and move the valve ahead. This action will continue until the eccentric crankpin reaches its forward dead-center position, at which time the pin *j* will have reached the limit of its movement on the arc *jo* as shown in Fig. 2. The valve has now moved the limit of its travel to the right and will have the front port wide open for the admission of steam and the back port wide open to the exhaust. The main crankpin is now on the bottom quarter and the piston has moved one-half stroke. As the main crankpin continues to turn to the back dead center and the eccentric crankpin to the bottom quarter, the pivot pin will move up on the arc *oj* until, with the main crankpin on the back dead center, the pin will be back again to the starting point *j* and the valve will be in mid-position as shown in Fig. 3. With the main crankpin turning to the top quarter position and the eccentric crankpin turning to the back dead center, the pivot pin will swing up to its limit of movement on the arc *jr*, Fig. 4, and cause the bell-crank to move the valve back and open the port fully. For the remainder of the stroke of the piston, the pivot pin will move down on its arc until with the main crankpin on the forward dead center, the valve gear and the valve will again be in the position shown in Fig. 1. Hence, for one turn of the driving wheel, the pivot pin will swing twice in the arc *or*.

5. Gear Not Reversible.—The valve gear just considered will impart the proper movement to the valve to keep the engine running forward, but it is deficient to the extent that the engine cannot be reversed and caused to run backward. For example, if the main crankpin *c*, Fig. 1, is turned backward, the point *j* will move upward on its arc and the action of the valve gear will be such as to cause the valve to move backward and open the front steam port to the exhaust, instead of moving forward and opening the port to steam. For the engine to run

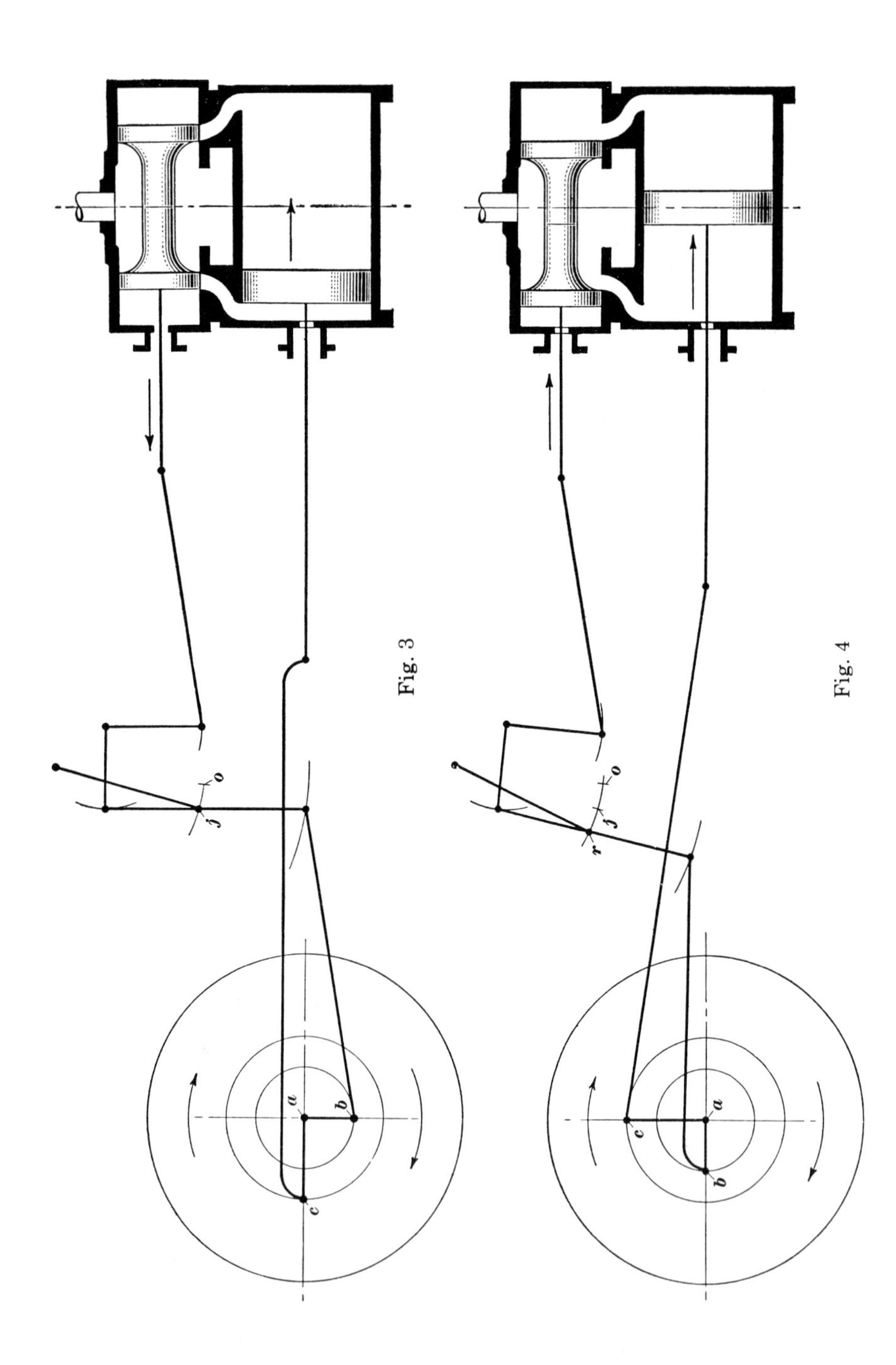

Fig. 3

Fig. 4

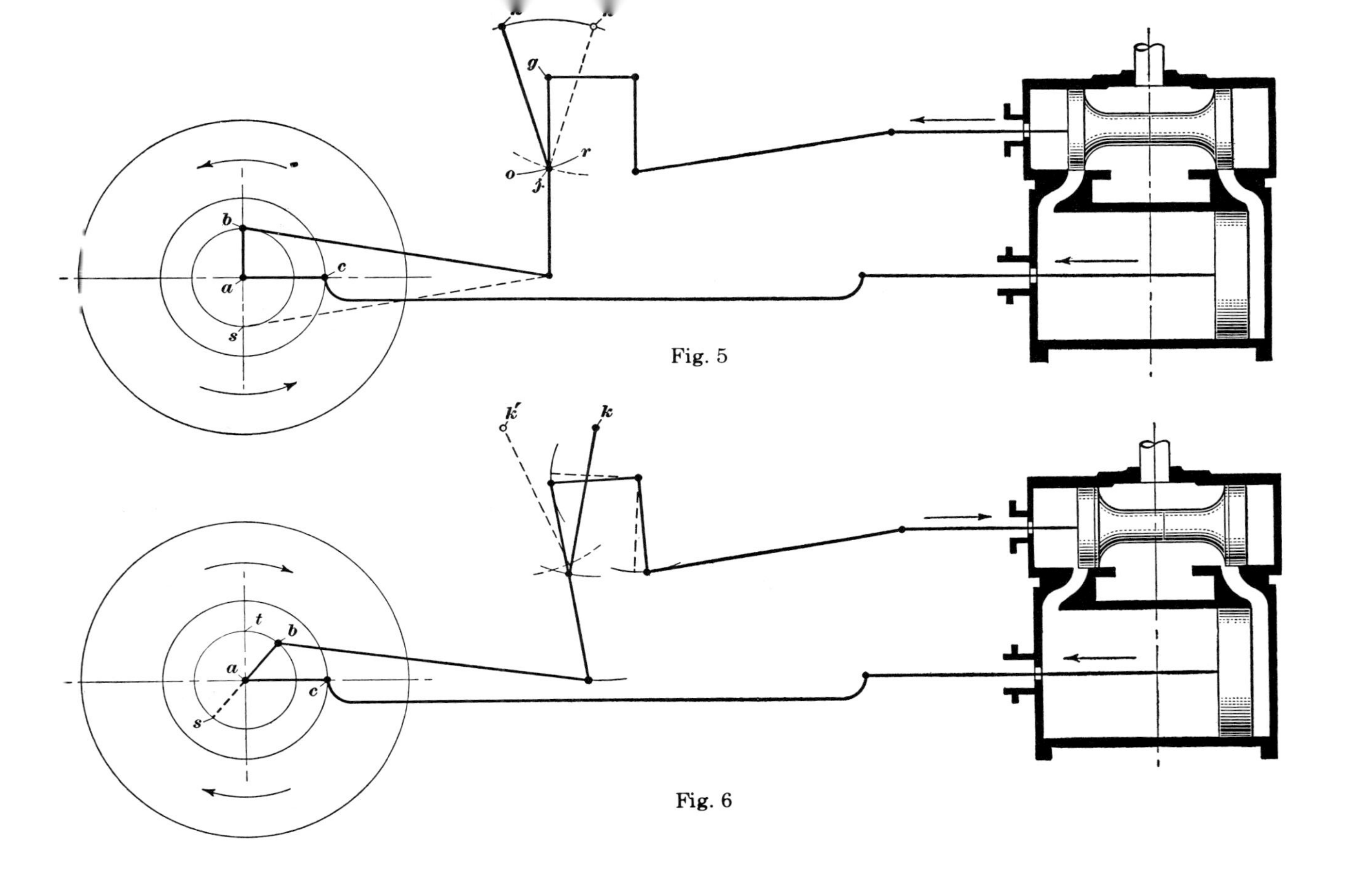

Fig. 5

Fig. 6

backward, the eccentric crankpin must be placed at *s*, because it is fundamental that the eccentric crank must be in the same position relative to the main crank when the engine is moving backward as when it is moving forward. That is, if the eccentric crank has to be set to follow the main crank in order that the engine may move forward, then it must be set to follow the main crank by the same amount so that the engine may move backward.

6. It would be impracticable to give the eccentric crank one-half turn each time it was desired to reverse, but exactly the same effect on the valve can be obtained by moving the pivot point *k*, Fig. 5, of the radius bar to the position *k'*. Such a movement will cause the pivot pin to follow the arc *or* and this will give the same movement to the valve when the driving wheel is turned backward as if the eccentric crank were given one-half turn from its forward-motion position to *s*. Thus, when the wheel is turned backward with the pivot point changed to *k'*, the pivot pin will begin to move downward on the arc *jo*, thereby pulling the end *g* of the bell-crank down and moving the valve ahead. Therefore, shifting the pivot point from *k* to *k'* has the same effect on the operation of the engine as if the pivot point were left in its original position at *k* and the eccentric crank shifted one-half turn, or 180 degrees.

7. Effect of Lap and Lead on Arrangement.—Owing to the back pressure that would develop on the return stroke of the piston, a steam engine cannot be operated at high speed unless provision is made to begin the exhausting of the steam before the piston completes its previous stroke. Neither can an engine be operated with economy unless provision is made for moving the piston by the expansive force of steam. Both of these conditions can be met by giving the valve steam lap. Steam lap not only causes an earlier release of the steam, but it also introduces an interval between the valve events, cut-off and release, instead of these events occurring simultaneously as with a valve with no lap. During this interval the steam expands in the cylinder and moves the piston. It will now be explained why the valve gear just considered cannot be used when a valve is given steam lap.

8. In Fig. 6 the valve gear already shown is employed to operate the valve when it has been given steam lap. This change in the valve requires the eccentric crank to be moved away from its position *at*; that is, at right angles to the main crank, to a position *ab*. The purpose is to displace the valve from mid-position the amount of the lap and the lead and have the front steam port open the amount of the lead when the piston is at the beginning of its backward stroke. If the position of the eccentric crank were not changed, the valve would be in mid-position with the piston at the end of either stroke. In such an event, the steam edges of both ports would be blanked by the amount of the steam lap, and steam would not be admitted to the cylinder until the piston had moved some distance on its stroke. This change in the position of the eccentric crank pulls the gear connecting-rod out of the position it occupied in the other illustration.

9. The valve gear will cause the valve to operate properly to keep the engine running forward, but it will not permit the motion to be reversed. A movement of the pivot point *k* of the radius bar to *k'*, as explained in Art. 6, has the same effect on the valve as if the eccentric crank were given one-half turn on the axle, and this would bring the crank to *s*. The valve would then be the amount of the lap and the lead to the left of its mid-position and would open the back steam port to steam; hence the piston could not move backward because of being blocked by the steam behind it. From this it follows that when one eccentric crank is employed for both forward and backward motions, it must be placed at right angles to the main crank, because this is the only position that will shift the effect of the eccentric crank to a similar or right-angle position on the other side of the main crank when the engine is reversed. Therefore, the right-angle setting of the eccentric crank must be retained and this requires that some means must be employed to bring the valve to its proper position with respect to the piston at the beginning of the piston stroke.

10. Arrangement for Obtaining Lap-and-Lead Displacement of Valve.—The right-angle setting of the eccentric crank can be adhered to and the lap-and-lead displacement of the valve from mid-position with the engine on the dead center can be obtained theoretically by the use of another crank set at right angles to the eccentric crank and acting on the valve independently of the eccentric crank. Thus, in Fig. 7, the lap-and-lead crank of a length equal to the lap and the lead is set 90 degrees in advance of the eccentric crank with the wheel turning forward. With the eccentric crank alone, the valve would be in mid-position with the main crank on the dead center, but the introduction of the lap-and-lead crank

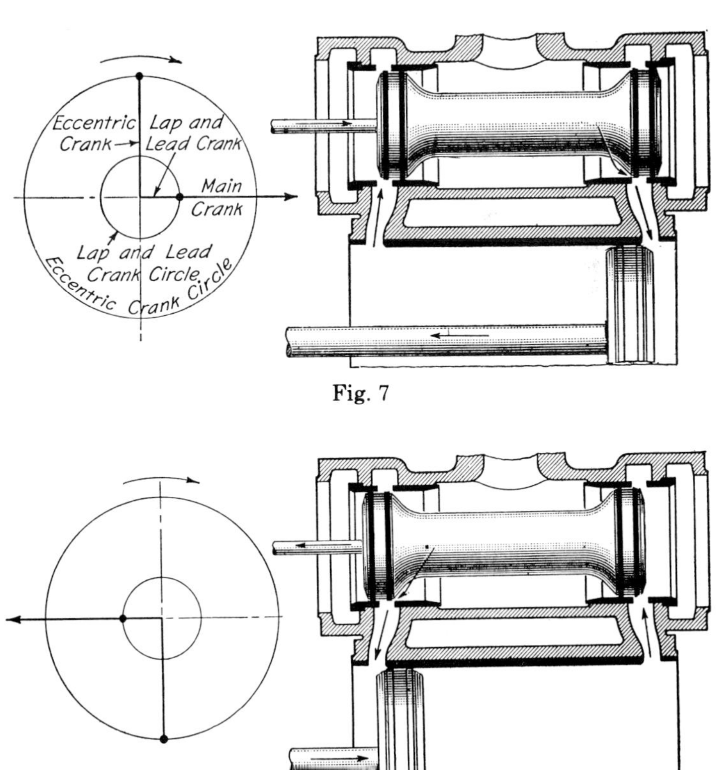

Fig. 7

Fig. 8

moves the valve the amount of the lap and the lead ahead of mid-position and opens the steam port in front of the piston the amount of the lead. If the cranks are given one-half turn, thus bringing the main crank to the back dead center, the lap-and-lead crank will then place the valve the amount of the lap and the lead to the rear of its mid-position as in Fig. 8, and open the back steam port the amount of the lead. The lap-and-lead crank, then, as it leads the eccentric crank one-quarter of a turn, always places the valve in its proper position relative to the piston at the beginning of each stroke.

11. Effect of Cranks on Movement of Valve.—The effect of two cranks of unequal lengths both acting independently on the valve is to cause it to move at a variable speed with respect to the piston, and it is due to this fact that the valve is always brought to the correct position relative to the piston at the beginning of each stroke. As the piston is nearing the end of its stroke, the two cranks are acting in conjunction and the speed of the valve increases with respect to the piston, thereby advancing the valve far enough ahead of the piston to open the port the amount of the lead, at the time the piston completes its stroke. The cranks are acting in opposition when the piston is beginning its stroke, so that the speed of the valve is less and the port opens slowly. Also, with the cranks acting in opposition to one another, the action of the lap-and-lead

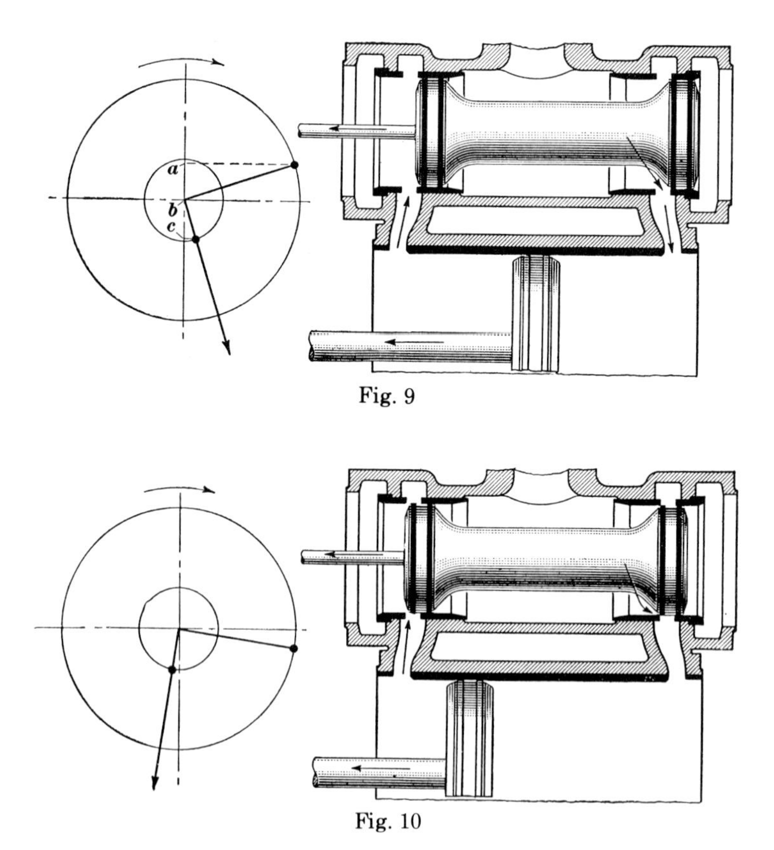

Fig. 9

Fig. 10

crank cases an earlier reversal of the valve and thereby decreases the valve travel imparted by the eccentric crank.

12. In the position of the cranks shown in Fig. 7, the action of one will begin to oppose the action of the other on the valve as soon as the cranks begin to move. The main crankpin is passing the front dead center, and the eccentric crank acts to move the valve forward and the lap-and-lead crank tends to pull the valve backward. However, the action of the longer crank predominates, but the valve will open the port slowly, owing to the retarding influence of the lap-and-lead crank. This action will continue until the cranks arrive in the position shown in Fig. 9. The effective length *ab* of the eccentric crank now becomes equal to the effective length *bc* of the lap-and-lead crank, and the valve has for the instant stopped. The eccentric crank is now moving the valve ahead at the same rate as the other crank is moving the valve back.

In Fig. 10, the two cranks are shown acting together to draw the valve back to cut-off at the front port, and the now higher valve speed will advance the valve the amount of the lap and the lead ahead of the piston at the time the main crankpin reaches the back dead center, Fig. 8. Were it not for the additional speed imparted to the valve at this time by the lap-and-lead crank, the valve would be in mid-position, or the amount of the lap and the lead from its correct position.

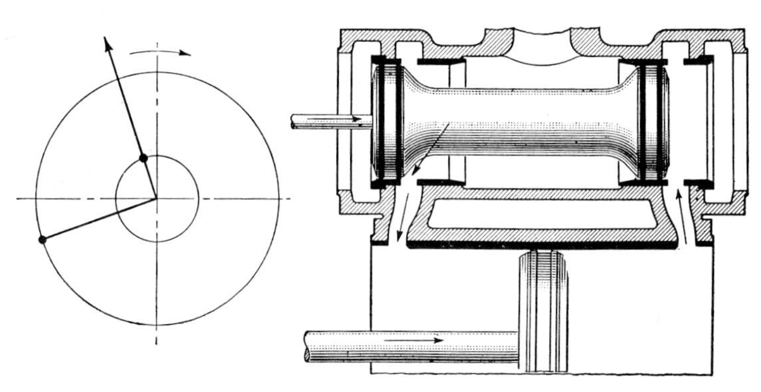

Fig. 11

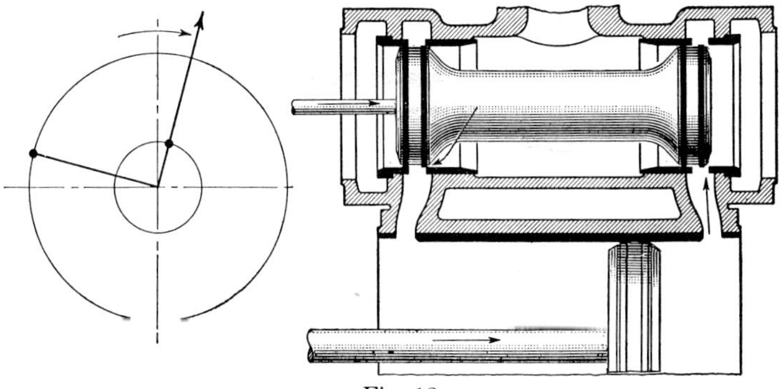

Fig. 12

In Fig. 11 the cranks are shown acting in opposition, the result being, as before, a slow movement of the valve in opening the back port.

In Fig. 12, the cranks are acting together to move the valve to cut-off at the back port, the speed of the valve now increasing to such an extent that it is brought to its proper position with repsect to the piston at the beginning of the backward stroke, Fig. 7. It will be noted that the lap-and-lead crank lies on the same line as the main crank so that if this crank were acting by itself the valve would always be moved in the same direction as the piston and close the port away from which the piston is moving.

13. Equivalent Lap-and-Lead Crank.—As it would be impossible to apply rods, without interference, to the cranks arranged as already shown, it is necessary to locate the lap-and-lead crank or its equivalent in some other position. The lap-and-lead crank lies in the same position and comprises a part of the length of the main crank, and the main crank obtains its movement from the piston and the crosshead. Hence, the crosshead can be employed to impart movement to a substitute lap-and-lead crank which, owing to the horizontal movement of the crosshead, must necessarily be a lever so arranged that its movement when transmitted to the valve will be equal to that of the lap-and-lead crank. This can be accomplished by a lever *ab*, which will be called the combination lever, arranged as shown in Fig. 13, with one end connected by a link *bc* to the crosshead, the other end connected to the valve rod *ad*, and the valve stem connected at the proper point *e* between the ends.

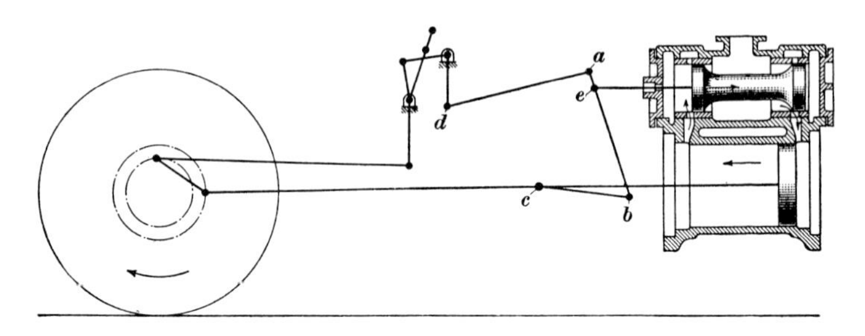

Fig. 13

With the main crankpin on the forward dead center, the valve would be in mid-position if the valve stem were connected directly to the valve rod. But with the valve stem connected to the combination lever lower down at *e* and the main crankpin in the same position as before, the valve will be displaced the amount of the lap and the lead ahead of its mid-position, and the front steam port will be open the amount of the lead. With the main crankpin on the back dead center, the valve will be the amount of the lap and the lead to the rear of its mid-position, and the back port will be open the amount of the lead. Thus, for one stroke of the piston the valve will move twice the amount of the lap and the lead when the combination lever is alone acting on the valve. Therefore, the combination lever serves the same purpose as the lap-and-lead crank and may be regarded merely as this crank placed in a different position. The distance the valve stem must be connected below the upper end of the combination lever to obtain the required valve movement if the lost motion in the connections is disregarded, is found by multiplying the total length of the lever by twice the lap and lead and dividing by the stroke of the piston.

14. The arrangement of the eccentric crank and the combination lever will produce practically the same action on the valve as the two cranks already described. That is, as the combination lever replaces the lap-and-lead crank, this lever, at certain times, increases the speed of the valve to such an extent that it is placed the amount of the lap and the lead in advance of the piston at the beginning of the stroke. At other times the combination lever acts not only to reduce the speed of the valve but also the valve travel. The variable speed imparted to the valve by the combination lever results in the port opening slowly after the lead opening has been obtained, and closing rapidly.

The movement imparted to the valve by the eccentric crank decreases as the reverse lever is drawn up until finally a point is reached on the quadrant where this movement becomes less than twice the lap-and-lead travel; the functions of the cranks then reverse. The total valve travel is now obtained from the combination lever, and the eccentric crank acts only to vary the speed of the valve. The valve travel now remains the same as the lever continues to be drawn back, even to mid-gear, but in the latter position the action of the eccentric crank in varying the speed of the valve is at a minimum.

With the combination lever all the way forward as with the crosshead at its extreme forward movement, Fig. 13, the equivalent lap-and-lead crank would be on its front dead center. With the lever all the way back, the crank would be on the back dead center, and, with the crosshead halfway in the guides, the crank would be on the quarter.

DESCRIPTION

15. Comparison With Walschaert Gear.—The Baker valve gear resembles the Walschaert valve gear in many respects. The gear is placd outside the frames and the valve derives its movement from an eccentric crank attached to the main crankpin and a combination lever connected at one end to the crosshead. Unlike the Walschaert valve gear a link is not used but a different arrangement that involves no sliding movement between the parts is employed to obtain a reversal of motion and a change in cut-off, these parts turning on pins equipped with roller bearings. The Baker valve gear is manufactured by the Pilliod Company.

16. Names of Parts.—The Baker valve gear is illustrated in Fig. 14, the gear being shown applied to a locomotive having inside-admission valves. The names of the parts are as follows: *1*, the eccentric crank; *3*, the eccentric rod; *38*, the gear connecting-rod; *19*, the reverse yoke; *16*, the radius bars; *10*, the bell-crank; *28*, the gear frame; *24*, the valve rod; *22*, the valve stem; *23*, the valve-stem crosshead; *4*, the combination lever; *27*, the union link; *30*, the gear reach rod; *37*, the reverse-shaft arm, connected to a reverse shaft which extends across the engine and connects to a similar arm on the other side; *31*, the reach rod. With the exception of the reach rod, all of the parts named are duplicated on the other side of the locomotive.

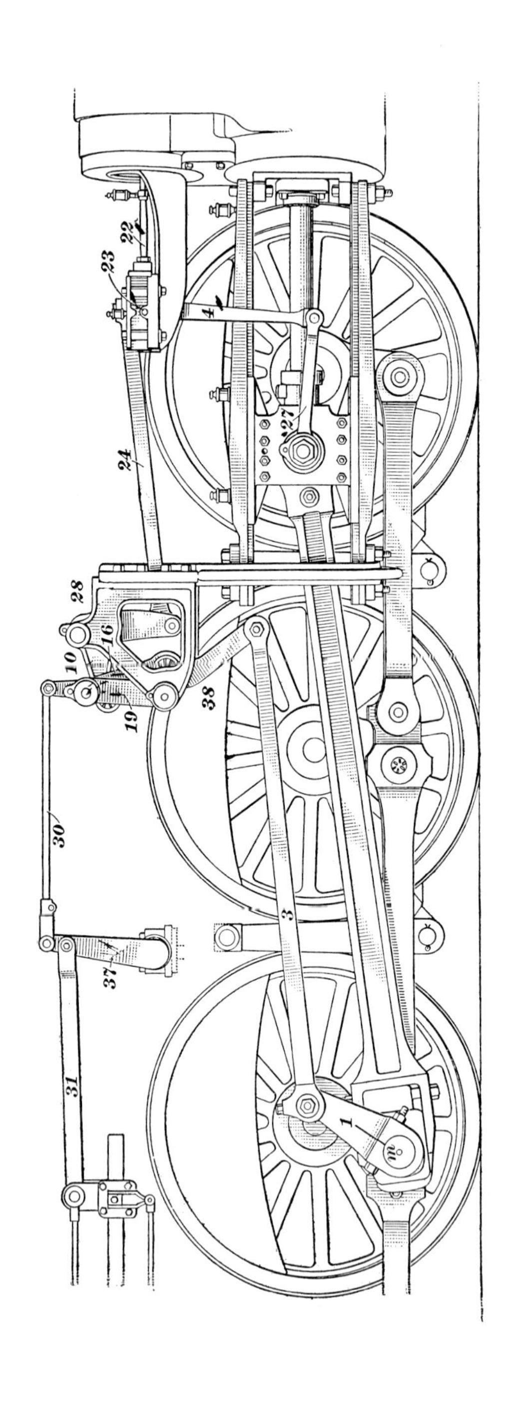

Fig. 14

17. Details of Arrangement.—It is difficult to understand the relation that exists between the different parts of the gear, particularly at the reverse yoke, from an outside view. Therefore, a view of a late type of reverse yoke and radius-bar assembly is shown in Fig. 15, and a rear view of the complete assembly at the reverse yoke, with one-half in section, is shown in Fig. 16. In brief, the arrangement comprises a gear connecting-rod *3*, with the lower end not shown, flexibly connected by pins to the bell-crank *6* and the radius bars *5*, the upper ends of which are suspended on pins in the reverse yoke *4*, which in turn is carried on pins in the gear frame *7*; all pins being equipped with roller bearings.

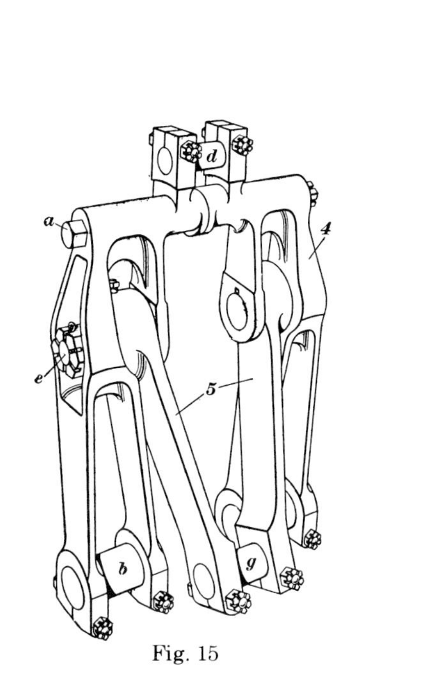

Fig. 15

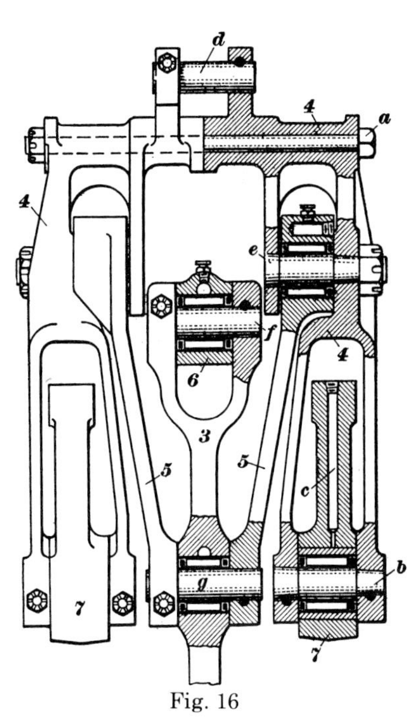

Fig. 16

18. The reverse yoke, which is made in two parts of cast steel, held together by the tie bolt *a* is carried in the gear frame *7* by the reverse yoke pins *b* with tapered ends, one on each side of the frame. Each pin is prevented from turning in the jaws of the reverse yoke by two pins and the wearing portion of a reverse yoke pin is equipped with a roller bearing to reduce the friction when the yoke is moved either way by the reverse lever. Lubrication is applied to the pins through oil holes *c* in the gear frame. The gear reach rod is connected to the reverse yoke by the reverse yoke and gear reach-rod pin *d*, which is held in its jaws by the two clamp bolts shown. The upper end of each radius bar hangs in the reverse yoke on a radius-bar pin *e* the ends of which have a taper of ¾ inch in 12 inches and are prevented from turning by a dowel. A roller-bearing assembly is also used at these pins as well as thrust washers to eliminate the necessity of building up the various parts with bronze when excess lateral develops. The connecting-rod pin *f*, which is also equipped with a roller bearing, is held in the forked end of the gear connecting-rod by clamp bolts and serves to connect this end of the rod to the bell-crank. The arrangement of the radius bar and connecting-rod pin *g*, which couples the radius bars to the gear connecting-rod, is similar to the top pin except that thrust washers are applied between the bars and the sides of the rod. At places where clamp bolts are used, this part of the gear is split for a draw of 1/8 inch.

When the locomotive is in motion, the reverse yoke is held at the inclination required for the cut-off, by the gear reach rod and the main reach rod.

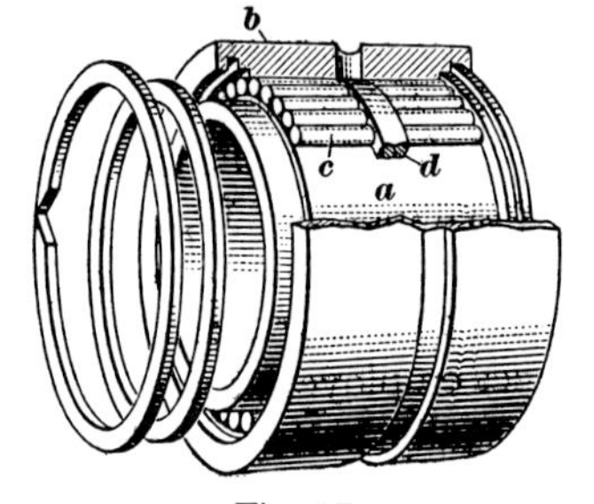

Fig. 17

19. Roller Bearing.—In Fig. 17 is shown the special type of roller bearing that is applied to certain of the connections in the reverse-yoke assembly. The inner race *a* is pressed on the pin and the outer race *b* into the adjacent part of the valve gear. The needle bearings *c* are maintained at the proper distance apart between the two races by the spacing ring *d* in which the bearings are retained. The snap rings shown, which spring into the ends of the outer race, prevent an endwise movement of the bearings. A hole through the top of the outer race is provided for lubrication.

DETAILS OF PARTS

20. Gear Frame.—Two kinds of gear frames are used, the Style No. 2, or short gear frame, and style No. 1, or the long gear frame; the one employed depends on whether the engine has a two-wheel or a four-wheel engine truck. The short gear frame requires only one support. The long gear frame requires two, one at each end. With a two-wheel truck, the guide yoke, owing to the position of the front driving wheel close to the cylinder, Fig. 18, can be located to the rear of this wheel and hence far enough from the steam chest to permit of a proper design of gear. The guide yoke can therefore serve as the single support required for a short gear frame. But with a four-wheel truck and the front driving wheel farther from the cylinder, Fig. 19, the guide yoke must usually be placed ahead of this wheel and too close to the steam chest for the application of a short gear frame, although an extension member is sometimes employed; therefore, the front end of the long gear frame is usually bolted to the guide yoke, and the rear end is connected by a gear-frame extension to one of the frame cross-ties. This type of gear frame locates the reverse yoke far enough away from the steam chest to permit of a proper design of valve gear. It will be noted that the guide yoke is used to support the guide-yoke end to which the guides are bolted.

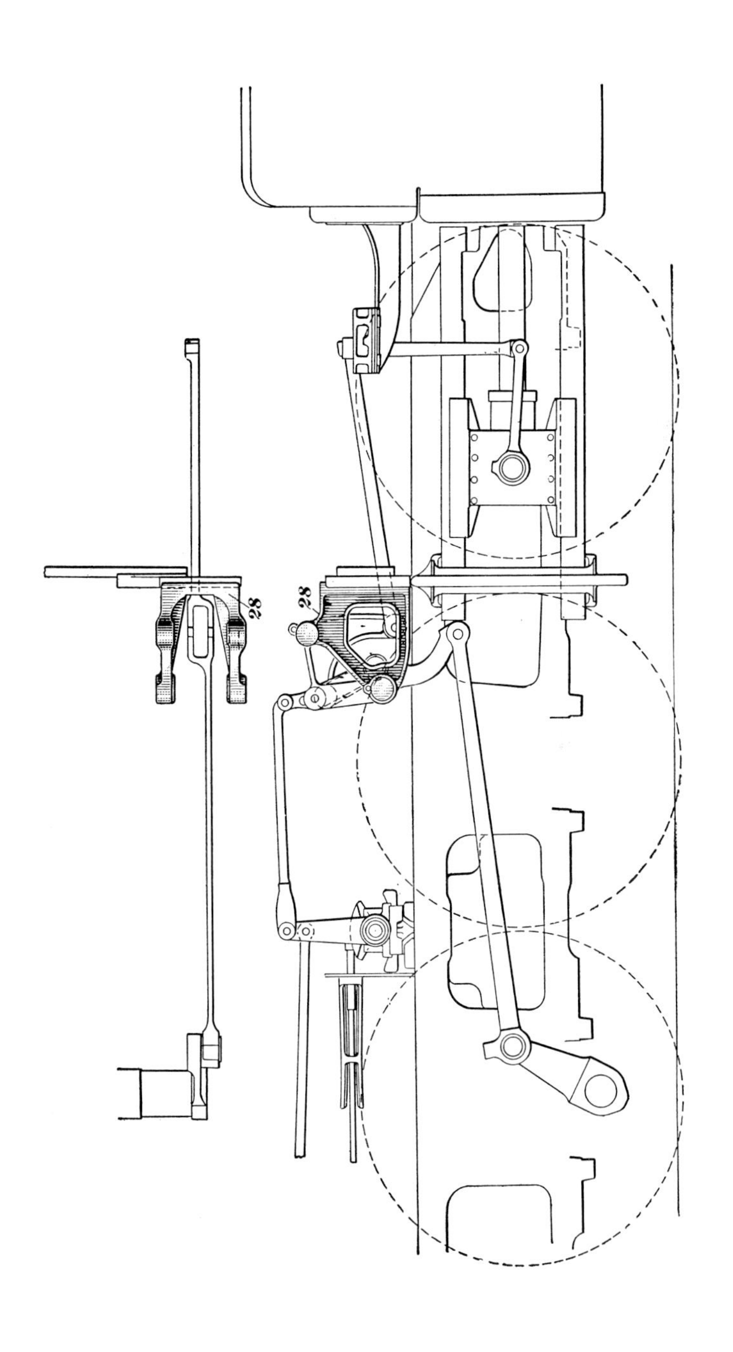

Fig. 18

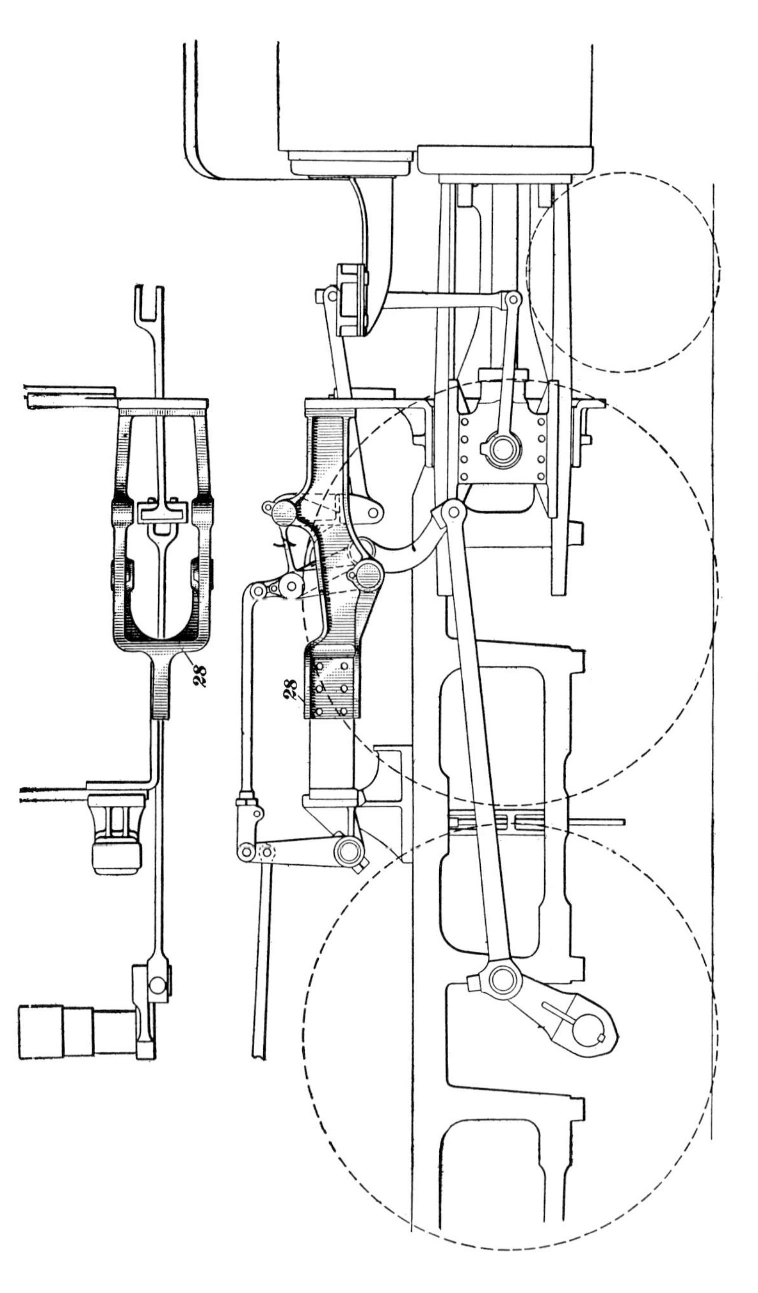

Fig. 19

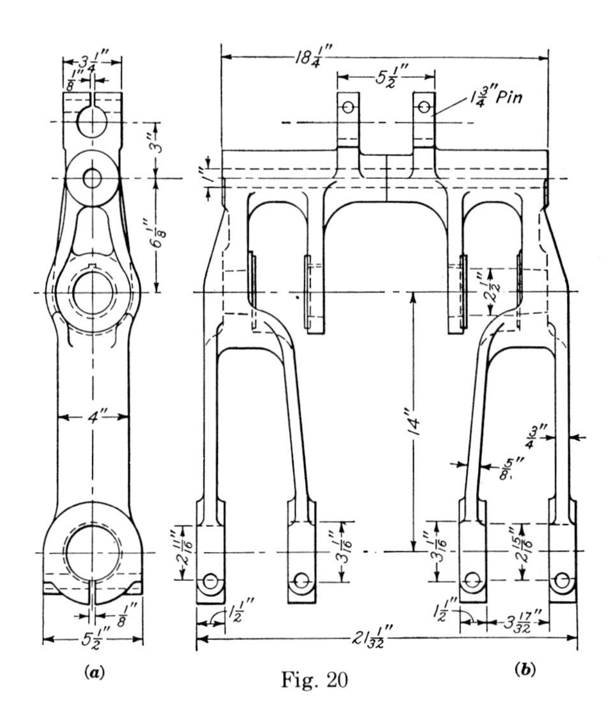

Fig. 20

21. Reverse Yoke.—A side view and a rear view of the reverse yoke are shown in Fig. 20 *(a)* and *(b)*, certain of the dimensions being given. The purpose of the reverse yoke is to bring about the reversal of the locomotive; also, it serves to vary the power developed by the locomotive by causing the valve to cut off the admission of steam to the cylinder at different points in the stroke of the piston. A movement of the reverse yoke from its inclined position forward to its inclined position backward produces an effect on the valves equivalent to one-half turn in the position of the eccentric crank on the axle, and hence causes the engine to move backward. As the inclination of the reverse yoke is increased, the arc scribed by the radius bar and the connecting-rod pin becomes more nearly vertical and the movement imparted by the bell-crank to the valve

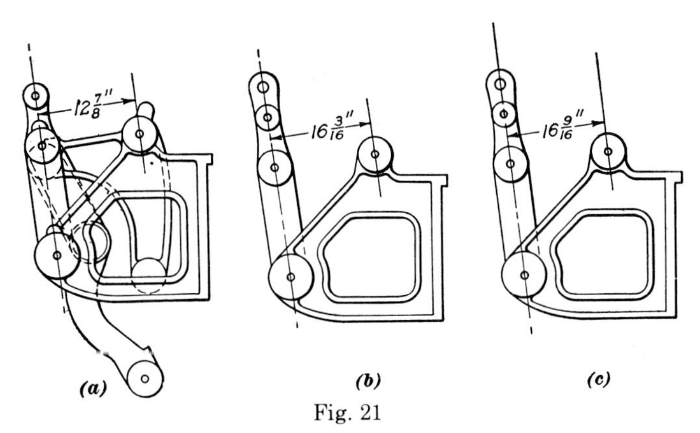

Fig. 21

increases, causing a later cut-off and the development of more power. A reverse action occurs when the inclination of the reverse yoke becomes less. In mid-position, the reverse yoke stands at right angles to a line drawn through the center of the main axle and the center of the pin that connects the eccentric rod to the gear connecting-rod. The movement of the eccentric rod will then be equal on each side of the mid-position of the reverse yoke.

With the standard valve gear, the distance between the center of the bell-crank pin in the frame and the center of the radius-bar pin for mid-gear position is shown in Fig. 21 *(a)*. The distance between these pins in mid-gear position for the long-travel valve gear is shown in view *(b)* and for the long-lap valve gear in view *(c)*.

22. Radius Bars.—The radius bars serve to suspend the gear connecting-rod in the reverse yoke and provide members on which the rod swings. The radius bar and connecting-rod pin that connects the radius bars to the gear connecting-rod swing in an arc, the inclination of which depends on the inclination of the reverse yoke. The length of the radius-bars with the standard gear between pinholes is 14 inches; their length with the long-travel gear is 14½ inches. The upper ends of the bars are supplied with bronze bushings. Side and front views of the radius bars with the more important dimensions given are shown in Fig. 22 *(a)* and *(b)*.

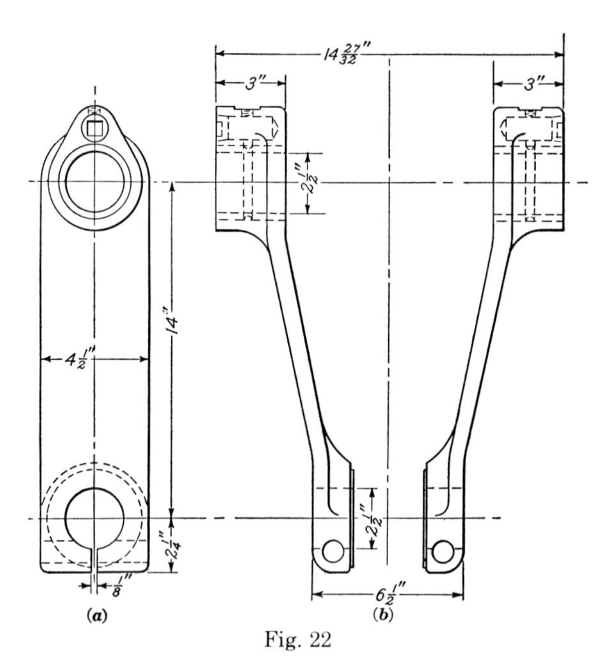

Fig. 22

23. Gear Connecting-Rod.—The purpose of the gear connecting-rod is to transmit the movement of the eccentric rod to the bell-crank. The center of the hole for the eccentric-rod pin is back-set 2 inches to the rear of the center of the hole for the radius bar and connecting-rod pin. As explained later, the back-set causes the gear connecting-rod to swing equally on each side of its mid-position and thereby equalizes the cut-offs. The bottom and the middle holes in the rod are provided with bronze bushings. The oil passages with a 15/16 inch drill at their outlets are shown by dash lines. Side and front views of the rod with the principal dimensions given are shown in Fig. 23 *(a)* and *(b)*.

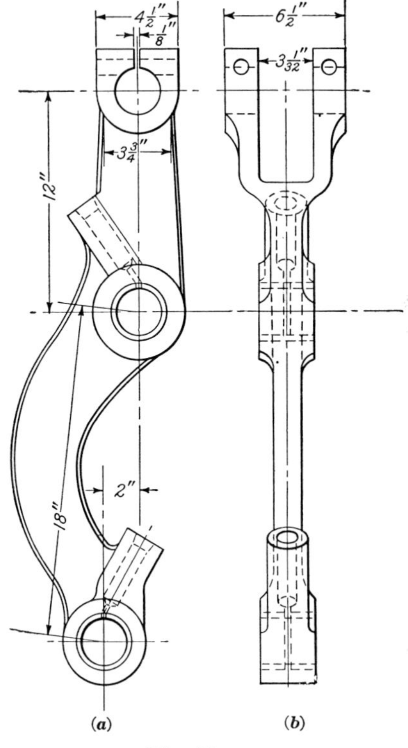

Fig. 23

24. Bell-Crank.—The purpose of the bell-crank is to transform the up-and-down movement of the upper end of the gear connecting-rod into a straight-line movement at the valve. A side view and an end view of the bell-crank are shown in Fig. 24 *(a)* and *(b)*.

The arm of the bell-crank to which the gear connecting-rod is pinned slants downward, thus bringing the pinhole farther down than the bell-crank frame pin. With the two pinholes on the same horizontal line, the connecting-rod pin and the arm of the bell-crank would move up higher from their mid-positions than they would move down. But by lowering the pinhole in the bell-crank the proper amount, the up-and-down movement will become equal.

The center of the pinhole for the valve rod is front-set 1¼ inches in advance of the center of the hole for the bell-crank frame pin. The purpose is to bring a line drawn through the center of these holes, when the valve gear is in mid-position, at right angles to the valve rod, which slants upward. The vertical arm of the bell-crank will then move the valve more

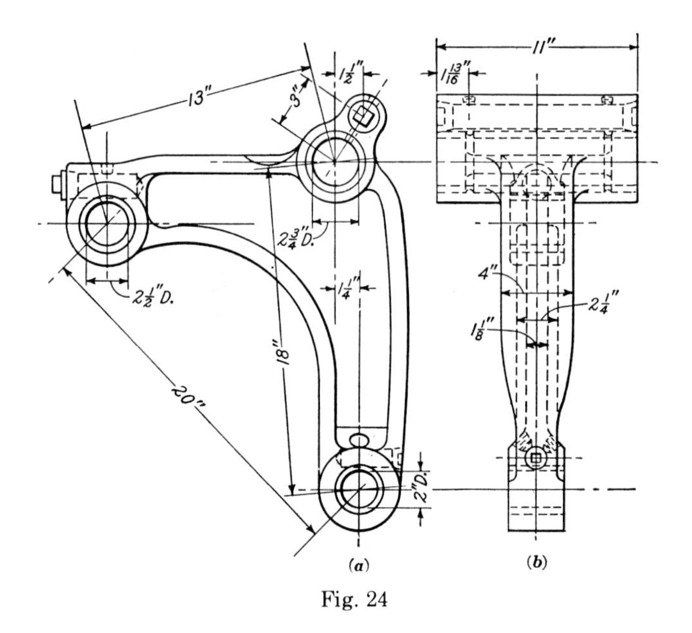

Fig. 24

nearly an equal distance on each side of its mid-position than otherwise. As explained farther on, the final correction to equalize the movement of the valve is made by front-setting the pinhole in the upper end of the combination lever.

The three pinholes are provided with waste cavities, supplied with oil holes and closed at the ends by pipe plugs.

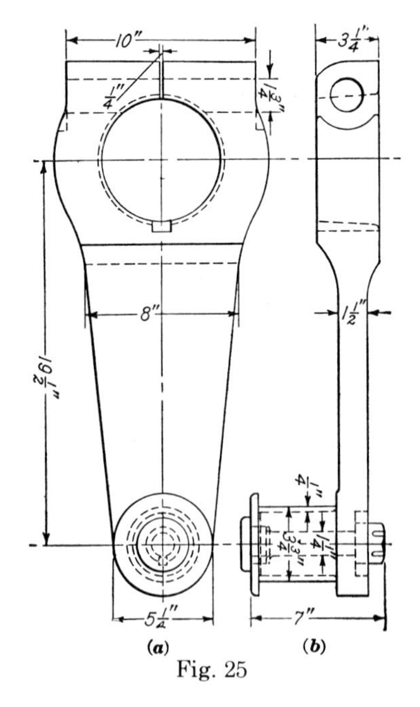

Fig. 25

25. Eccentric Crank.—The purpose of the eccentric crank is to impart to the valve at maximum cut-off the travel specified in the design. The movement transmitted to the valve by the eccentric crank can be varied by the reverse yoke. Eccentric cranks are either cast steel or forged. Some have a trunnion cast integral with the crank for driving the eccentric rod; others have a pin pressed in and riveted over. Where the trunnion is cast integral with the crank, a hardened sleeve is pressed over it to insure a smooth bearing. In Fig. 25 *(a)* and *(b)* are shown views of an eccentric crank in which the pin is pressed in. The pin is hollow and the eccentric rod is held on the pin by a bolt and a washer, the nut being on the inside of the crank. The end of the eccentric crank is split, with a draw of 1/8 inch, and is clamped on the end of the main crankpin by a 1 3/4 inch bolt. A tapered key prevents the eccentric crank from turning on the main crankpin.

26. The eccentric crank is not in itself a crank, but its design and its position on the axle give the effect of a crank having a length equal to the distance between the center of the axle and the center of the eccentric-rod pin. Thus, in Fig. 26 the length of the eccentric crank is 19 7/8 inches, which, owing to its position, gives a crank circle of a diameter of 20 1/4 inches and a crank length of 10 1/8 inches. In designing the valve gear, the eccentric crank is designed with a throw that will give the valve its specified travel, 7 1/2 inches, 8 inches, or whatever it may be. Its throw must always exceed the valve travel because, with the standard Baker valve gear, the eccentric crank moves the eccentric rod 4 inches for each inch of valve movement; that is, the gear ratio is 4 to 1. This is due chiefly to the distance, or 18 inches, between the point at which the eccentric rod is connected to the gear connecting-rod and the radius bar and connecting-rod pin. Allowance must also be made, when designing the crank, for the reduction in valve travel caused by the action of the combination lever, as well as the reduction in travel due to the valve rod being connected to this lever above the valve stem. Thus, in Fig. 26, the valve travel is 7 inches and this travel, on account of the factors just mentioned, requires a crank with a throw of 20 1/4 inches.

As shown in Fig. 27, the eccentric crank is so set on the crankpin that the actual crank makes an angle of 90 degrees with the center line of motion of the eccentric rod, which position places it slightly more than this amount behind the main crankpin with the engine moving ahead. This position would insure, were it not for the angularity of the eccentric rod, that the gear connecting-rod would swing equally on each side of its mid-position, but this error is to be taken care of by back-setting the lower end of this rod.

If the gear connecting-rod were of such a length as to bring the center of its bottom pinhole on the center line of the axles, the center line of motion of the eccentric rod would coincide with the center line of the axles; hence the eccentric crank would occupy a position 90 degrees, or a quarter turn, from the main crank. However, a gear connecting-rod of such a length would require an eccentric crank of an unreasonable throw.

With the Baker valve gear, the eccentric crank always follows the main crankpin with the engine moving forward.

27. Combination Lever.—The combination lever, when moved by the crosshead, serves the same purpose as a crank with a length equal to the lap and the lead and set at 90 degrees from the eccentric crank. Its

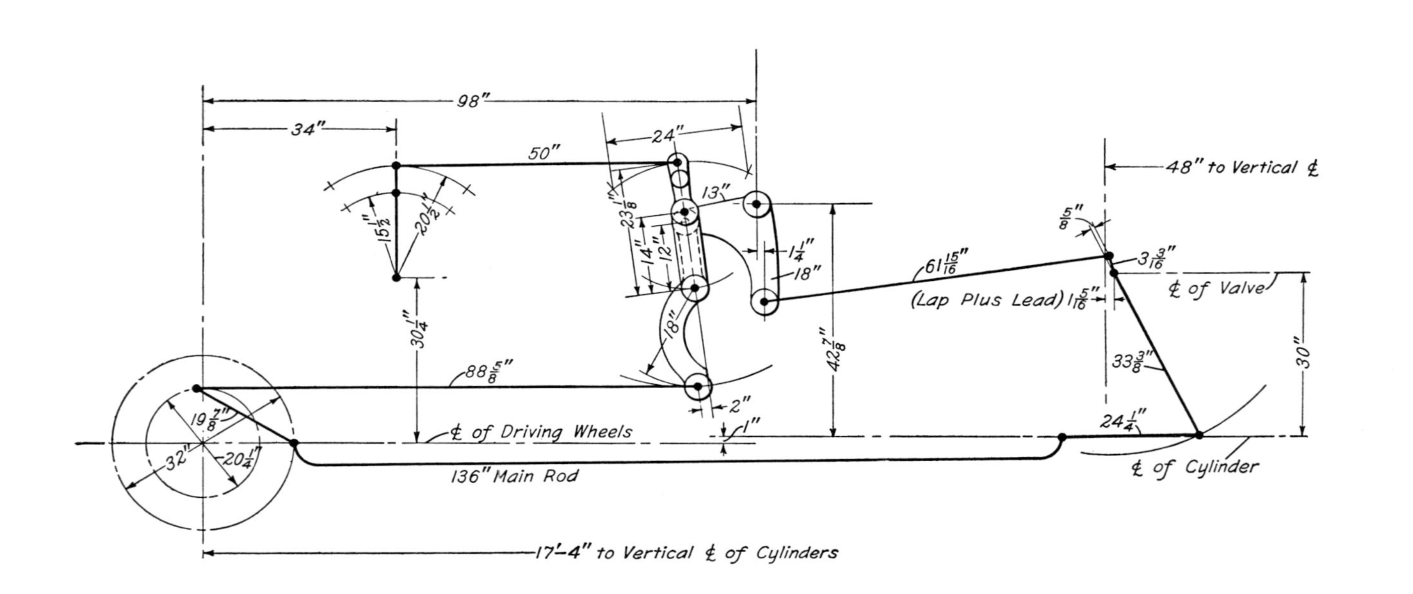
98"
34"
50"
24"
13"
48" to Vertical ℄
$20\frac{1}{2}$"
$15\frac{1}{2}$"
$23\frac{1}{8}$"
14"
12"
$1\frac{1}{4}$"
18"
$61\frac{15}{16}$"
$\frac{5}{8}$"
$3\frac{3}{16}$"
(Lap Plus Lead) $1\frac{5}{16}$"
℄ of Valve
$30\frac{1}{4}$"
18"
$42\frac{7}{8}$"
$33\frac{3}{8}$"
30"
$88\frac{5}{8}$"
2"
$19\frac{7}{8}$"
℄ of Driving Wheels
1"
$24\frac{1}{4}$"
32"
$20\frac{1}{4}$"
136" Main Rod
℄ of Cylinder
17'-4" to Vertical ℄ of Cylinders

Fig. 26

action, except at very short cut-offs, is to so vary the speed of the valve as to bring it to its proper position relative to the piston at the beginning of the stroke. Its arrangement subtracts from the valve travel, thereby requiring an eccentric crank of a somewhat greater throw than otherwise.

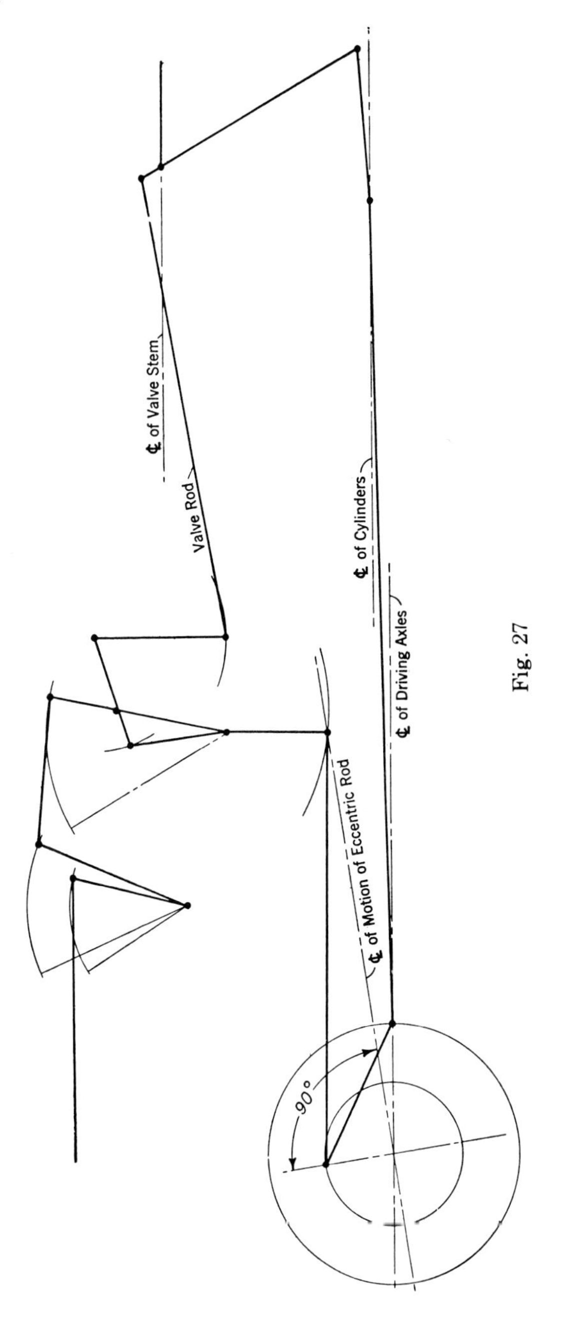

Fig. 27

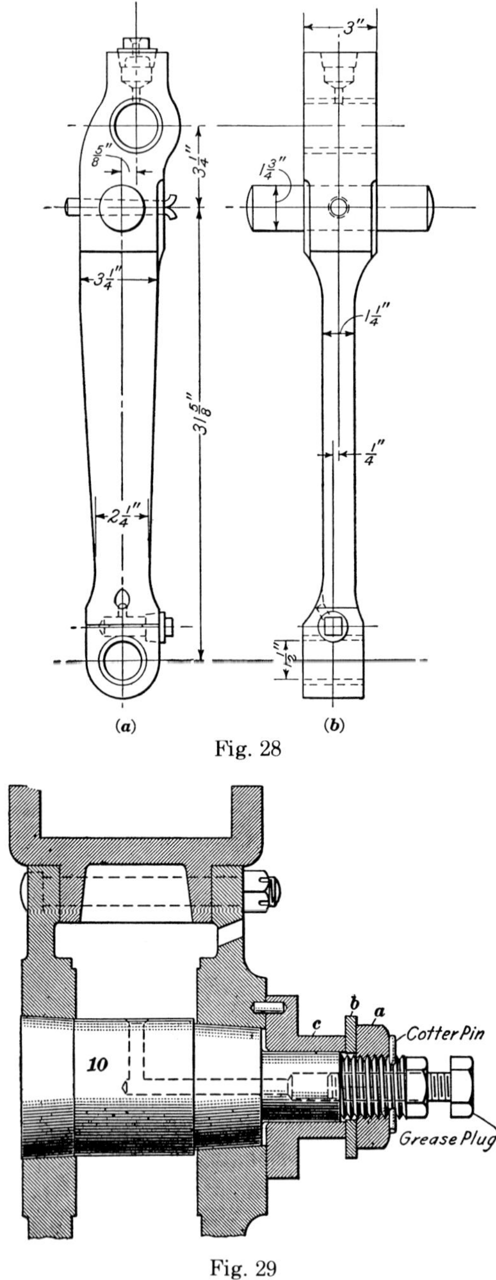

Fig. 28

Fig. 29

Two views of the combination lever are shown in Fig. 28 *(a)* and *(b)*. The rod is connected to the valve-stem crosshead by a 1¾ inch pin held in the rod by a ½ inch tapered pin, split at the end. Waste cellars closed by pipe plugs and provided with oil holes are located at the points shown.

The center of the upper pinhole has a front set of 5/8 inch, the reason for which is as follows:

Owing to the inclination of the valve rod, the top of the combination lever will move farther to the rear of its mid-position than to the front for an equal swing of the bell-crank on each side of its mid-position, and this inequality of movement will be transmitted to the valve-stem crosshead and to the valve. Front-setting the top of the lever, in this instance 5/8 inch, acts to equalize the travel by adding to the movement to the front of the mid-position point and subtracting from the movement to the rear.

The union link serves to connect the lower end of the combination lever to the crosshead pin of the locomotive. The method of connecting the union link to the crosshead pin is shown in Fig. 29. The arrangement of the nut a, the washer b, and the collar c on the end of the crosshead pin will be evident. The collar provides a wearing surface for the link, it prevents the washer from binding on the link when the nut is tightened up and it assists in holding the pin tight in the crosshead.

28. Proportioning Combination Lever.—The formulas used by the Pilliod Company for proportioning or calculating the lengths of the long and short ends of the combination lever used with an inside admission valve are given below, the proper allowances being made for lost motion in the pins that connect the valve rod, the valve-stem crosshead, and the union link to the lever.

$$\text{Long end} = \frac{B\,(S - 2L + 1/16)}{S} - 1/8$$

$$\text{Short end} = \frac{B\,(2L + 1/16)}{S} + 1/8$$

In these formulas, as shown in Fig. 30,

L = the lap and lead
B = the whole length of the lever
S = the stroke of the piston.

EXAMPLE.—The total length of a combination lever is 32½ inches, the lap and lead is 1 5/16 inches, and the stroke is 28 inches. Find the long end of the lever.

SOLUTION.—Substituting the values of L, B, and S in the formula, the length of the long end of the lever will be found to equal

$$\frac{32\tfrac{1}{2}\,(2\tfrac{5}{8} + \tfrac{1}{16})}{28} - \tfrac{1}{8} = 29\tfrac{1}{4} \text{ in.}$$

The length of the short end is then equal to $32\tfrac{1}{2} - 29\tfrac{1}{4} = 3\tfrac{1}{4}$ in.

EXAMPLE.—The total length of a combination lever, Fig. 26, with an offset of 5/8 inch, is 36 9/16 inches, the lap and lead is 1 5/16 inches, and the stroke is 32 inches. Find the length of the short end of the lever.

SOLUTION.—From the formula, the length of the short end equals

$$\frac{36\tfrac{9}{16}\,(2\tfrac{5}{8} + \tfrac{1}{16})}{32} + \tfrac{1}{8} = 3\tfrac{3}{16} \text{ in. Ans.}$$

The length of the long end will then be $36\tfrac{9}{16} - 3\tfrac{3}{16} = 33\tfrac{3}{8}$ in.

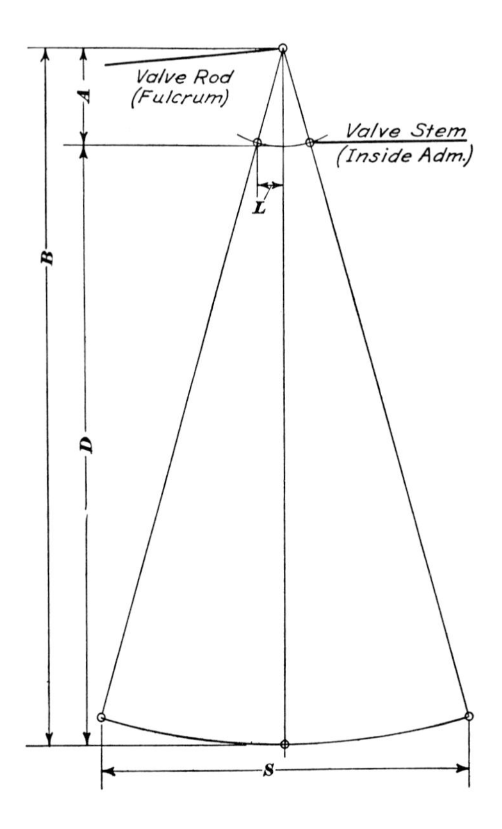

Fig. 30

29. The reason for the addition of ⅛ inch in the formula for calculating the short end of the lever is to lengthen it and thus compensate for the reduction in valve travel that would be caused by wear in the valve-rod pin at the upper end of the lever. But, to make up for the reduction in valve travel caused by wear in the union-link connections, ⅛ inch must be subtracted when calculating the length of the long end of the lever. Shortening the lever as shown in Fig. 31 makes it swing beyond the positions it would assume were the rod longer, thereby compensating for the reduction in movement caused by wear in the connections. The 1/16 inch given in both formulas compensate for wear in the valve-stem crosshead pin.

30. Changing the Lead.—To increase the lead, the short end of the combination lever must be made longer; to decrease the lead, this end must be made shorter. The formula used to find the length of the short

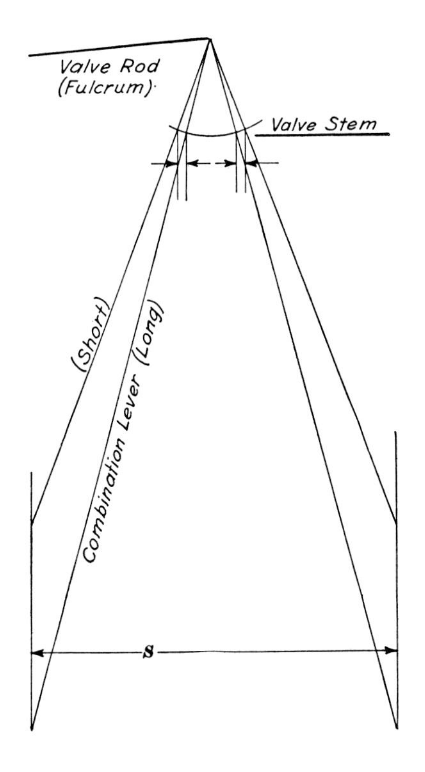

Fig. 31

end of the lever can also be employed when necessary to calculate a required change in the lead.

EXAMPLE.—Let it be assumed that it is required to decrease the lead of the valve gear shown in Fig. 32 from 1/4 to 3/16 inch. Find the new length of the short end of the lever.

SOLUTION.—Substituting the new value for the lead in the formula $\frac{B\,(2L + 1/16)}{S} + 1/8$ the formula becomes $\frac{45\,11/16\,(3\,10/16 + 1/16)}{30} + 1/8 = 5\,11/16$. Therefore, the distance between the centers of the top and the middle pinholes is to be shortened 1/4 in.

Any alteration ordinarily requires a new combination lever, and a change of this nature must, of course, be sanctioned by the proper authority.

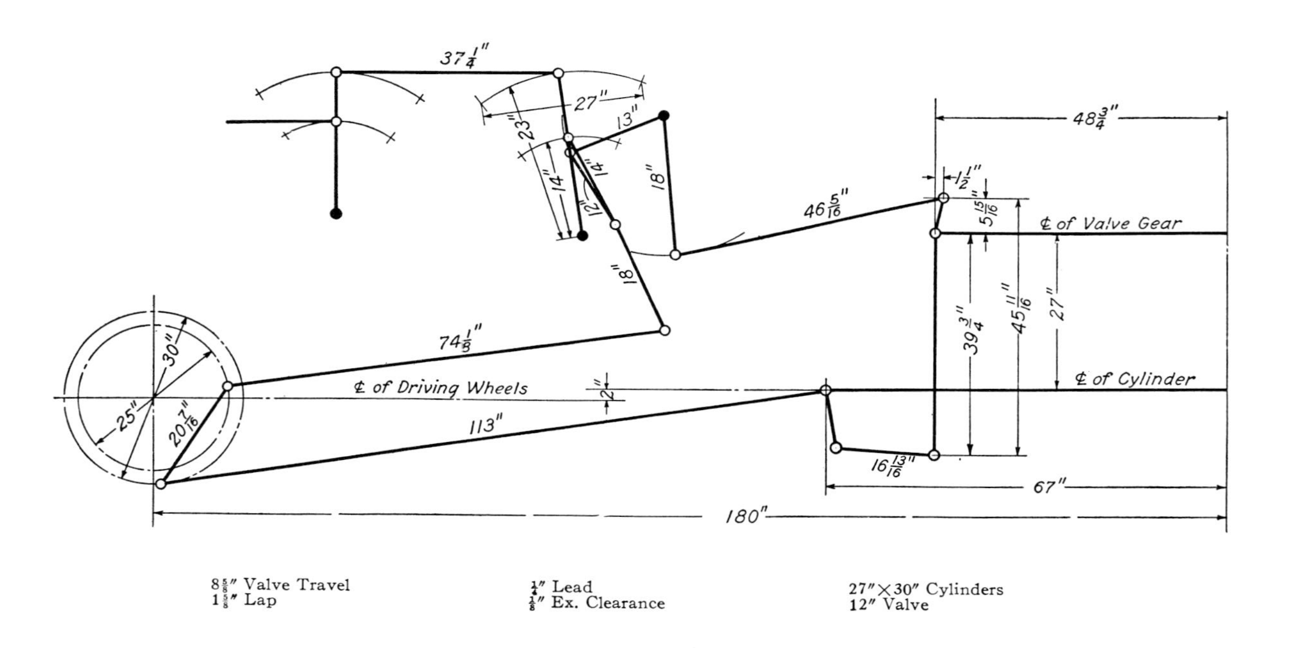
$37\frac{1}{4}''$
$27''$
$23''$
$13''$
$14''$
$14''$
$12''$
$18''$
$18''$
$46\frac{5}{16}''$
$48\frac{3}{4}''$
$\frac{1}{2}''$
$5\frac{15}{16}''$
℄ of Valve Gear
$39\frac{3}{4}''$
$45\frac{11}{16}''$
$27''$
℄ of Cylinder
$74\frac{1}{8}''$
$30''$
$25''$
$20\frac{7}{16}''$
℄ of Driving Wheels
$2''$
$113''$
$16\frac{13}{16}''$
$67''$
$180''$
$8\frac{5}{8}''$ Valve Travel
$1\frac{5}{8}''$ Lap
$\frac{1}{4}''$ Lead
$\frac{1}{8}''$ Ex. Clearance
$27'' \times 30''$ Cylinders
$12''$ Valve

Fig. 32

OPERATION OF GEAR

MOVEMENT OF THE PARTS

31. Movement of Parts at Reverse Yoke.—The movement of the parts of the gear at the reverse yoke will be explained by considering the diagrammatic view of the gear shown in Fig. 33, in which one radius bar and one leg of the reverse yoke are shown. The parts are shown in two positions, the dash lines showing the position of the parts in full gear, and the full lines, with the reverse lever drawn up. Let it be assumed that the reverse yoke has been inclined forward until the radius-bar pin is at *a*, the radius bar being shown by dash lines. Then, as the bottom of the gear connecting-rod *b* is moved to-and-fro by the action of the eccentric crank and the eccentric rod, the point *c*, the center of the radius bar and connecting-rod pin, will be constrained to move in an arc *de*, as shown by the dash lines. This movement will cause the top of the rod to pull the arm of the bell-crank down on the downward swing of the point *c*, and push it up on the upward swing, thereby transmitting, through the other arm of the bell-crank and the valve rod *f*, a to-and-fro movement to the valve.

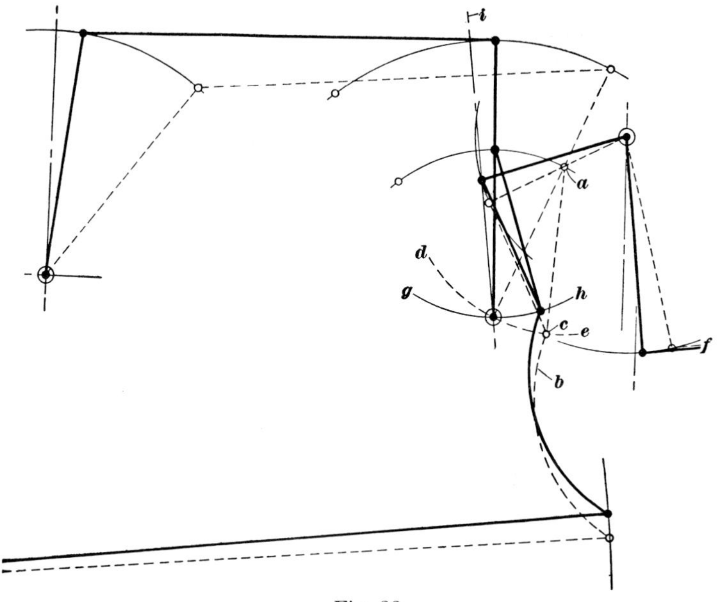

Fig. 33

By moving the reverse yoke back, thus drawing the radius bar back to its full line position, the arc *gh* now scribed by *c*, the point of attachment of the radius bar to the gear connecting-rod, will become flatter than before, resulting in a lesser movement of the bell-crank and a shorter travel of the valve.

The arcs become progressively flatter as the reverse yoke is drawn back until in mid position, indicated by the line *i*, the ends of the arc become horizontal. The gear connecting-rod will still impart a small movement to the bell-crank for the reason that the length of the upper end of the rod is 2 inches, less, with the standard gear, than the length of the radius bars. The effect of this inequality in length is to cause the point *c* to swing in an arc about 4½ inches on each side of the reverse-yoke pin, and impart a small movement to the bell-crank and the valve. If both parts were of the same length, there would be no movement of the valve with the reverse yoke in mid-gear.

With the reverse yoke in back gear, the arc scribed by the point *c* will be tilted in a direction opposite to that shown.

32. When the engine is standing on either dead center, there will be no movement imparted to the valve when the reverse yoke is moved from one gear to the other because, as shown in Figs. 20 and 22, the distance between the centers of the radius-bar pin and the reverse-yoke pin with the standard gear is 14 inches, and this corresponds to the length of the radius-rod hanger.

33. Variation in Cut-Off.—Any change in the position of the reverse lever causes the eccentric crank either to increase or to decrease the valve travel, but the action of the combination lever on the valve remains unaltered and is independent of the position of the reverse lever.

As already pointed out, the combination lever may be considered as comprising a short portion of the main crank, hence the lever is always moving the valve in the same direction as the piston, and would close the port away from which the piston is moving, were it not for the eccentric crank.

When considering a variation in cut-off, it is convenient to regard the combination lever as a port-closing crank, and the eccentric crank as being a port-opening crank. It then follows that, as the effect of the eccentric crank on the valve is decreased, owing to the reverse lever being drawn up, the action of the combination lever, which remains unchanged to close the port, will overcome earlier in the stroke of the piston the action of the eccentric crank to open the port. Lowering the reverse lever has an opposite effect, as this increases the action of the eccentric crank on the valve and delays the port-closing action of the combination lever in moving the valve to cut-off. However, the eccentric crank and the combination lever are not always acting in opposition when the valve is moving to cut-off, except when the cut-off is short; at long cut-offs the crank and the lever act together.

This is illustrated in Fig. 34, which shows the parts of the valve gear when operating at a cut-off of approximately 75 per cent. The action of

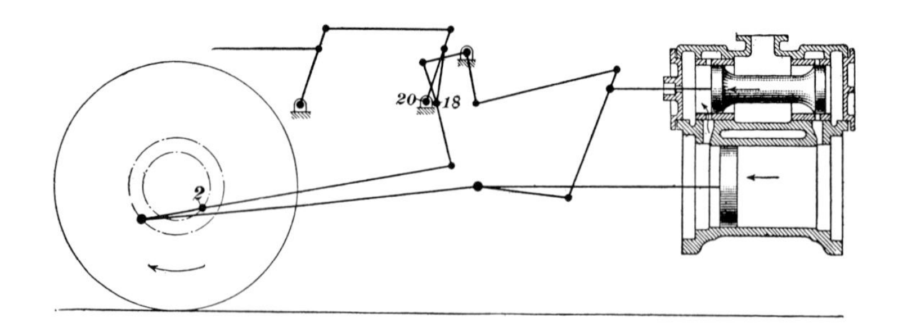

Fig. 34

the combination lever in moving the valve to cut-off is now assisted by the eccentric crank, which is near the bottom forward eighth. However, in Fig. 35, which shows the valve gear operating at a cut-off of about 33 per cent, the combination lever has moved the valve to cut-off against the opposition of the eccentric crank, which is still in a position to move the valve forward.

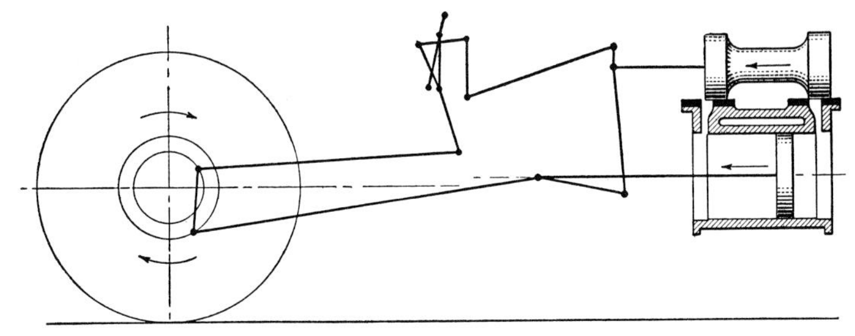

Fig. 35

Therefore, an earlier cut-off is obtained with the Baker valve gear as well as with the Walschaert valve gear by reducing the action of the eccentric crank on the valve without changing the action of the combination lever.

34. Principle of Reverse Arrangement.—To obtain reversal of motion it is fundamental that the true eccentric crank, the distance between the eccentric crankpin and the center of the axle, should occupy the same position relative to the main crank in one gear as in the other. This implies that the valve gear must be so designed as to produce the same effect as if the eccentric crank were shifted to the opposite side of the axle, or the equivalent of one-half turn when the reverse lever is moved from one gear to the other. With the Walschaert valve gear, a movement of the link block from one side of the link trunnion to the other produces an effect equivalent to one-half a turn in the position of the eccentric crank. With the Baker valve gear, the same effect is produced by moving the reverse yoke from one side of its mid-position to the other side. The eccentric crank with the Baker valve gear always follows the main crankpin when the engine is running forward; hence, when the engine is reversed, the effect on the valve will be the same as if the eccentric crank were shifted across the axle to follow the pin when the engine is running backward.

35. In Fig. 36, with the reverse yoke in forward gear the radius bar *16* will be inclined forward as shown, and the radius bar and connecting-rod pin *18* will begin to move down in an arc *ab* when the wheel is turned forward, the bell-crank will push the valve ahead, and the piston will move back. If the eccentric crank *2* and the eccentric rod *3* are shifted to their dotted-line positions, *2′* and *3′*, and the wheel is turned backward, the valve will receive the same movement. However, such a shift is unnecessary, because the valve can be caused to move ahead by shifting the reverse yoke to its dotted-line position. Then, when the wheel is turned backward, the radius bar and connecting-rod pin will move down in the arc *a′b′* and the valve will move ahead as before. Shifting the reverse yoke then produces reversal of motion, because it reverses the position of the arc scribed by the radius bar and connecting-rod pin.

The front dead center position of the main crankpin is selected when explaining the reason for reversal of motion, because the operation is not complicated by any movement of the valve and valve gear except the swinging of the reverse yoke and the radius bars. The fact that the valve moves in the same direction, or forward, when the wheel is turned in either direction is proof that reversal occurs. On the opposite side, the main crankpin is on the top quarter so that moving the reverse yoke to back gear is equivalent to a change in the position of the eccentric crank from one dead-center position to the other; the valve will be moved accordingly and will open the front steam port. When an engine is reversed, the valves move to admit steam to the opposite side of one or both pistons, depending on the positions of the main crankpins.

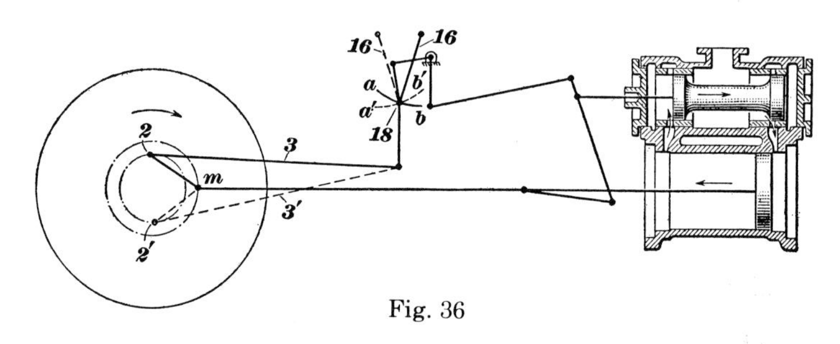

Fig. 36

36. While reversal of motion produces an effect equivalent to giving the eccentric crank one-half turn on the axle, yet in practice the position of the theoretical crank with respect to the main crank in back gear is not the same as the actual crank in forward gear. The reason is that the eccentric crank is set at right angles to the center line of motion of the eccentric rod, which position places the eccentric crank slightly more than a right angle from the main crank when running forward, and the theoretical crank slightly less than this amount from the main crank when running backward.

37. Gear Ratio.—The standard Baker valve gear has a gear ratio of 4 to 1, this meaning that the eccentric rod must be moved 4 inches in order to move the valve 1 inch with the valve gear in full gear. The Baker long-travel valve gear has a gear ratio of 3 to 1, and the Baker long-lap gear has a gear ratio of 5 to 1 at a 50 per cent cut-off.

38. Direct and Indirect Gears.—A valve gear is said to be direct when the eccentric rod and the valve rod both move in the same direction, and indirect when these parts move in opposite directions. When a valve gear is direct in forward motion, it necessarily follows that it will be indirect in backward motion, for the valve must move in the same relation to the piston in both motions, but the eccentric rod must move in the opposite direction, as the motion of the drivers is reversed. In Fig. 36 it is shown that the Baker valve gear is direct in forward motion, as the forward movement of the eccentric rod causes the valve to move in the same direction. When the wheel is turned in the reverse direction, the gear becomes indirect, as the eccentric rod is then moving to the left and the valve is moving in the same direction as before, or to the right. When the Baker valve gear operates an outside-admission valve, the gear is indirect in forward motion and direct in backward motion.

39. Admission at Back Port.—Conventional views of the Baker valve gear operating an inside-admission valve at a cut-off of about 86 per cent of the stroke are given in Figs. 37 to 42. These show the positions assumed by the parts of the gear at admission, cut-off, and release for the forward and backward strokes of the piston, or while the driving wheel is making practically one complete revolution. In Fig. 37, the main crankpin *m* is on the back dead center, and the back steam port is open an amount equal to the lead. The eccentric crankpin is at *2,* approximately on the bottom quarter. The radius bar *16* stands directly in front of the reverse yoke, pivoted at *20,* and cannot be seen until the bar is moved away from the yoke by the backward movement of the gear connecting-rod. It will be noted that only one radius bar and one leg of the reverse yoke are shown.

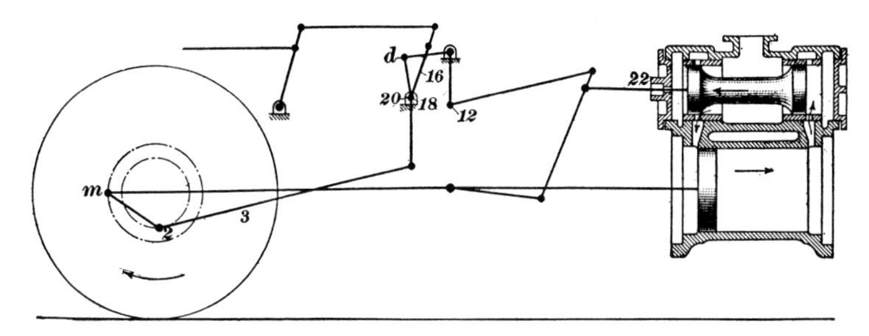

Fig. 37

40. As the driving wheel turns in the direction indicated by the arrow, the action of the eccentric rod *3,* Fig. 37, will draw the lower end of the gear connecting-rod to the left. As the radius bar *16* is connected at its lower end to the gear connecting-rod by the pin *18*, the rod will be compelled to move in the arc described by the radius bar as it swings in the reverse yoke, and, on account of the inclination of the radius bar at this time, the arc will be tilted somewhat toward the vertical. The pin *18* and the gear connecting-rod will then begin to follow an upward arc, causing the pin *d* and the upper arm of the bell-crank to move in the same direction. The pin *12* and the lower arm of the bell-crank move to the left, carrying the valve rod and the valve stem *22* in the same direction, and the valve opens the back steam port. As the piston is now moving forward, the lower end of the combination lever is being moved to the right with the piston and crosshead, which would tend to carry the valve in the same direction. However, any tendency of the combination lever to move the valve forward is overcome by the eccentric crank in pulling it backward, on account of the small movement that can be imparted to the crosshead by the main crankpin *m,* as compared with the eccentric crankpin *2.* The horizontal movement transmitted by the pins *m* and *2* decreases as either approaches dead center, and therefore the upper end of the combination lever is being pulled backward faster by the action of the eccentric crank than the lower end of the rod is being moved forward by the crosshead, and thus a backward movement is imparted to the valve.

41. Cut-Off at Back Port.—In Fig. 38 the valve has just closed the back steam port and has cut off steam to the back end of the cylinder. In bringing about this valve movement, the gear connecting-rod has been carried first in an upward arc, thus moving the valve backward, after which the pin *18* has reversed its movement and has come back to the position shown. The pin, as it moves in its downward arc, carries the gear connecting-rod and the upper arm of the ball-crank downward, thus

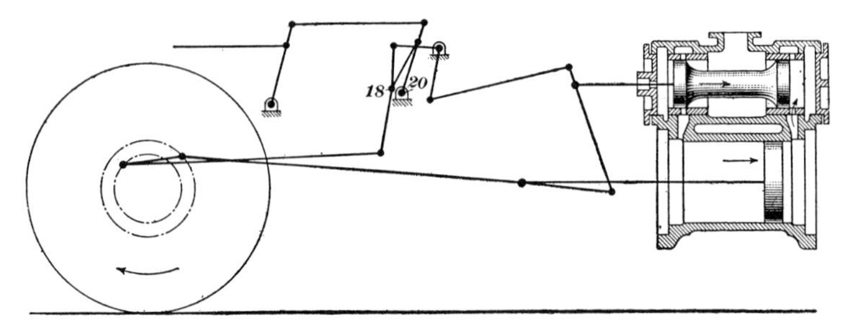

Fig. 38

moving the lower arm of the bell-crank, valve rod, valve stem *22,* and valve ahead to the point of cut-off. The valve has been moved to cut-off by the combined movement of the eccentric crank and combination lever, as the movements imparted by these parts are now in the same direction, or forward.

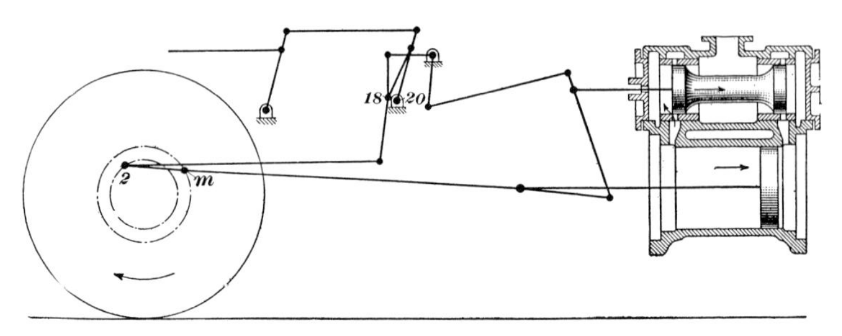

Fig. 39

42. Release at Back Port.—In Fig. 39, steam is shown exhausting from the back end of the cylinder. The valve is still being moved to the right by the combined action of the eccentric rod and combination lever, as the movement of the pin *18* is still in a downward arc. The extent of the valve movement from cut-off to release is equal to the steam lap, and the piston during this time is being moved by the expansive power of steam.

43. Admission at Front Port.—In Fig. 40 the main crankpin is shown on the forward dead center. The valve has been moved to the right until it has opened the front steam port for the admission of steam by an amount equal to the lead. This movement of the valve has been brought about by the combined action of the eccentric crank and combination lever, as the pin *18* and the lower end of the radius bar are still moving in a downward arc, thus imparting a forward movement to the valve rod. The movement of the combination lever has also been forward. However,

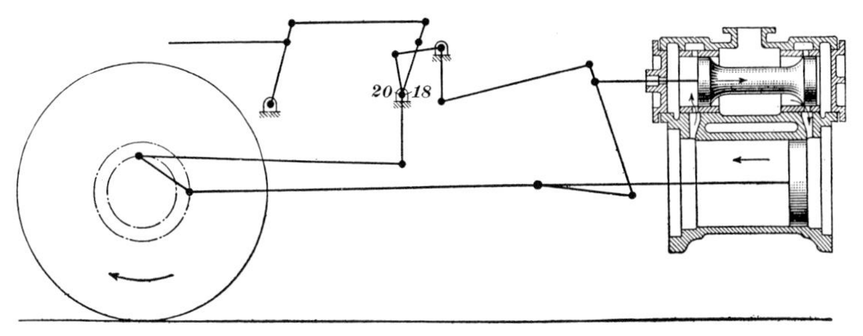

Fig. 40

as the main crankpin nears the dead center the movement transmitted to the lower end of the combination lever by the crosshead will be slight in comparison to that given by the eccentric crank, which is approaching the top quarter.

44. Cut-Off at Front Port.—In Fig. 41 the valve is shown at cut-off at the front steam port. The pin *18*, from the time the main crankpin moves from the front dead center to cut-off, moves first in a downward arc, then reverses the movement and moves in an upward arc. The movement in a downward arc causes the upper arm of the bell-crank to move down and the valve to move forward, thus opening the front steam port wider. The upward movement raises the gear connecting-rod and the

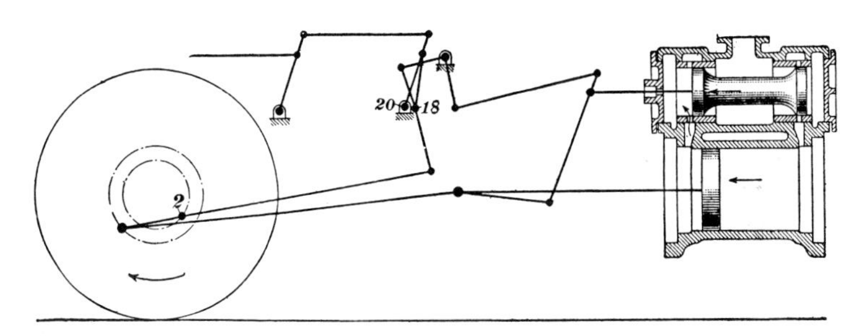

Fig. 41

bell-crank upper arm, and pulls the valve back to cut-off. During the opening movement of the valve, the upper end of the combination lever and the valve move ahead, although the crosshead is moving the lower end of the lever back. The movement imparted to the valve to open the port may then be considered as being due to the action of the eccentric crank only; for the eccentric crank, owing to its position, affects the valve more than the crosshead, as the main crankpin is near the dead center.

45. Release at Front Port.—In Fig. 42 the valve is just about to open the front steam port and permit the steam in the front end of the cylinder to escape. The movement of the valve is still backward, and the crosshead and the eccentric crank combine to produce this movement. The pin *18* is still moving in an upward arc, which moves the upper arm of the bell-crank upward and the lower arm backward to the left. As the valve moves from cut-off to release, or the amount of the steam lap, the

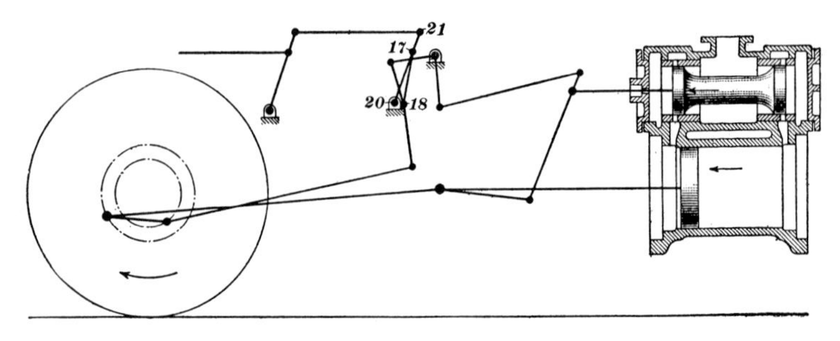

Fig. 42

piston is being moved by the expansion of steam in the cylinder. When the main crankpin reaches the back dead center, the valve will have the back steam port open an amount equal to the lead.

46. It will be noted that, while the main crankpin is moving from the back dead center in Fig. 37 to the front dead center in Fig. 40, the gear connecting-rod swings on the radius-bar pin *18* first in an upward arc, and then reverses and moves in a downward arc, the position of the gear being the same in the latter case as in the former, except that the position of the combination lever is reversed. As the main crankpin moves from the forward dead center to the back dead center in Fig. 42, the radius-bar pin *18* moves in a downward arc, and then reverses and moves in an upward arc until the gear assumes the same position on the back dead center as on the front dead center, with the exception that the combination lever is then inclined in the reverse direction. The point in the stroke of the piston at which the pin *18* reverses the direction of its movement will depend on the position of the reverse lever. Raising this lever causes the reversal of movement to occur earlier and lowering the reverse lever will cause the pin to reverse its movement later in the stroke of the piston.

EFFECT OF ANGULARITY OF RODS ON CUT-OFF

47. Effect of Angularity of Main Rod.—In Fig. 43 the main crankpin will move from its front dead center to *a* when moving the crosshead 25 per cent of its backward stroke, and from the back dead center to *b* when moving 25 per cent of its forward stroke. Although the distance moved by the crosshead is equal on both strokes, the arcs described by the crankpin for the crosshead movements are unequal, the arc *c* being shorter than the arc *d*. The difference in the lengths of these arcs for an equal movement of the crosshead is due to what is commonly called the angularity of the main rod.

It would be possible to set the valves so that cut-off would occur for the same length of crankpin arc as measured from one dead center or from the other, and it could then be said that the error due to the angularity of the rod was compensated for. But equal arcs would not imply equal movements of the crosshead and the piston, so that, if the valves were set to cut-off at equal crankpin arcs, the unequal movement of the piston would cause a greater volume of steam to be admitted to one end of the cylinder than to the other. The engine would then sound somewhat out of square and unequal turning impulses would be transmitted to the crankpins. In practice no account is taken of the angularity of the main rod when setting the valves. The valves are set to cut off the steam for equal movements of the crosshead; that is, if the valves are set for a cut-off of 25 per cent, then cut-off will occur when the piston has moved this part of the stroke from either end of the cylinder with the reverse lever in the 25 per cent notch in the quadrant. The error introduced in the position of the piston by the angularity of the rod is small. With a main rod 10 feet long, and a 30-inch stroke, the maximum error in the displacement of the piston due to the angularity of the rod is less than 1 inch and this is taken care of by a slight adjustment of the eccentric rod when setting the valves.

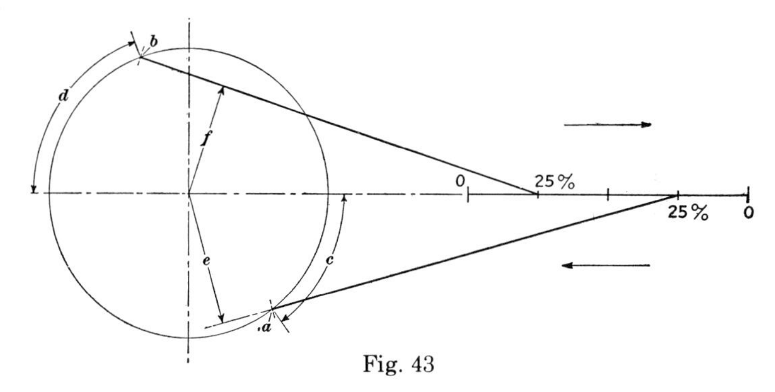

Fig. 43

48. It would seem that the revolution of the crankpin through unequal arcs during the admission of steam to the cylinder would be an undesirable condition. However, this is not the case, the reason being that the length of the effective crank arm is greater while the crankpin is moving from the forward dead center to *a*, Fig. 43, than its length when it is moving from the back dead center to *b*. The effective lengths of the crank arms with the crankpin at *a* and *b* are indicated by the lines *e* and *f*. During any part of the revolution, a larger crankpin arc for a certain movement of the crosshead is always accompanied by a shorter effective crank arm and a shorter arc by a longer effective crank arm. A long effective crank arm and a short arc, and a short effective crank arm and a long arc compensate for each other, so that the turning effort on the crankpin remains about the same.

49. Compensating for Angularity of Eccentric Rod.—With a properly designed constant lead valve gear, the gear connecting-rod or the link is midway of its swing with the locomotive on either dead center; this condition is brought about by back-setting the lower ends of these parts, and so compensating for the angularity of the eccentric rod.

In the absence of a back-set, the eccentric crankpin, in moving from its back center to the top quarter, moves through one-half of its upper arc, yet owing to the angularity of the eccentric rod, the gear connecting-rod has made less than half its swing. For the same reason, during the movement of the eccentric crankpin through the other half of its arc to the front center, the gear connecting-rod moves more than half its swing. From the front center to the bottom quarter, the gear connecting-rod moves more than half its swing and less than half its swing from the bottom quarter to the back center. The valve will then move unequally on each side of its mid-position, and the result will be unequal cut-offs. The unequal swing of the gear connecting-rod caused by the angularity of the eccentric rod, can be remedied only by introducing a back-set in the lower end of the gear connecting-rod. It might seem that back-setting the end of the gear connecting-rod would merely decrease the forward swing by the amount of the back-set and increase the backward swing by the same amount and thereby leave the condition the same as before. However, this is not the case, for while the forward swing is restricted to an extent practically equal to the amount of the back-set, yet the backward movement is increased in excess of the back-set. The reason is that the back-set causes the angle between the eccentric rod and the gear connecting-rod as it is approaching the end of its rearward swing, to assume more nearly a straight line, thereby causing, for any certain movement of the eccentric rod, a greater swing of the gear connecting-rod to the rear than would otherwise be the case. The effect is the same as if the rod were progressively shortened during the latter part of the rearward swing of the gear connecting-rod. The forward swing of this rod is not materially affected owing to the angle at this time remaining more nearly normal. The eccentric rod is of course shortened the amount of the back-set.

As shown in Fig. 44, the back-set of the gear connecting-rod is 2 inches; that is, the center of the pinhole at the lower end of the rod is 2 inches to the rear of a vertical line drawn through the centers of the other two pinholes. It will be noted that the back-set begins at the middle hole and increases gradually about halfway down, then it decreases to the amount shown. This construction gives the lower portion of the gear connecting-rod a curved shape, which is necessary in order to provide clearance between the rod and the bell-crank.

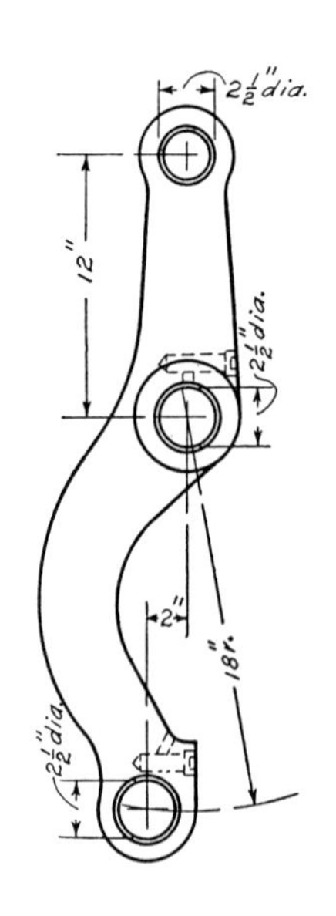

Fig. 44

50. Broken Eccentric Crank, Eccentric Rod, or Gear Connecting-rod.—Should any of the parts mentioned break, take down the eccentric rod, and block the bell-crank, if of the single-arm type, by the blocking *b*, Fig. 45, with the lower bearing 1¼ inches ahead of the upper

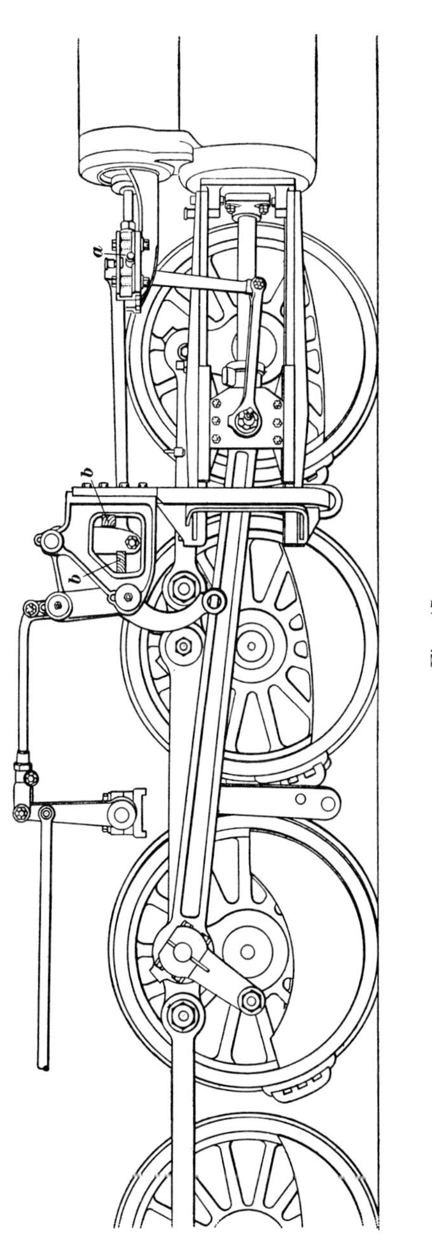

Fig. 45

bearing. If the bell-crank is of the two-arm type, block it in a vertical position. When the bell-crank is so blocked, the valve on the disabled side will receive a movement from the combination lever equal to twice the lap and the lead. Care must be taken not to stop with the main crankpin on the disabled side on either quarter, because the valve on this side will be in mid-position, and the main crankpin on the other side will be on a dead center, thus making it impossible to start the locomotive.

It may be found difficult to block the bell-crank with the long type of gear frame, in which event, in addition to taking down the eccentric rod, remove the union link, tie the combination lever out of the way, and secure the valve with the port opened slightly. However, this method of disconnecting will render the mechanical lubricator inoperative if operated from the combination lever on the disabled side.

Either one of the procedures here outlined should also be followed, should the radius bars or the reverse yoke break. A locomotive can almost always be run in without disconnecting any parts, should only one radius bar or one side of the reverse yoke break.

51. Broken Vertical Arm of Bell-Crank or Valve Rod.—Should either one of these parts break, remove the valve rod, and secure the valve by means of the setscrew *a* leaving the other parts of the gear intact as shown in Fig. 46. This will permit the mechanical lubricator to be operated should it be connected to the combination lever on this side.

52. Broken Combination Lever or Union Link.—In the event of either one of these parts breaking, remove the eccentric rod and the union link, tie the combination lever forward out of the way of the crosshead, as shown in Fig. 47, and secure the valve in the position desired. To disconnect the union link from the end of the crosshead pin *10,* Fig. 29, remove the cotter pin, the nut *a*, and the washer *b;* the link can then be taken off the collar *c*. The nut and the washer must be replaced to prevent the crosshead pin from working loose.

53. Broken Gear Reach Rod.—With a broken gear reach rod, remove the broken parts and block or tie the reverse yoke in the position to obtain the cut-off desired. The locomotive must not be reversed without changing the position of the reverse yoke on the disabled side. Should the main reach rod fail, block both reverse yokes at the desired cut-off.

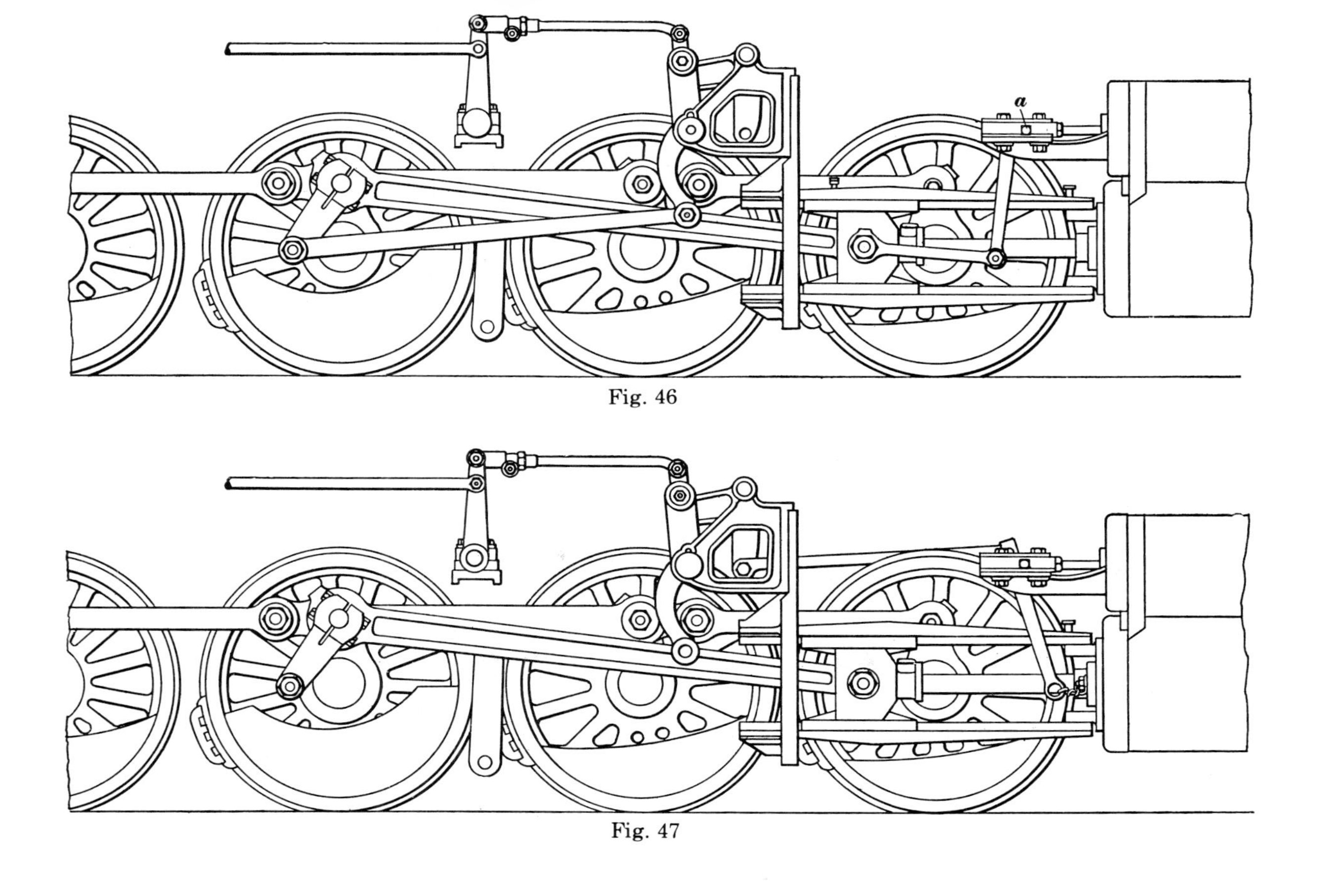

Fig. 46

Fig. 47

BAKER LOCOMOTIVE VALVE GEAR

EXAMINATION QUESTIONS

Notice to Students.—*Study the Instruction Paper thoroughly before you attempt to answer these questions.* **Read each question carefully and be sure you understand it;** *then write the best answer you can. When your answers are completed, examine them closely, correct all the errors you can find, and* **see that every question is answered.**

(1) How does the Baker valve gear differ from the Walschaert valve gear?

(2) Name the principal parts of the Baker valve gear.

(3) Explain the arrangement of the gear at the reverse yoke.

(4) What is the purpose of the reverse yoke?

(5) What is the purpose of: (a) the gear connecting-rod? (b) the radius bars? (c) the bell-crank?

(6) Explain how the movement of the eccentric crank is imparted to the valves.

(7) How is the eccentric crank set in relation to the main crankpin?

(8) Explain why moving the reverse yoke forward increases the movement imparted to the bell-crank by the gear connecting-rod.

(9) With an inside-admission valve, is the Baker valve gear direct or indirect when the engine is moving ahead?

(10) What should be done in the event of a broken eccentric crank, eccentric rod, or gear connecting-rod?

(11) What should be done if the gear reach rod should break?

(12) What should be done if the main reach rod should break?

(13) What should be done if the valve rod should break?

(14) What should be done if the combination lever or union link should break?

Southern Locomotive Valve Gear

By
J. W. Harding

2073-2 Printed in U.S.A. Edition 1

SOUTHERN LOCOMOTIVE VALVE GEAR

CONSTRUCTION AND OPERATION

INTRODUCTION

1. The Southern locomotive valve gear, like the Walschaert and Baker valve gears, is of the radial-gear type. The chief difference between it and these gears is that the Southern gear has no lap and lead lever, the valve deriving its total movement from the eccentric crank.

It was invented by William Sherman Brown, of Knoxville, Tenn., a locomotive engineer employed on the Southern Railway, and was first applied to a locomotive in February, 1913.

CONSTRUCTION

2. Arrangement and Details of Parts.—The arrangement of the Southern valve gear as applied to a locomotive having inside admission valves is shown in Fig. 1, in which *1* is the eccentric crank; *2*, the eccentric rod; *3*, the radius hanger; *4*, the transmission yoke; *5*, the link support; *6*, the link block; *7*, the bell-crank; *8*, the valve rod; *9*, the auxiliary reach rod; *10*, the main reach rod; *11*, reversing shaft; *12*, reversing-shaft arm; *13*, the reverse lever; *14*, the bell-crank bracket; *15*, the link.

Fig. 2 illustrates the gear as applied to a locomotive having outside admission valves. For clearness, the parts of the gear are indicated by heavy lines representing center lines.

The eccentric crank *1*, which is similar to that used by other single-eccentric valve gears, is set theoretically at right angles to the main crank. When an inside admission valve is used, the eccentric crank *1* follows the main crank when the engine is running forwards, as in Fig. 1, whereas with an outside admission valve the eccentric crank leads the main crank, as in Fig. 2. The lower ends of the radius hanger *3* and transmission yoke *4* are both connected to the eccentric rod *2*. The other end of the radius hanger *3* is attached to the link block *6*, and the transmission yoke *4* connects at its other end to the bell-crank *7*, which turns in the bell-crank bracket *14* on the bell-crank center pin. The valve rod *8* connects to the lower arm of the bell-crank *7*. The link is stationary and is fastened to the link support *5*. The movement of the link block *6* in the link is controlled by the auxiliary reach rod *9*, main reach rod *10*, and reverse lever *13*.

3. Radius Hanger.—Fig. 3 shows two views of the radius hanger. The lower end of the hanger is forked and connects to the eccentric rod by pin *a*. The upper end is also forked, and straddles the link, link block, and auxiliary reach rod. The link-block pin *g* connects the link block to the radius hanger.

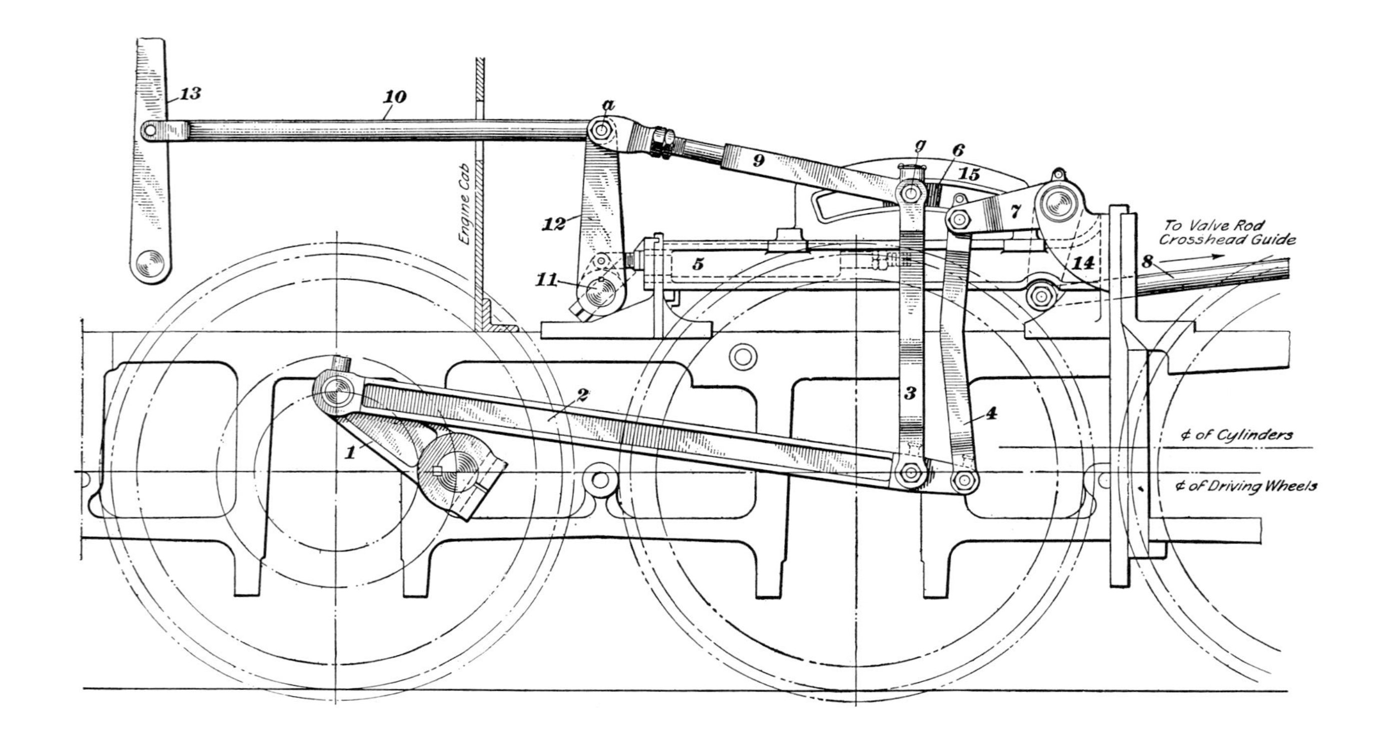
13
10
a
9
g
6
15
7
Engine Cab
12
5
11
To Valve Rod
Crosshead Guide
14
8
2
3
4
1
℄ of Cylinders
℄ of Driving Wheels

Fig. 1

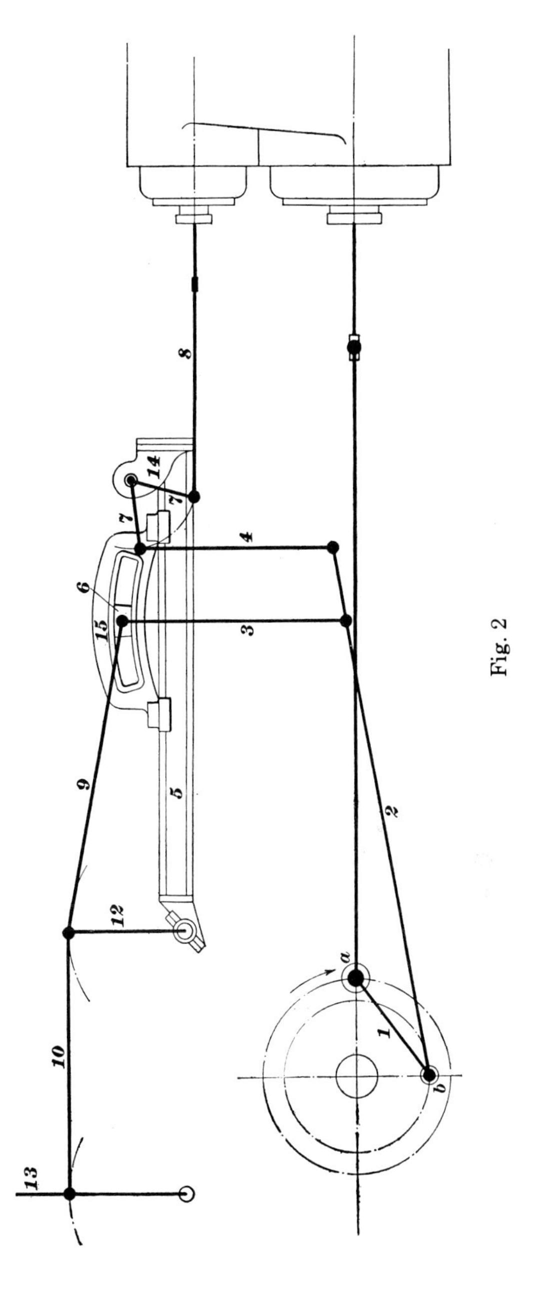

Fig. 2

The radius hanger performs three duties. It is connected to the eccentric rod at such a point as to cause the eccentric crank to transmit the lap and lead movement to the valve; its inclination, which is governed by the position of the link block in the link, determines the extent of the movement transmitted to the transmission yoke, bell-crank, and valve, and it also permits of the gear being reversed.

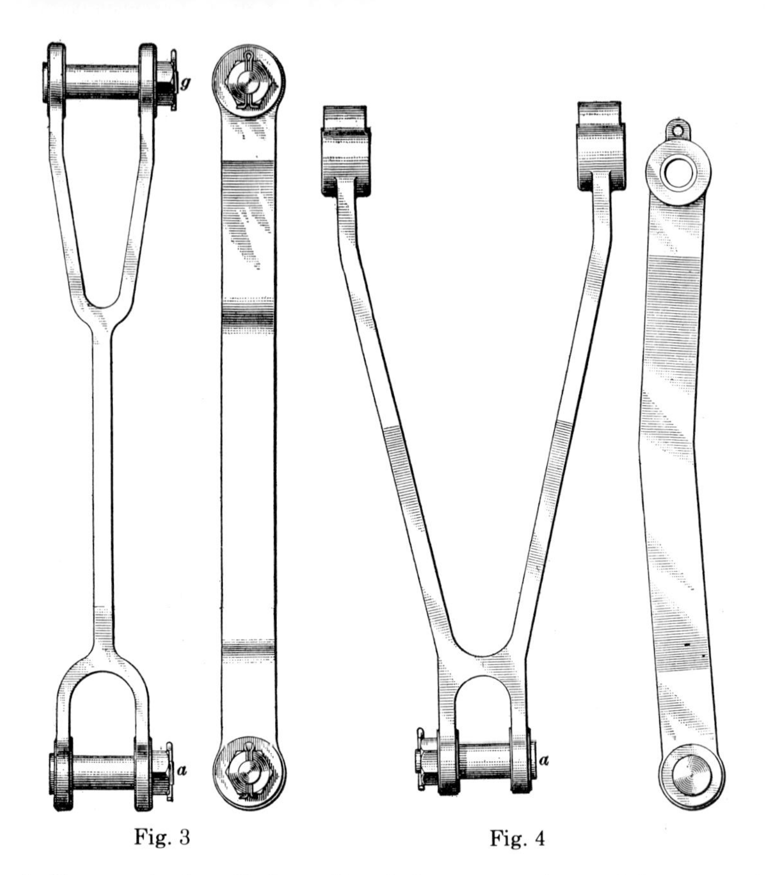

Fig. 3

Fig. 4

4. Transmission Yoke.—Fig. 4 shows two views of the transmission yoke. The lower end is attached to the eccentric rod by pin *a,* and the upper ends connect to the upper arms of the bell-crank. The duty of the transmission yoke is to transmit the movement of the eccentric crank and eccentric rod as influenced by the radius-hanger connection, to the bell-crank.

5. Bell-Crank.—Fig. 5 shows three views of the bell-crank; view *(a)* illustrates it as seen from the side; view *(b)* shows it as viewed from the top; and view *(c)* shows it as it would appear viewed from the front looking toward the rear of the engine. The part that projects on the left in view *(c)* is the portion of the forked end and nut of the upper bell-crank arm of view *(b)* as seen when the bell-crank is viewed from the front. The upper ends of the transmission yoke are connected to the upper arms *7* of the bell-crank by the pins *a*; the pin *b,* view *(c),* connects the valve rod to the lower bell-crank arm 7. The bell-crank is hollow and turns on a bell-crank center pin, which connects it to the bell-crank hanger, or bracket, *14*, Figs. 1 and 2.

6. Auxiliary Reach Rod.—Fig. 6 shows two views of the auxiliary reach rod, both ends of which are forked. Pin *a* connects the rear end to the reversing-shaft arm *12,* Fig. 1. The link and link block lie between the ends of the two arms *c,* which are placed between the two upper arms of the radius hanger, Fig. 3. The link block, auxiliary reach rod, and radius hanger are secured together by the link-block pin *g,* Figs. 1 and 3.

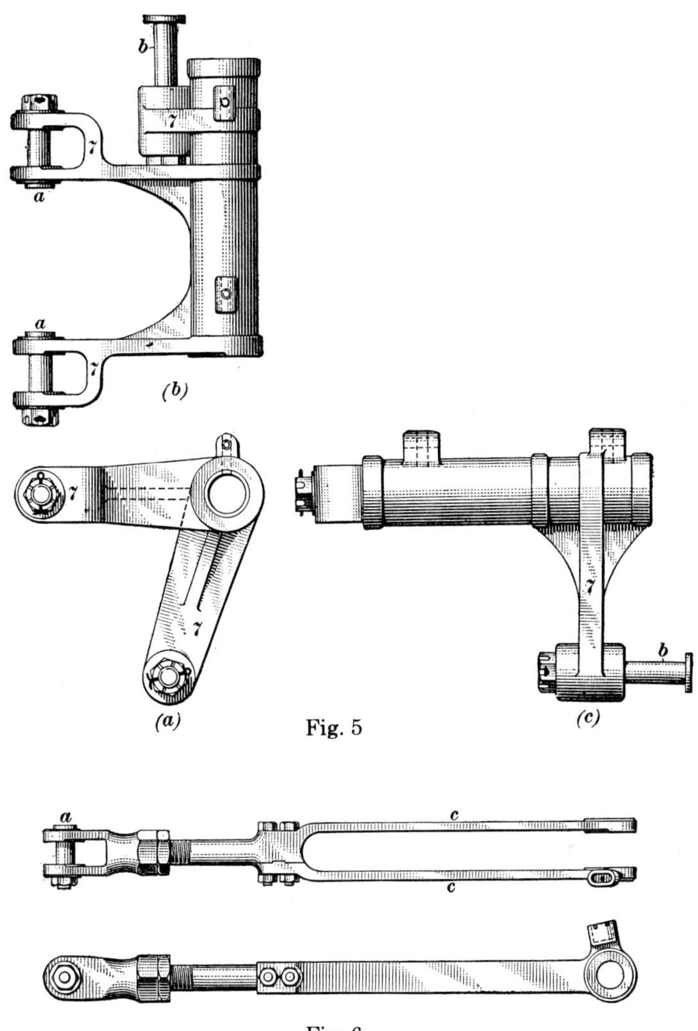

Fig. 5

Fig. 6

OPERATION

7. Movement Imparted by Eccentric Crank.—In Fig. 7 the radius hanger *3* is shown connected to the eccentric rod *2* at *c*, much farther from the transmission yoke than is actually the case in practice. The purpose of doing this is to show more plainly how the radius-hanger connection affects the movement of the front end of the eccentric rod with the engine in motion. With the main pin at *a* on the forward dead center and the radius hanger *3* in the position *g c*, the circular movement of the eccentric crank *1* causes the point *c* to move in the arc *d c b′* when the pin *a* is turned in the direction indicated by the arrow. The first movement of point *c* will be downwards toward *b′*. Constraining the point *c* to move in the arc *d c b′* causes point *e′* to move in the path shown by the dotted figure *f*, which

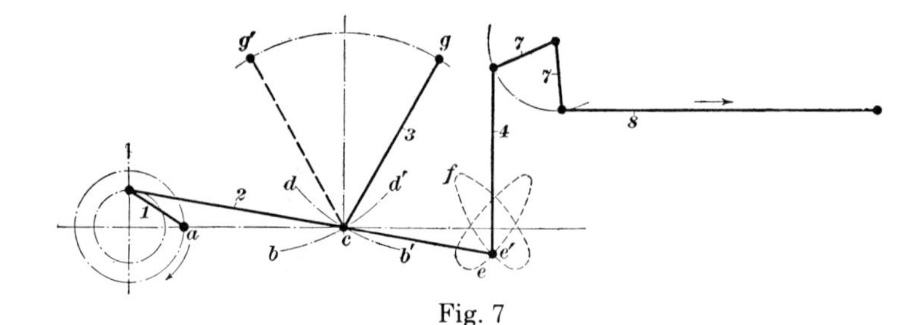

Fig. 7

indicates the movement of this point for one-half turn of the drivers when the engine is running in forward gear. An up-and-down movement is then imparted by the forward end of the eccentric rod *2* to the transmission yoke *4* and the upper arm of the bell-crank *7*. This movement is transmitted to the valve through the lower arm of the bell-crank *7* and the valve rod *8*, causing it to move back and forth across the ports. With the radius rod *3* in the position *g′ c*, and the engine moving backwards, the point *c* will move in the arc *b c d′*. The point *e′* will then describe the path shown by the dotted figure *e*.

8. Extent of Valve Travel.—The extent of the movement imparted to the bell-crank *7* by the transmission yoke *4* and the consequent valve travel depends on the inclination of the radius hanger *3*, and this in turn is governed by the position of the link block in the link. As the point *g*, Fig. 7, is moved toward the center of the link the arc *b′ c d* becomes less vertical and the figure *f* described by the point *e′* becomes flatter. The up-and-down movement of the transmission yoke and bell-crank *7* is reduced, resulting in a decreased valve travel and shorter cut-off. When the link-block pin *g* is in the center of the link the figure described by the forward end of the eccentric rod becomes horizontal and the movement then imparted to the bell-crank *7* by the transmission yoke *4* causes the valve to move twice its lap and lead for each half turn of the drivers.

The cut-off is changed by changing the inclination of the radius hanger *3*, the link and link block merely serving to guide its upper end when the reverse lever is moved. The link also prevents any up or down movement of the radius hanger when the engine is in motion.

As the upper end of the radius hanger *3* conforms to the curvature of the link with the main crankpins on dead centers, the lead therefore remains constant for all positions of reverse lever.

9. Actual Movement of Gear.— Fig. 8 (on page 518A) shows the position assumed by the valve gear for two positions of the main pin. The full lines show the position of the valve gear with the main crankpin at *a* on the forward dead center, while the broken lines show the change in position after the pin has moved to *a′*; view *(a)* shows the position of the valve when the main crankpin is at *a*, and view *(b)* shows its position when the main pin has moved to *a′*. View *(a)* shows the valve with the front steam port open the lead, while in view *(b)* this port is shown wide open. The movement of the parts of the valve gear will now be traced while it is moving the valve from lead opening to full port opening.

As the main crankpin *a* moves from the front dead center, the forward movement of the eccentric crankpin *b* carries the eccentric rod *b d* forwards. The link-block pin being at *f*, the lower end of the radius hanger

3 moves in a downward arc towards its dotted-line position *c'*. This causes the transmission-yoke pin *d* to move forward and downwards toward its full port opening position *d'*, and pulls the upper arm of bell-crank *7* from its starting position *e* toward its full port opening position *e'*. The other arm of the bell-crank moves the valve rod and valve ahead until the valve fully opens the forward steam port. With the link block in the position shown, the length of the arc described by the lower end of the radius hanger is such that a further movement of the eccentric crank from *b'* results in the transmission yoke *4* moving upwards. The valve will then move backwards toward cut-off.

The movement of the valve gear will next be traced while it is moving the valve from cut-off at the front steam port, Fig. 9 (on page 518A), view *(a)*, to release at the same port, view *(b)*. The distance traveled by the valve between cut-off and release is equal to the steam lap provided the valve is line and line on the exhaust side. The full lines indicate the position of the valve gear at cut-off, and the broken lines indicate its position at release.

The lower end of radius hanger *3*, Fig. 9, is moving in an upward arc from *c* toward its dotted-line position *c'* and carries the transmission yoke *4* upwards from *e* toward *e'*. This movement, transferred through bell-crank *7* and valve rod *8*, causes the valve to move backwards until it connects the front steam port with the exhaust passage *k*. The movement of the valve continues backwards and when the main pin *a'* is on the back dead center and the eccentric crankpin *b'* on the bottom quarter, the back steam port will be open the amount of the lead.

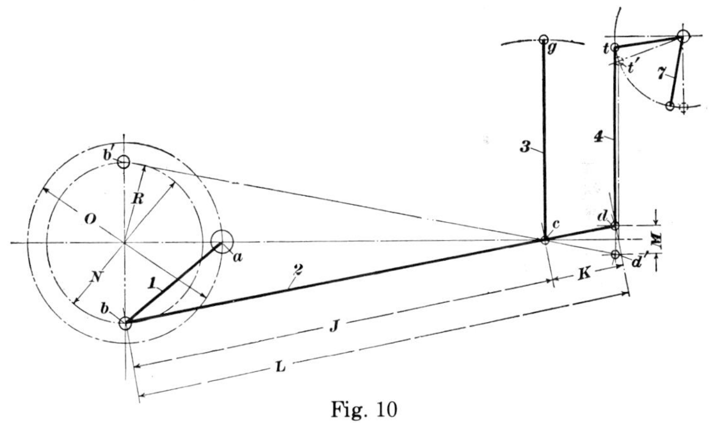

Fig. 10

10. Ratio of Eccentric Rod.—Fig. 10 shows the link-block pin *g* in the center of the link and the main crankpin *a* on the forward dead center. Moving the main pin to the back dead center causes the eccentric crankpin *b* to move one-half a turn from *b* to *b'*. The transmission-yoke pin *d* moves from *d* to *d'*, the point *t* of the bell-crank from *t* to *t'*, and the movement imparted to the valve is equal to twice its lap and lead *M*.

If desired, the main crankpin may be assumed stationary. Detaching the eccentric rod from its crankpin *b* and moving it upwards to *b'* causes the transmission yoke *4* and bell-crank *7* to transfer movement to the valve equal to twice its lap and lead *M* provided the radius-hanger pin *c'* is properly located on the eccentric rod.

The formula for calculating the ratio of the eccentric rod is as follows:

$$J = \frac{L \times R}{R + M}$$

$$K = \frac{J \times M}{R}$$

or

$$K = L - J$$

$$M = \frac{K \times R}{J}$$

in which

J = long end of eccentric rod;
K = short end of eccentric rod;
L = total length of eccentric rod;
M = lap plus lead;
R = radius of eccentric-crank circle.

EXAMPLE.—The length of the eccentric rod is 7 feet, the lap and the lead is 1½ inches, and the radius of the eccentric-crank circle is 5½ inches. Find the length of the long and the short ends of the eccentric rod.

SOLUTION.—From the formula, J, or the length of the long end is

$$\frac{L \times R}{R + M} \text{ or } \frac{7 \times 5\tfrac{1}{2}}{5\tfrac{1}{2} + 1\tfrac{1}{2}} = 5\tfrac{1}{2} \text{ ft.}$$

The length of the short end is 7 - 5½ + 1½ ft. Ans.

REVERSING

11. General Explanation.—A locomotive with the Southern valve gear can be run backwards by moving the reverse lever to the back corner of the quadrant, thereby placing the link blocks in the rear ends of the links. The position assumed by the radius hanger *3′* and the reversing-shaft arm *12′* when the locomotive is reversed is shown in Fig. 11. As with any locomotive valve gear, the reverse lever must be moved all the way in the quadrant to insure reversal, and even then the starting impulse in the opposite direction is frequently imparted to one side only. The starting impulse varies with the positions of the main crankpins, and in certain positions of the pins the slack in the train has to be taken in order to obtain a more advantageous position from which to start.

12. Principle Involved in Reversing.—The principle involved in the reversing of a locomotive will be explained by referring to the simplified arrangement of parts shown in Fig. 12. With the main crankpin c on the bottom quarter, and the eccentric crankpin b on the back dead center in which position it is ahead of or leads the main crankpin, the engine can be made to run backwards by moving the pin b to the front dead center at b'. The reason is that this change places the eccentric crankpin ahead of the main crankpin the same amount when running backwards as when running forwards, and opens the back steam port the same amount as the front port was opened before reversal. The fact that it should be necessary for the eccentric crankpin to be in the same relative position with respect to the main crankpin with the engine running in one direction as in the other should be almost self-evident.

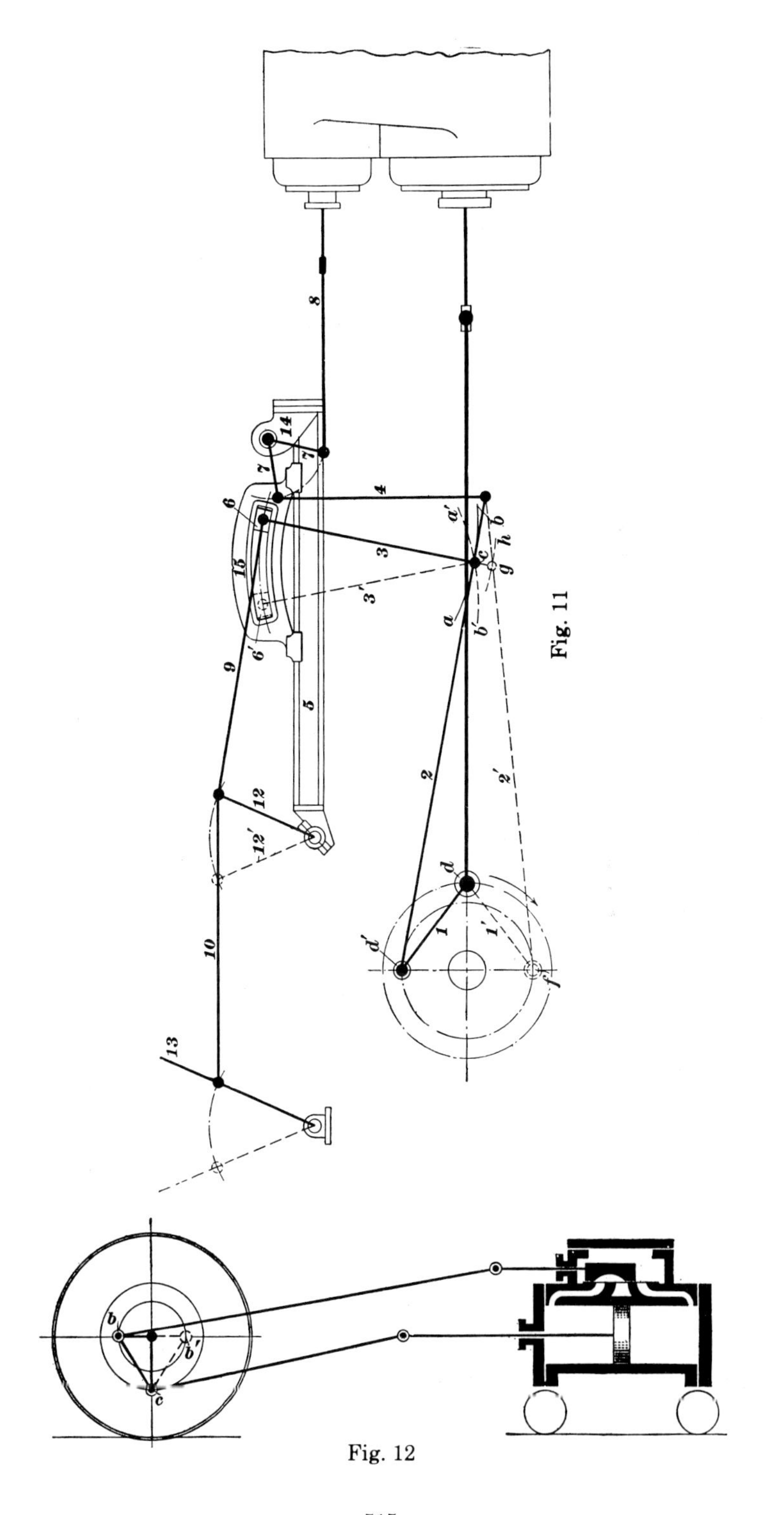

Fig. 11

Fig. 12

The following fundamental principle can be derived from the foregoing: Reversal of motion requires eccentrics or eccentric cranks, or an effect equivalent to eccentrics or eccentric cranks at the same distance from the main crankpins when the locomotive is moving backwards as when moving forwards. All locomotive valve gears are designed to produce this effect when the reverse lever is moved from either corner of the quadrant to the other. With the Stephenson valve gear the back-up eccentrics fulfill the requirement of eccentrics at the same distance from the main crankpins when running backwards as when running forwards. With the Southern valve gear and but one eccentric crank for each gear, the same effect is produced when the reverse lever is moved from the front to the back corner of the quadrant as if each eccentric crank were given one-half of a turn in the proper direction.

13. Although each eccentric crank is fixed on the main crankpin, yet each one can be considered as occupying the same position with respect to the main crankpin when the locomotive is running backwards as when running forwards. That is, if the eccentric crankpins are set to follow the main crankpins when running forwards, the action of the eccentric crankpins when the gear is reversed is the same as if these crankpins were also following the main crankpins when the locomotive is running backwards. In Fig. 11, the eccentric crankpin *f* follows the main crankpin *d* when the locomotive is running forwards. When running backwards, the eccentric crankpin leads or is ahead of the main pin, but the effect on the valve when the gear is reversed is the same as if the eccentric crank were at *1'*, and the eccentric crankpin at *f* following the main pin.

The effect of eccentric cranks at the same distance from the main crankpins when moving back as when moving ahead, places the valves in the proper position to start the locomotive backwards, and also gives them the proper movement for the reverse direction.

14. Reversing Arrangement.—When considering the reversing arrangement of the Southern valve gear, it will only be necessary to explain that the required change in the position of the eccentric crankpins imparts the same movement to the valves as is produced by the movement of the link blocks to the rear of the links. If it is found that the actions obtained are similar, it can then be assumed that the valves when the locomotive is reversed will be moved and will continue to be moved in the proper manner to cause the locomotive to run backwards.

In Fig. 11, the locomotive is shown on the forward center on the right side. The eccentric crankpin *d'* is placed one-quarter of a turn from the main crankpin *d,* therefore an inside admission valve is used, and the front steam port with the locomotive on the front center is open the amount of the lead. If the main crankpin is now turned in the direction of the arrow with the reverse lever *13* in forward gear as shown, the valve will be moved forwards, and will open the front steam port wider. The valve when the locomotive starts backwards must necessarily move in the same direction or forwards in order to admit steam into the front end of the cylinder and thereby move the piston backwards.

15. The reversing of the locomotive will first be considered by assuming that the main crankpin *d* is held fixed and with the eccentric rod *2* uncoupled, that the eccentric crank *1* is turned on the main pin from its present position at *1* to *1'*, thereby bringing the eccentric crankpin to *f*. The radius hanger *3* is next lengthened by the amount *c g*, so that the transmission yoke *4* will not be moved when the eccentric rod *2'* is coupled up again. This new position places the eccentric crankpin *f* in the same relation to the main crankpin *d* when the locomotive is running backwards as when running forwards; that is, the eccentric crankpin follows the main crankpin in each case. According to the principle already laid down, the valve with the eccentric crankpin at *f* should now receive a movement in the proper direction to cause the locomotive to move backwards. When the main crankpin *d* is turned backwards, the eccentric crankpin *f* will move the eccentric rod *2'* forwards, and the point *g* will follow the arc *g h* and will move downwards. The transmission yoke will also move downwards and the bell-crank *7* and the valve rod *8* will move the valve forwards or in the proper direction for the backward rotation of the main crankpin. However, the same movement of the valve can be obtained by leaving the eccentric crank at *1* and the eccentric rod at *2*, and by simply moving the link block *6* to *6'*, thereby moving the radius hanger from *3* to *3'*.

16. In this event when the main crankpin *d* is turned backwards the point *c* will begin to move downwards in the arc *c b'* with the result that the transmission yoke *4* will be pulled downwards and the valve will move forwards as before. Therefore, by moving the radius hanger *3'* to the position shown the same movement is given to the valve as if the hanger were left in position *3* and the eccentric crankpin turned to position *f*. In brief, the movement of the radius hanger by the reverse lever is equivalent to giving the eccentric crank one-half turn to position *1'*, thereby placing the eccentric crankpin at *f*. The gear then conforms to the principle governing the reversal of motion given in Art. 12. On the left side of the locomotive, the main crankpin is on the top quarter and the valve on this side will have the back steam port open. When the locomotive is reversed, the valve will move forwards and open the front port, therefore for the instant the starting impulse backwards develops on the left side. However, the valve on the right side does not move when the locomotive is reversed with the main crankpin on the dead center. When running forwards, the point *c* follows the arc *a b* and when running backwards the arc *a b'*. The preceding can be summarized as follows: To reverse a locomotive it is necessary for the valve gear to be so designed that when the reverse lever is shifted an effect will be produced equivalent to placing the eccentric crankpins in the same position relative to the main crankpins when running in one direction as in the other. Such an action not only causes the valves to admit steam to the opposite end of one or both cylinders but also gives the valves the proper movement to keep the locomotive in motion.

17. Valve Movement During Reversal.—The valve movement during reversal depends on the position of the main crankpins and is the greatest on the side on which the main crankpin is on the quarter and is the least on the side with the main crankpin on the center. On account of practically the infinite number of positions in which the main crankpins can be placed, it is impossible to study the movement of the valves in all positions of the crankpins when the locomotive is reversed. However, in Figs. 13, 14, 15, and 16 are shown the changes in the positions of the valves when a locomotive with inside admission valves is reversed with the main crankpins in certain representative positions. The positions of the main crankpins are shown in views *(a)*, the positions of the valves before reversal in views *(b)* and *(c)*, and their positions after reversal in views *(d)* and *(e)*. The crankpins and the cylinders on the left side are marked *l* and on the right side they are marked *r*. As shown by the valve stems, the front of the locomotive points to the left. The arrows in the ports show the direction of the flow of steam, and the arrows on the pistons indicate their direction of movement.

18. It is assumed that the reverse lever is in full forward gear before reversal and that it is in full backward gear after reversal. The lever is usually moved all the way back when reversing so that the maximum valve movement and the greatest port opening possible will be obtained. With the main crankpins in the positions shown in Fig. 13 *(a)*, the valve on the right side, view *(b)*, has the front steam port open the amount of the lead, and the valve on the left side, view *(c)*, has the back steam port about wide open for the admission of steam. When the locomotive is reversed, the valve on the left side, view *(e)*, opens the front steam port for the admission of steam the same amount as the back port before reversal while the valve on the right side, view *(d)*, does not move. In this case for the instant the locomotive starts backwards from the left side.

19. The valve movement during reversal with the main crankpins as shown in Fig. 14, view *(a)*, is as follows: The valve on the right side, view *(b)*, is about at cut-off at the front steam port, and the front steam port on the left side, view *(c)*, is open fully for the admission of steam. When the locomotive is reversed, the valve on the right side, view *(d)*, opens the back steam port fully for the admission of steam, and the valve on the left side, view *(e)*, moves to cut-off at the back steam port. In this instance the locomotive starts backwards from the right side.

With the crankpins in the positions shown in Fig. 15 *(a)*, the valve on the right side has the back steam port opened fully for the admission of steam, and the valve on the left side, view *(c)*, is about at cut-off at the front port. When the locomotive is reversed, the valve on the right side, view *(d)*, moves to cut-off at the front port. The valve on the left side, view *(e)*, opens the back port for steam fully, and the locomotive starts back from this side.

With the crankpins in the positions shown in Fig. 16 *(a)*, the valve on the right side, view *(b)*, has the back steam port opened wide, and the valve on the left side is about to open the front steam port and exhaust the steam. When the locomotive is reversed the valve on the right side, view *(d)*, moves so as to open partly the front steam port, and the valve on the left side, view *(e)*, opens the back steam port fully. Therefore, the locomotive starts backwards from both sides.

The foregoing shows that the locomotive when reversed usually receives a starting impulse from one side only. Generally a locomotive starts backwards from both sides when one crankpin is near the dead center as shown in Fig. 16. With both of the pins some distance above or below the dead-center positions, the movement in the reverse direction is generally received on one side.

Figures 13, 14, 15, and 16 appear on page 518B.

20. Broken Eccentric Crank or Eccentric Rod.—If the eccentric crank or the eccentric rod should become broken, disconnect the rod from the crank and from the radius hanger and transmission yoke, then secure the hanger and yoke to clear any moving parts. If no other provision is made to lubricate the cylinder, the valve should be clamped to open the steam port slightly for this purpose.

21. Broken Radius Hanger.—If the radius hanger breaks, disconnect it and the transmission yoke from the eccentric rod and remove the rod. Clamp the valve in central position, or to slightly open the steam port as conditions may require. If the rules permit, the eccentric rod and transmission yoke may be left up and carried by the bell-crank.

22. Broken Transmission Yoke or Horizontal Arm of Bell-Crank.—If the transmission yoke or horizontal arm of the bell-crank should become broken, disconnect the yoke from the eccentric rod and secure the valve as already described. The eccentric rod will then swing on the radius hanger.

23. Broken Vertical Arm of Bell-Crank.—If the vertical arm of the bell-crank should become broken, clamp the valve in the position desired.

24. Broken Main Reach Rod, Auxiliary Reach Rod, or Reversing Shaft Arm.—If the main reach rod, auxiliary reach rod, or reversing shaft arm should become broken, block both link blocks in the same position in the links to give the cut-off desired. If one auxiliary reach rod breaks, clamp the link block on that side in the position desired, and secure reverse lever so that the engine cannot be reversed.

SETTING THE VALVES

25. Set the link supports so that the dimensions conform to the figures on the erecting card. Then connect the gear as shown on the erecting card.

26. Links.—Set the link *15* temporarily so that dimension *M* conforms to that on erecting card. (See Fig. 17 for this as well as for the operations that follow.)

27. Main Reach Rod.—Set the reverse lever *13* in center of the quadrant and adjust the main reach rod so that reversing shaft arms *12* will stand in a vertical position.

28. Auxiliary Reach Rods.—Adjust the auxiliary reach rods *9*, so that the link block *6* will be in the center of the link when the reverse lever *13* is in the center of quadrant.

29. Checking Length of Eccentric Crank.—Check the length of the eccentric crank by using a tram with an adjustable long leg and a fixed short leg. With a hollow crankpin or with one having a large lathe center, bridge the hole temporarily and obtain an accurate center. Also establish an accurate center on the eccentric crankpin. Then by

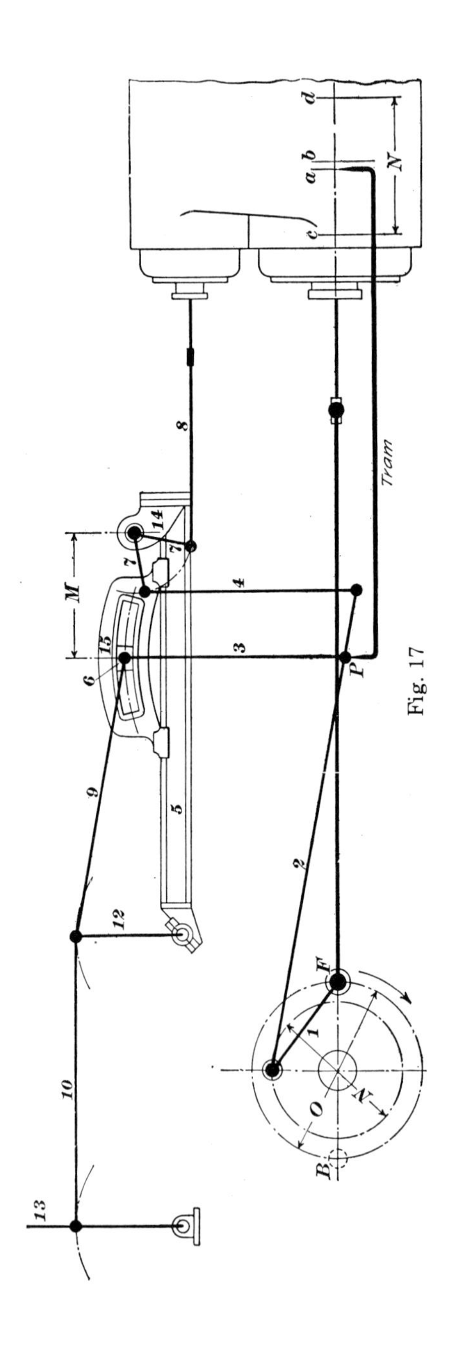

Fig. 17

adjusting the tram to these centers, the length of the eccentric crank can be checked. A new crank must be applied if the old one is found to be of an improper length. If the proper tram is not available, measure the distance from the center of the main crankpin to the base of the eccentric crank pin and add one-half the diameter of the eccentric crankpin.

Set a new eccentric crank temporarily in its correct position as nearly as possible. For outside admission, the eccentric crank leads the main crankpin and for inside admission it follows the main crankpin, with the locomotive running forwards in both cases.

30. Setting Eccentric Crank.—To set the eccentric crank, first find the dead centers in the usual way. Then with the engine on the front dead center *F*, tram from the center of the radius-hanger pin *P* to a point on the cylinder casing or the guide yoke and scribe the line *a*, taking care to hold the tram as nearly horizontal as possible. Next place the engine on the back dead center and scribe the line *b* from the center *P*. If the lines *a* and *b* coincide, the setting of the eccentric crank is correct. If they do not, knock the eccentric crank toward or away from the center until they do.

The eccentric crank can also be set by means of a gauge; still another way is to set it so as to give the proper throw as explained in Art. 31.

31. Checking Length of Eccentric Crank.—If an eccentric crank of the proper length is properly set, it cannot help but give the proper throw. On the other hand, if the eccentric crank is known to be of the proper length and gives the proper throw, then it must be properly set. As a check against possible errors in calculating the length of the crank, its length can be checked by checking its throw in the following manner: Revolve the wheel and move the engine one full turn, tram in the same center *P*, and scribe lines *c* and *d*, thereby marking the extreme movement that the eccentric crank carries the point *P* in either direction. If the full travel as shown on the cylinder is within $\frac{1}{16}$ inch of the diameter of the eccentric crank circle *N*, as shown on the erecting card, it is the correct length and is also properly set. If there is a difference of as much as $\frac{1}{8}$ inch in the full travel of the eccentric crankpin, change the eccentric crank one-fourth of the difference and reset it as already described. For example, if an eccentric circle of a diameter of 21 inches is required and the distance from *c* to *d* is found to be only $20\frac{1}{2}$ inches, then the eccentric crank must be lengthened $\frac{21 - 20\frac{1}{2}}{4}$ or $\frac{1}{8}$ inch. If, on the other side of the engine, the distance *cd* is found to be $21\frac{1}{2}$ inches, the eccentric crank is too long and must be shortened $\frac{21\frac{1}{2} - 21}{4}$ or $\frac{1}{8}$ inch. The reason for taking one-quarter of the difference is that for every $\frac{1}{16}$ inch added to the length of the crank, about four times that amount, or $\frac{1}{4}$ inch, is added to the diameter of the crank circle. This, however, does not appear until after the crank is reset.

32. Adjustment for Still Valve.—Place the engine on the front and back dead centers *F* and *B* and move the reverse lever the entire sweep of the quadrant. If the valve moves in the same direction as the link block, move the link ahead; if it moves in the opposite direction, move the link back until a still valve, or a point where the valve does not move, is found. When the link is correctly set, the auxiliary reach-rod *9* must be adjusted to bring the tumbling shaft arm plumb when the link-block is in mid-position.

33. Equalizing the Cut-Off.—Next, run the engine over, and measure the cut-off in the usual working notch which corresponds to a cut-off of 25 or 30 per cent. The cut-offs are equalized by altering the valve rods and when calculating the change the best plan is to determine by actual experiment the ratio between the valve travel and the stroke of the piston in the working notch. To do this, ascertain how far the valve moves in the vicinity of cut-off while the piston is moving 1 inch. If this is found to be 1/16 inch, this is the ratio required. Then, for each inch that it is desired to change the cut-off, alter the valve stem 1/16 inch.

For example, assume a difference of 2 inches in the cut-offs at the front and the back ends of the cylinder, the cut-off at the front end being 6 inches and at the back end, 8 inches. To find the alteration to be made to square the cut-offs, first add them together and divide by two; thus, in this case, 7 inches will be the cut-off desired. Then lengthen the valve stem 1/16 inch, thus keeping the front port open longer or while the piston is moving another inch. This alteration results automatically in the back port closing earlier so that the cut-off is shortened 1 inch at the back port.

34. Back Gear.—In this method of valve setting no attention has been given to back gear, as it is the general practice to favor forward gear only. To make a locomotive function equally well in both gears calls for many refinements in the valve setting, but when only one motion is considered many of these may be overlooked. Hence, with the setting described, the locomotive cannot be expected to operate efficiently when running in back gear.

(No questions are included with the original text for this section.)

Locomotive Boilers Part 1

By

J. W. Harding

1967A Printed in U.S.A. Edition 3

1945 Edition

LOCOMOTIVE BOILERS
(PART 1)

CONSTRUCTION AND DETAILS

BOILER SHELL

GENERAL DESCRIPTION

1. Definition.—The locomotive boiler is a steel shell for containing water, which, when converted into steam by the heat of the fire in the firebox, furnishes the energy to move the locomotive. Locomotive boilers are of the internal-firebox, straight-firetube type and are made up of a cylindrical portion, which contains the tubes and flues, a back end, which is enlarged and shaped to accommodate the firebox, and a smokebox at the front, on which the stack is placed.

2. Description.—An exterior view of a locomotive boiler is shown in Fig. 1 (on page 540A). A sectional view taken lengthwise of the boiler is given in Fig. 2 *(a)* (on page 540A), and a cross-sectional view is shown in *(b)*. The boiler shell is made up of a number of sheets and courses, namely, the back head, an outside firebox sheet, one on each side, the roof sheet, the throat sheet, the dome course, the taper, or conical, course, the first course, and the smokebox. However, the names of the courses vary according to the location of the steam dome. Should the dome be placed on the course next to the smokebox, as is the usual practice with smokebox throttles, the first course would become the dome course. Also, the taper course would become the dome course should the dome be located on it. In the absence of the steam dome on the course next to the roof sheet, this course would become the combustion-chamber course because the combustion chamber of the firebox projects partly into it. The rear end of the boiler is shaped as shown in order to conform to the shape of the firebox on the inside, the firebox being rectangular. The rear end of the boiler is riveted to the foundation ring *o*, Fig. 2 *(a)*, which serves as a base to secure this part of the boiler to the frame of the locomotive.

The firebox, which comprises the firebox proper and the combustion chamber, is made up of a crown sheet, two side sheets, a door sheet, and a back tube-sheet. The back tube-sheet and the front tube-sheet are connected by flues *a* and tubes a_1, which serve to carry the smoke and gases from the firebox to the smokebox. The arch tubes *b* serve to promote the circulation of the water around the firebox as well as to support the brick arch b_1.

The purpose of the steam dome *d* is to collect and hold dry steam, from where it is conveyed through the throttle pipe and the dry pipe to the superheater. In the older types of locomotives, the throttle valve was located in the throttle pipe in the dome, but with the throttle in the smokebox and incorporated with the superheater header, the dome houses the dry pipe only. The four openings shown along the bottom of the boiler are the belly washout holes, and the angles shown at the rear of these holes are the waist-sheet angles to which the waist sheets are riveted. The purpose of the waist sheets is to secure to the frame of the locomotive the particular part of the boiler to which they are attached.

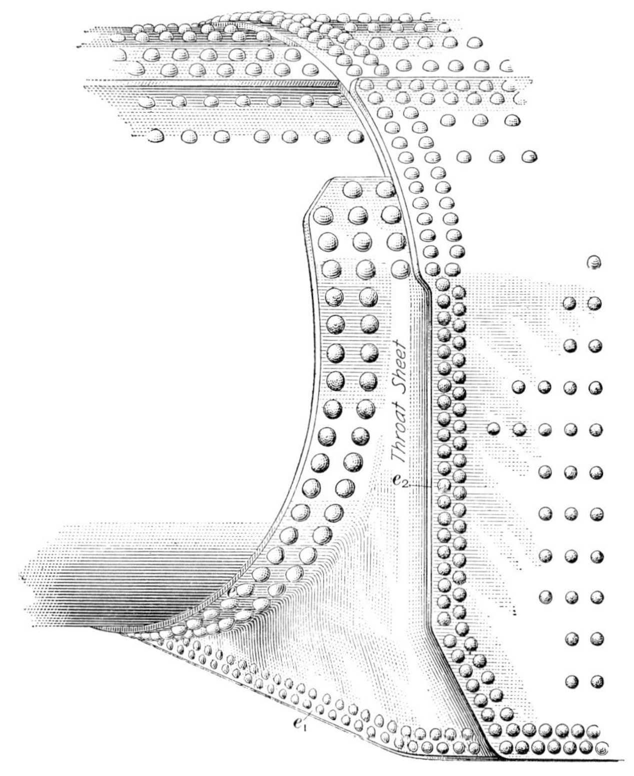

Fig. 3

3. The throat sheet, Fig. 3, must be of a special shape because it serves to connect the cylindrical part of the dome course to the outside firebox sheets. One part of the sheet is flanged to conform to the cylindrical part of the dome course that rests on it, the remainder of the sheet, except where it is flanged to join the outside firebox sheets at e_2, being flat. At the lower edge e_1, the throat sheet is riveted to the foundation ring.

The smokebox, Fig. 1, the bottom surface of which is bolted to the cylinders of the locomotive, is the part of the boiler from which the products of combustion, after passing from the firebox through the tubes and the flues, are discharged to the stack. Certain draft appliances are installed in the smokebox principally to prevent the emission of live sparks of such size as to cause fires, but also to make the smokebox selfcleaning. The smokebox is the only part of the boiler that is rigidly connected to the locomotive frame, the connections at the other points being flexible to permit the free expansion and contraction of the boiler lengthwise due to heating and cooling. An exhaust pipe through which the exhaust steam from the cylinders is discharged to the smokebox projects upward into it through the opening shown. The heavy ring shown riveted inside the front edge of the smokebox, Fig. 2 *(a)*, is the smokebox front ring and its purpose is to form a bearing for the attachment of the smokebox front, which is a cast-iron or pressed-steel plate usually somewhat curved and which, with the smokebox door, closes the end of the smokebox. The smokebox door, which permits access to the interior of the smokebox, is hinged to the smokebox front and is held closed usually by nuts screwed onto studs set in the front. The smokebox bottom liner is riveted on the inside and at the bottom of the smokebox shell to strengthen it where the steam pipe opening is cut out and also to prevent wear on the smokebox shell by the abrasive action of the cinders.

CONSTRUCTION OF BOILER SHELL

4. Definition.—The boiler shell comprises all the exterior sheets of the boiler and therefore consists of the cylindrical courses in front of the firebox, the roof sheet, the back head, the throat sheet, the front tubesheet, and the outside firebox sheets. However, boilermakers refer to the part between the firebox and the smokebox as the boiler shell and the other exterior sheets as the outer firebox sheets.

5. Lap Joints.—The boiler shell forward of the firebox is made up of circular courses, which vary in number according to the length of the boiler and are jointed together at the ends by lap joints. With this type of joint, the end of one course laps over the end of the course next to it and the two are riveted together. One, two, or three rows of rivets may be used, depending on the size of the boiler. When two rows of rivets are used, the joint is referred to as a double-riveted lap joint. The dome course and the first course, Fig. 1, are connected to the conical course by double-riveted lap joints. A triple-riveted lap joint is one in which three rows of rivets are used.

6. Details of Lap Joint.—In Fig. 4 *(a)* is shown an exterior view of a triple-riveted lap joint, and in view *(b)*, a sectional view. One course *a*, view *(b)*, fits on the inside of the course *b*, and the two are held together by three rows of rivets, *c*, which are driven to a full head on each side. When there are two or more rows of rivets, it is preferable to place the rivets in alternate rows, as shown in view *(a)*. At *d*, view *(b)*, is shown a recess in the edge of the plate, which is made by calking it to insure a steam-tight joint. The calking should be done with a blunt round-nose tool held in a pneumatic hammer, care being taken that the plates are not sprung apart in the process of calking. Both the inside and the outside edges are sometimes calked to insure a more uniform job.

7. Butt Joints.—The courses that comprise the cylindrical part of a boiler are usually made in one piece. When the ends of a course are connected, the seam will come lengthwise of the boiler, as shown in Fig. 1. The kind of joint which is used to connect the ends of the course is called a butt joint, because the ends butt together and do not overlap as in the girth joints.

8. Details of Butt Joint.—An exterior view of a butt joint is shown in Fig. 5 *(a)*, and a sectional view in view *(b)*. As shown in view *(b)*, the ends *a* of the course are brought together but not overlapped, and strips,

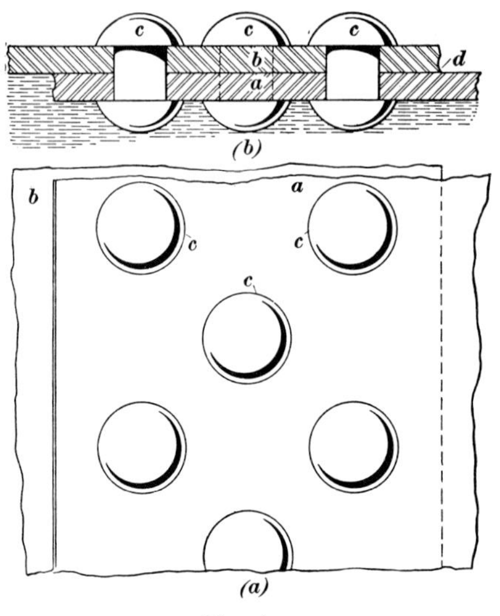

Fig. 4

or welts, b and c, which cover the joint on the inside and the outside, are then riveted to the plate by rivets that pass all the way through.

The width of the inside welt is the distance between x and y, view *(a)*, and the width of the outside welt is the distance between x_1 and y_1. The dotted line x_2y_2 shows where the ends of the course meet. The name applied to a butt joint depends both on the number of rows of rivets used on each side of the junction of the ends of the course and on the number of welts. In this case, three rows of rivets are placed on each side of the line x_2y_2 and two welts are used, so the joint is referred to as a triple-riveted, double-welt butt joint. A quadruple-riveted, double-welt butt joint is one in which four rows of rivets and two welts are used, and a quintupler riveted, double-welt butt joint is one that employs five rows of rivets on each side of the seam and has two welts.

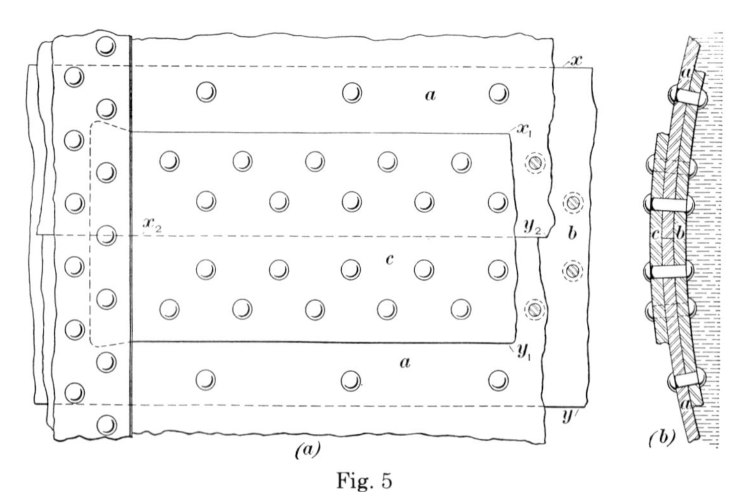

Fig. 5

In the construction of a butt joint, the object should be to obtain a joint whose strength equals that of the solid plate as nearly as possible. The form of joint shown in Fig. 5 has a strength equal to about 90 per cent of the solid plate; it is therefore said to have an efficienty of 90 per cent.

Welding is not permissible on the cylindrical part of the boiler because there are no additional means of strengthening the joint. On the other hand, the firebox is further supported by staybolts, so welded joints or seams are allowed.

9. Relative Stresses on Lap and Butt Joints.—Calculations show that a seam running lengthwise of a boiler course, such as a butt joint, is subject to nearly twice the stress that acts on a lap joint in the same course; therefore, a joint running lengthwise of the boiler will withstand only one-half the boiler pressure that a lap joint of the same construction will withstand. For this reason, care should be taken that butt joints be properly constructed to withstand the pressures and stresses to which they are subject. A lap joint, besides being subject to only half the stress of the boiler pressure, also is strengthened by the staying properties of the tubes and flues and by the stays that connect the front and back heads of the boiler.

10. Back Head.—The back head, Fig. 1, is connected to the roof sheet and the outside firebox sheets by a single riveted lap joint, and to the foundation ring by a double row of rivets. In Fig. 6 is shown a detail of the joint at the back head *a* and the roof sheet *b* with the sheet calked at *c*. A

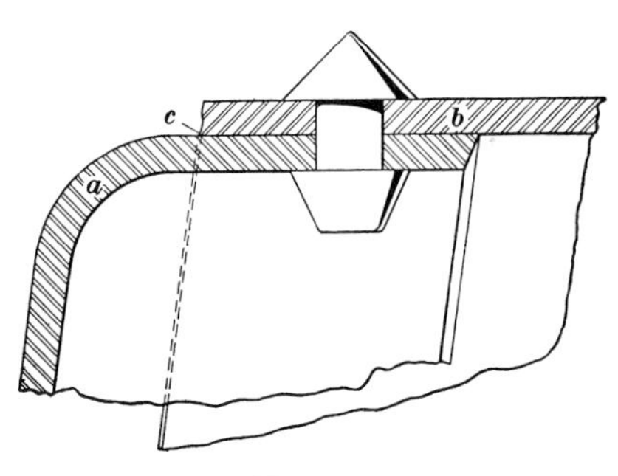

Fig. 6

single row of rivets has been found sufficient to carry the strain on the back head because of the way the head is stayed at other points. The portion of the backhead below the crown sheet is rigidly stayed to the door sheet and the foundation ring. Above the crown sheet, the back head is stayed to the roof sheet by diagonal braces, one type being shown in Fig. 2 *(a)*. Therefore, the greater part of the load on the back head is carried by the stay-bolts and the braces thereby relieving the stress on the rivets of the single riveted lap joint.

11. Door Hole.—The joint at the back head *b*, Fig. 7, and the door sheet f_2 of the firebox to form the door hole is constructed in a number of different ways. As shown in view *(a)*, the door sheet surrounding the door opening is flanged to a large radius and extends through to the back head, the two sheets being riveted together at *o*. In view *(b)*, the door sheet

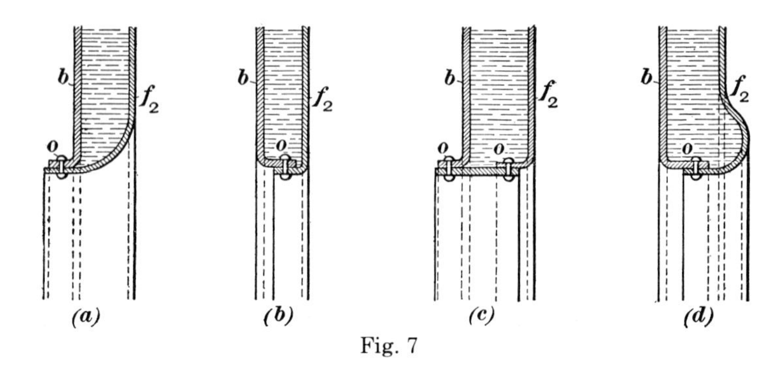

Fig. 7

is shown flanged outwards and the back head is flanged inwards, and the two are riveted together. View *(c)* shows what is called the sleeve door, in which both sheets are flanged outwards, and a separate piece is rolled to the shape of the door hole and is welded and then riveted to the sheets. Owing to the continual expansion and contraction caused by the alternate heating and cooling of the inner sheet around the fire-door as the door is opened and closed, the sheet, when flanged in a single curve as already shown, is liable to crack. A method of overcoming this is shown in view *(d)*, where the door sheet is flanged to an ogee curve. This construction allows considerable movement of this sheet without causing any rupture or distortion. The door hole is the only point at which the firebox is riveted to the boiler shell.

12. Dome Course and Roof Sheet.—The dome course, or the combustion chamber course, as the case may be, is connected to the roof sheet in the same way as the cylindrical courses are connected; that is, if the cylindrical courses are connected by double-riveted lap joints, the same construction is used at the junction of the dome course with the roof sheet. The joint extends down, as shown in Fig. 3, until it meets a similar type of joint in the throat sheet and the side sheets.

13. Throat Sheet and Side Sheets.—As shown in Fig. 3, the throat sheet is connected to the side sheets and the cylindrical course by a double-riveted lap joint, but in some cases the connection is made by a triple-riveted lap joint.

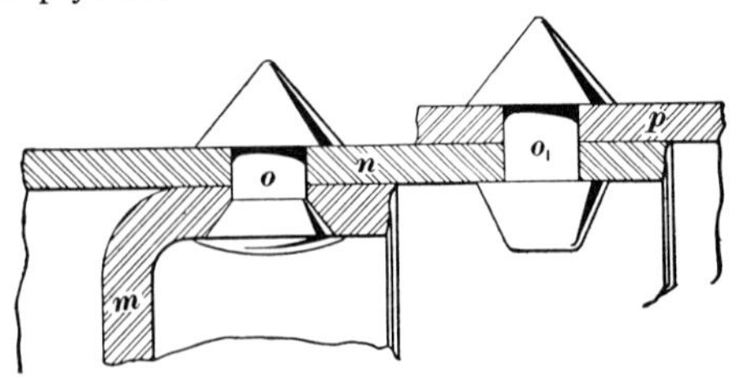

Fig. 8

14. Smokebox and First Course.—As shown in Fig. 8, the smokebox p is connected to an extension of the first course n beyond the tube-sheet m by a single or double row of rivets o_1. The smokebox may be riveted directly to the course n, as shown, or a wrought iron n_1, Fig. 9, called the smokebox ring, may be inserted between the course n and the smokebox. The ring provides a stopping point for the insulation that is applied to the outside of the boiler. Two rows of rivets o_1 are generally used with the ring n_1.

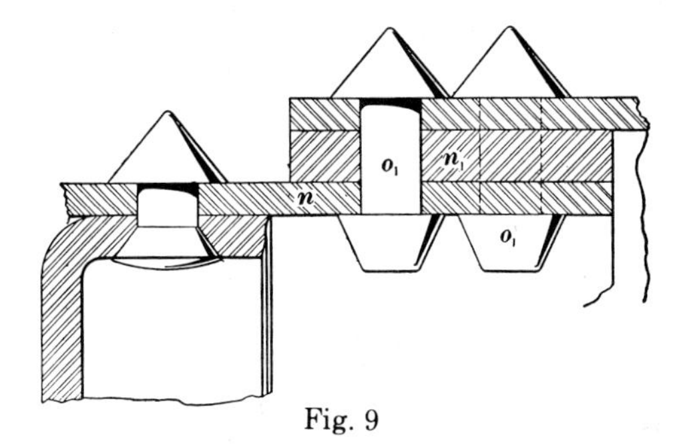

Fig. 9

15. Tube Sheets, Tubes, and Flues.—The end of the boiler course next to the smokebox is closed by the front tube-sheet, whereas the back tube-sheet forms the front end of the firebox. The tubes and the flues extend through the boiler from one tube sheet to the other and serve not only to convey the smoke and the gases from the firebox to the smokebox but also to act as stays for the tube sheets. At the firebox end, the tubes and the flues are electrically welded to the tube sheet all the way around to prevent leaks. The reason for the use of the flues, which are usually 3½ inches in diameter, is to accommodate a series of steam pipes known as the superheater units in which the steam, in its passage to the cylinders, absorbs additional heat. Hence the flues have to be made larger than the tubes, which ordinarily are 2¼ inches in diameter. In modern boilers, there are generally three or four times as many flues as tubes.

In Fig. 8 is shown how the front tube-sheet m is connected to the first course n by a single row of rivets o, which pass through the flange of the sheet and the outer course. The rivets o are sometimes driven slightly flattened, as shown, and sometimes with a flat head on both sides. The first method, which makes it easier to use the tube expanders on the side rows of tubes and also to calk the rivets should they leak, is the better one. In Fig. 10 is shown how the part of the front tube-sheet above the tubes and flues is braced. The front ends of the braces c are connected by the pins d with the **T** irons e riveted to the tube-sheet. The back end of the brace is flattened out sufficiently to allow two or three rivets to be used to connect it with the first course.

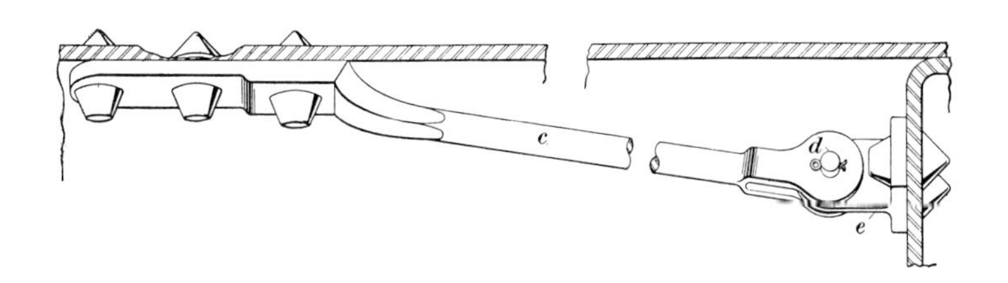

Fig. 10

16. Minimum Net Gas Area.—The minimum net gas area of a flue is the cross-sectional area of the inside of the flue after the area occupied by the superheater unit has been deducted. This area decreases with the higher steam pressures because the flue then has to be made thicker while it maintains the same outside diameter, or 3½ inches. A flue with an outside diameter of 3½ inches, used with a boiler carrying up to 200 pounds pressure per square inch, has, when the area of the superheater unit is deducted, a minimum net gas area of 6.1319 square inches. For the same flue and with pressures up to 250 pounds, the minimum net gas area is 5.9898 square inches; and for pressures from 250 pounds to 310 pounds, the net gas area is 5.8476 square inches.

17. Gas Area Through Tubes and Flues.—The net internal gas area through the tubes and flues varies between 15 and 17 per cent of the grate area and is one of the most important factors in the design of the boiler. If the gas area is too small, the size of the nozzle must be reduced, thereby increasing the cylinder back pressure in order to draw the gases through the tubes and flues. The maximum gas area that can be put into a boiler is influenced largely by the area of the front tube-sheet. The diameter of the front end of the boiler should be large enough for the front tube-sheet to take the maximum number of flues that can be installed in the back tube-sheet.

18. Foundation Ring.—The foundation ring, also called the firebox ring or the mud-ring, is a casting or forging that serves as a base to connect the rear end of the boiler with the frame of the locomotive and that also serves to space the firebox correctly with respect to the boiler shell. The rectangular part of the firebox fits within the foundation ring with the edges of the sheets flush with the bottom surface of the ring. The outside firebox sheets, the back head, and the throat sheet are similarly placed around the outside of the ring, and the complete assembly is riveted together. The space formed by the separation of the sheets by the mud-ring is called the water-leg.

The foundation ring, as viewed from the top, is shown in Fig. 11 *(a)*, and, as viewed from the side and the end, in views *(b)* and *(c)*, respectively. A section taken through the ring and the firebox sheets is shown in Fig. 12 *(a)*, and the sheets are riveted to the ring by two rows of zigzag rivets, headed over on each side, is shown in view *(b)*. At their junction at the mud-ring, the edges of adjacent sheets are scarfed, chamfered, or beveled so that their total thickness will be the same as one sheet; hence the sheets will fit tight against the side of the ring. This is the only place where this type of joint is employed in a boiler.

Foundation rings are made from 5 to 7 inches wide and from 2¼ to 4½ inches thick, depending on the width of the water-leg and whether one or two rows of rivets are used.

19. Water Leg.—The water-leg *i*, Fig. 12 *(a)*, is the space between the side sheets and the door sheet of the firebox and the boiler shell, with a width at the bottom equal to that of the mud-ring. It is a very important part of the water space of the boiler because, containing water, it prevents the firebox sheets from becoming overheated by the extreme heat of the fire. The water evaporates very rapidly in the water-leg and this evaporation causes a continuous circulation of water around the firebox. The water is at its greatest depth in the water-leg, so the pressure at the bottom of the water-leg exceeds that at any other part of the boiler.

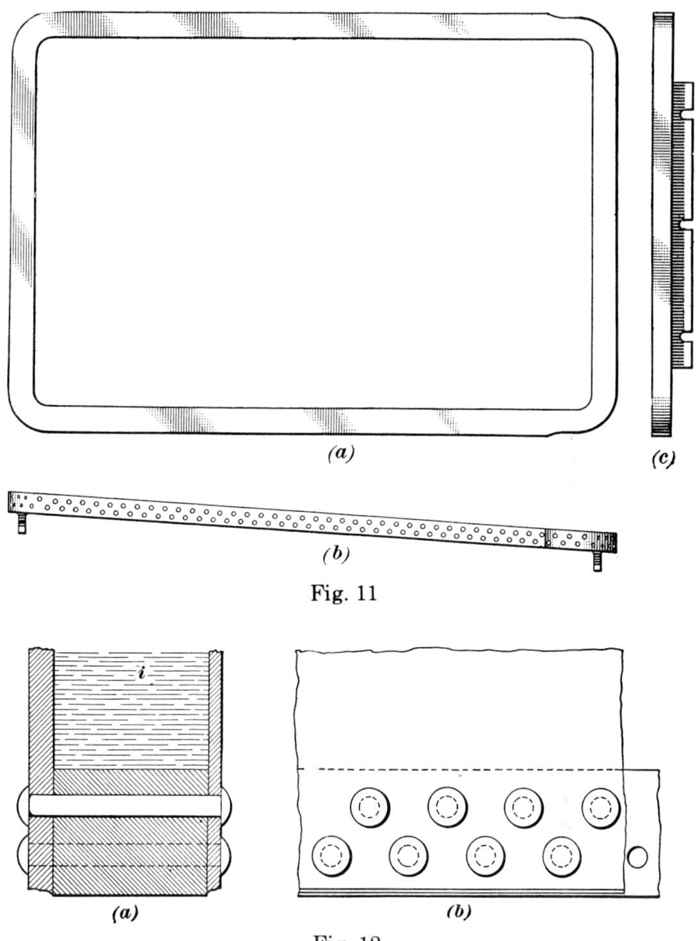

Fig. 11

Fig. 12

SMOKEBOX DETAILS

20. Names and Purpose.—In Fig. 13 is shown an arrangement of the smokebox details, known as the Master Mechanics' front-end design, with the smokebox partly broken away to make the arrangement clear. The design consists of an exhaust pipe, or stand, *a* with a round-bore exhaust nozzle *b*, a smokestack *c* with a stack extension *d* bolted to it, a diaphragm *e*, a table plate *f* supported by the exhaust pipe and attached to the diaphragm and the sides of the smokebox, an adjustable diaphragm apron, or damper, *g* attached to the front of the table plate, and a sloping smokebox netting *h* attached to the table plate and the interior of the smokebox. The purpose of the diaphragm, the table plate, the damper, and the netting is to offer an obstruction to the flow of the sparks and the cinders in their passage to the stack, thereby breaking them up into such small pieces that they will cool quickly after leaving the stack and hence reduce the fire hazard.

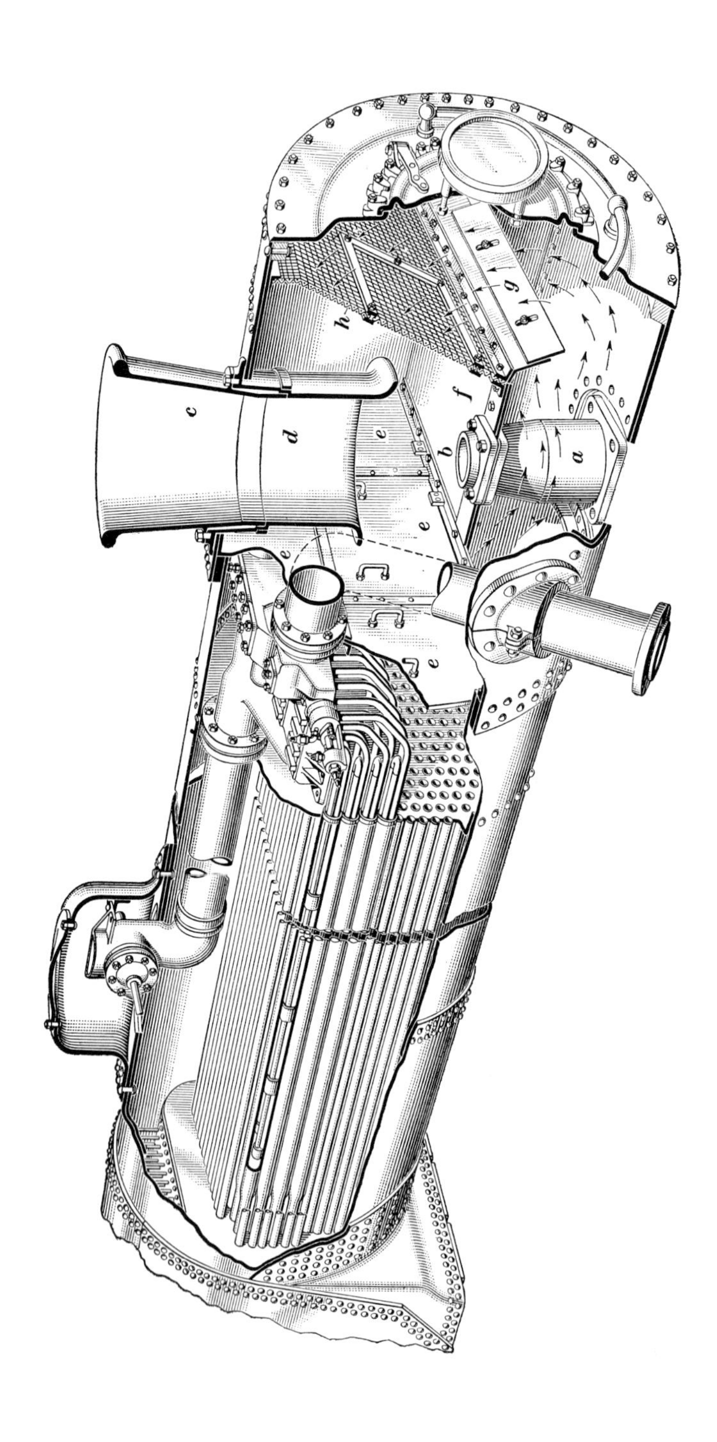

Fig. 13

21. The diaphragm conforms in shape to a part of the circumference of the smokebox and hence is semicircular. It is applied either vertically or with a slight slope about 30 inches ahead of the front flue-sheet and introduces a partition in the smokebox down to the junction of this plate with the table plate. The edge of the diaphragm is bolted to angle irons on the interior wall of the smokebox, the fit of the plate being such as to be spark-tight. The two flanges of the superheater header to which the steam pipes are connected pass through the diaphragm, and here, also, the plate fits the flanges close enough to prevent the escape of sparks. The diaphragm is made in several sections to facilitate removal and application when making repairs to the tubes, the flues, and the superheater units.

The pulsating draft, produced by the partial vacuum developed in the smokebox at each discharge of the exhaust steam, carries the sparks and the cinders against the diaphragm and causes them to break up into smaller pieces.

The table plate is rectangular in shape and is applied horizontally. It is connected both to the diaphragm and to the wall of the smokebox by angle irons. The exhaust pipe projects through an opening in the table plate, the edges of the plate being caught between the exhaust pipe and the exhaust nozzle when these parts are bolted together. The partial vacuum in the smokebox that follows each exhaust acts to lift the sparks and cinders up against the table plate and causes them to be broken up.

The diaphragm apron, or damper, also assists in breaking up the solid products of combustion, but its real function is to regulate the intensity of the draft through the fire and prevent it from being torn by the heavy exhaust at low speeds.

If there were no damper at all, the gases would rush into the smokebox as rapidly as the resistance of the tubes and the arch would permit them to flow, and, as the resistance of the grates and the fuel bed would allow the air to enter the firebox, the result would be a draft of sufficient intensity to tear the fire at low speeds. However, when an adjustable damper is used, the space through which the products of combustion have to pass to enter the front of the smokebox can be changed as desired. If the draft through the fire is insufficient, a higher adjustment of the damper will increase the space beneath it and so increase the effectiveness of the partial vacuum in the smokebox in producing a draft through the fire. If the draft is too strong, a lower adjustment will lessen the space and thus will decrease, or dampen, the effectiveness of the smokebox vacuum in causing air to be drawn through the fire and will reduce the draft. Hence, raising the damper will increase the total draft through the fire, and lowering it will decrease the draft.

22. It was formerly thought that the draft could be equally distributed through the fire by adjusting the damper, but carefully conducted experiments have shown that raising or lowering the damper has no effect on an equal distribution of the draft. With an arch fitted close against the flue sheet, the draft is always stronger through the rear grates because the arch, being open at the rear, offers no obstruction to the flow of the air.

Some railroads secure an equal distribution of air through the fire by using grates of a smaller opening at the position of greater draft, or at the rear, and grates of a greater opening at the point of less draft, or at the front.

23. Finding Height of Table Plate.—The various heights at which the table plate must be placed to obtain certain gas areas beneath it for different smokebox diameters are given in Table I. The table can be used also to obtain the height of the diaphragm apron, or damper.

The preferred maximum and minimum areas under the table plate are 95 per cent and 85 per cent, respectively, of the minimum net gas area through the tubes and the flues, and the preferred area under the diaphragm apron is 75 per cent.

The following example illustrates the use of the table:

EXAMPLE.—It was found that 95 per cent of the minimum net gas area through the tubes and the flues was 1,543 square inches and that 75 per cent of the gas area was 1,218 square inches. What should be the height of the table plate and the diaphragm apron if the inside diameter of the smokebox is 84 inches?

SOLUTION.—By following the column under 84 in Table I down to the number 1,543, it will be found that the number opposite 1,543 in the column headed Height of Table Plate is 27 in. Ans.

To find the height of the diaphragm apron for an area not given in the table, such as 1,218, select under the proper smokebox diameter the nearest gas area and height of diaphragm apron, which in this instance are 1,235 and 23 respectively, then multiply 1,218 by 23 and divide by 1,235. Thus, $\frac{1218 \times 23}{1235} = 22\frac{5}{8}$ in. Ans.

TABLE I

HEIGHT OF TABLE PLATE

Height of Table Plate Inches	Smokebox Diameters, Inches										
	66	68	70	72	74	76	78	80	82	84	86
	Areas Under Table Plate, Square Inches										
15	586	595	605	613	625	636	646	655	663	674	682
16	642	653	663	672	685	697	708	719	728	739	784
17	699	711	722	733	747	760	772	784	793	806	816
18	758	771	783	795	810	824	837	850	861	874	886
19	817	831	845	858	874	890	904	918	929	944	957
20	877	893	908	922	939	957	972	986	999	1,015	1,028
21	939	955	972	987	1,006	1,024	1,040	1,056	1,070	1,087	1,103
22	1,001	1,019	1,037	1,053	1,073	1,093	1,110	1,127	1,142	1,160	1,177
23	1,063	1,083	1,102	1,120	1,141	1,163	1,181	1,199	1,216	1,235	1,253
24	1,126	1,148	1,169	1,187	1,210	1,233	1,252	1,272	1,290	1,310	1,330
25	1,190	1,213	1,235	1,256	1,279	1,304	1,325	1,346	1,365	1,387	1,408
26	1,255	1,279	1,303	1,325	1,350	1,376	1,398	1,421	1,441	1,465	1,487
27	1,319	1,345	1,371	1,394	1,420	1,449	1,472	1,497	1,518	1,543	1,566
28	1,384	1,412	1,439	1,464	1,492	1,522	1,547	1,573	1,585	1,622	1,646
29	1,450	1,479	1,508	1,535	1,564	1,595	1,622	1,649	1,674	1,702	1,728
30	1,561	1,547	1,578	1,606	1,636	1,669	1,698	1,727	1,753	1,782	1,809

24. Netting.—The netting, Fig. 13, extends upward from the point where the table plate and the apron meet to the top and front of the smokebox. Its purpose is to prevent sparks that have not been sufficiently reduced in size by the other smokebox appliances, from being pulled through from the firebox and thrown out of the stack by the action of the exhaust. The netting consists of a network of wires about ⅛ inch in diameter and with a mesh about 3/16 to ¼ inch square, or it may consist of a perforated steel plate. The netting is held in position by being bolted to the table plate and to angle irons on the sides and the top of the smokebox, or it may be bolted to a piece of plate which extends down a short distance from the top of the smokebox. The preferred total area of the openings in the netting is 130 per cent of the total minimum net tube-and-flue-area of the boiler.

The netting obstructs the flow of gases to such an extent that the size of the nozzle must be decreased slightly to produce the same amount of draft that would be obtained if no netting were used. Experiments have shown that the obstruction to the gas flow caused by the netting is equal to the combined effect of the diaphragm, the table plate, and the damper. For this reason, a smokebox arrangement has been designed on the centrifugal principle and no netting is used.

25. Smokestack.—The smokestack, or stack, is a cast-iron pipe that is bolted to the top of the smokebox. Its purpose is to carry off the products of combustion from the smokebox to a point above the engine and the train. Fig. 14 is a half-outside and a half-sectional view of a tapered stack

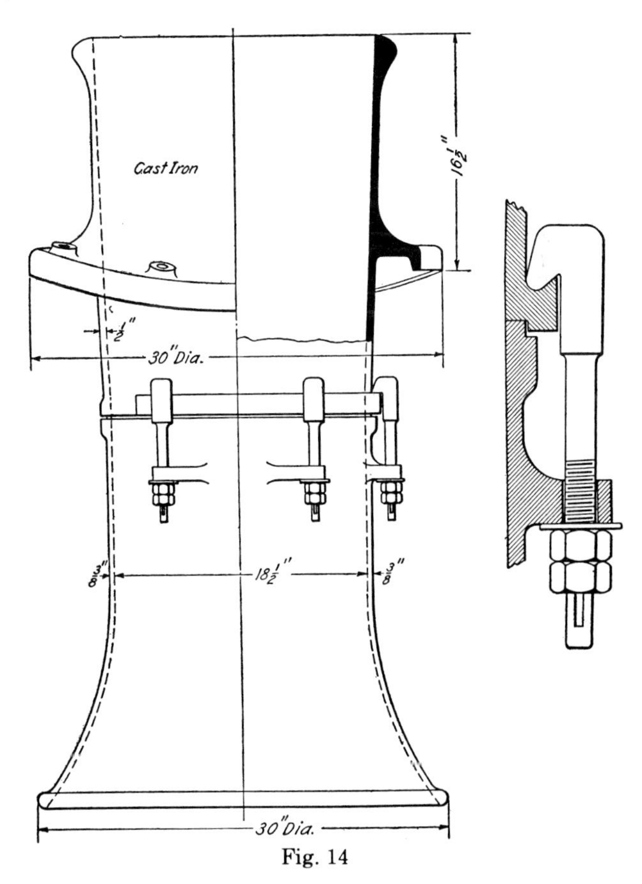

Fig. 14

of cast iron, which is used on boilers of large diameter. This stack is largest at the top, from where it tapers to the top of the stack extension, which is connected to the base of the outside stack, as shown, and projects down into the smokebox close to the nozzle. The purpose of the stack extension is to make the stack of such height that, when the engine is running at moderate speed, one exhaust is entering at the bottom of the stack before the preceding one has escaped at the top, thus preventing a rush of cold air down the stack to fill the vacuum in the smokebox.

With large boilers, the stack, owing to clearance limitations, must be made short and, since it is not of sufficient height to prevent the entry of air, it is accordingly lengthened by the stack extension. The smokestack must be set in an exactly vertical position on the boiler, and its center must line up accurately with the center of the exhaust nozzle.

26. Exhaust Pipe.—The exhaust pipe is made of cast iron and is used to conduct the exhaust steam from the cylinders and auxiliaries to the proper point in the smokebox. The intensity of the draft is governed by the exhaust nozzle on the top of the exhaust pipe.

A print of an exhaust pipe is shown in Fig. 15. The top is machined and drilled for the exhaust nozzle, Fig. 16, and the base is machined and drilled where it rests on the cylinders. The base must make a steam-tight joint and both surfaces are spotted in with a surface plate. To do this, the surface plate is coated with red lead and then rubbed on the base of the pipe and its seat on the cylinders; the high spots are filed and scraped after each application of the plate until the surfaces are flat. The surfaces are then ground into each other by using a grinding compound and rubbing one over the other. The exhaust nozzle is ground into its seat in practically the same way.

The exhaust pipe is secured to its base by twelve studs, and the nozzle is recessed into the end of the exhaust pipe and is held in place by four studs.

In the application of an exhaust pipe, the opening in its base must coincide exactly with the opening in the cylinders. The pipe then serves as an extension of the exhaust passages. The opening through the nozzle should be of such size as to cause the steam that passes through it to produce the desired draft on the fire. If the exhaust nozzle is unduly contracted, the force of the exhaust steam will be extremely violent and holes will be torn in the fire. A contracted nozzle will also increase the back pressure in the cylinders. If the nozzle is too large, the draft will be too weak and the required amount of fuel will not be burned.

27. Diameter of Nozzle.—One rule for finding the diameter of a single circular exhaust nozzle is to use 22½ per cent of the diameter of the cylinder for the diameter of the nozzle. In Table II is shown the correct nozzle sizes calculated on this basis for cylinders from 20 to 32 inches in diameter. If a bridge is used in the nozzle, it must be made larger so as to retain the net area given in the table.

28. Draft.—Draft in a locomotive is the term applied to the rapid flow of air through the grates and the fire that occurs when the locomotive is working. Draft is divided into two classes, natural and forced. With natural draft, no apparatus or force is used to induce the flow of gases other than the difference between the weight of the hot gases in the firebox and the stack and that of the cooler air outside. The draft through a locomotive boiler, when the engine is standing and the

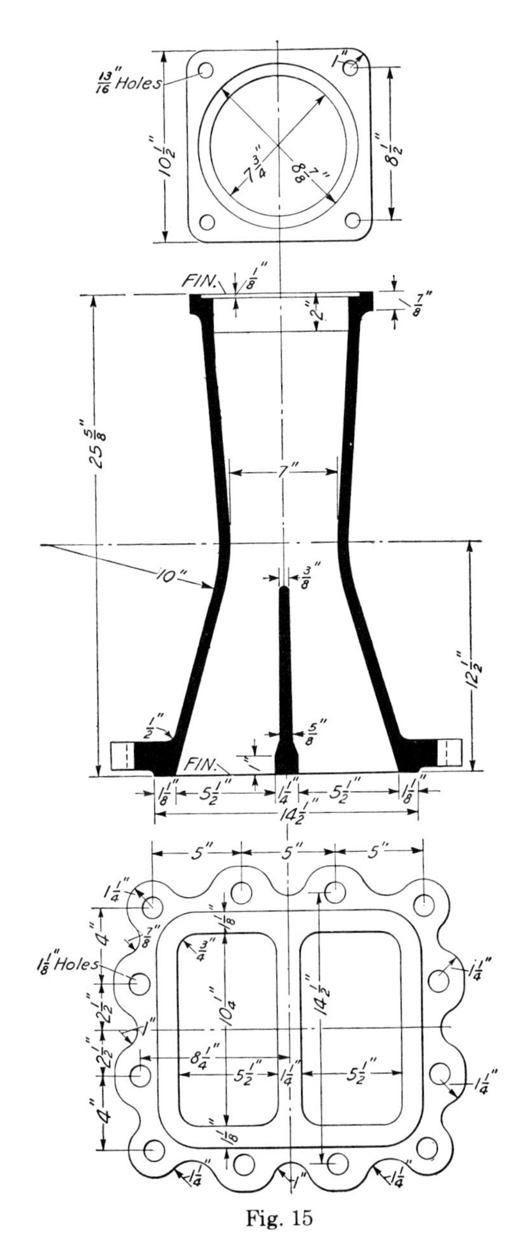

Fig. 15

TABLE II
EXHAUST-NOZZLE SIZES—SIMPLE, TWO-CYLINDER

Boiler Details	Diameter Sizes Inches												
Cylinder	20	21	22	23	24	25	26	27	28	29	30	31	32
Exhaust Nozzle	4½	4¾	5	5³⁄₁₆	5⁷⁄₁₆	5⅝	5⅞	6⅛	6⁵⁄₁₆	6½	6¾	7	7¼

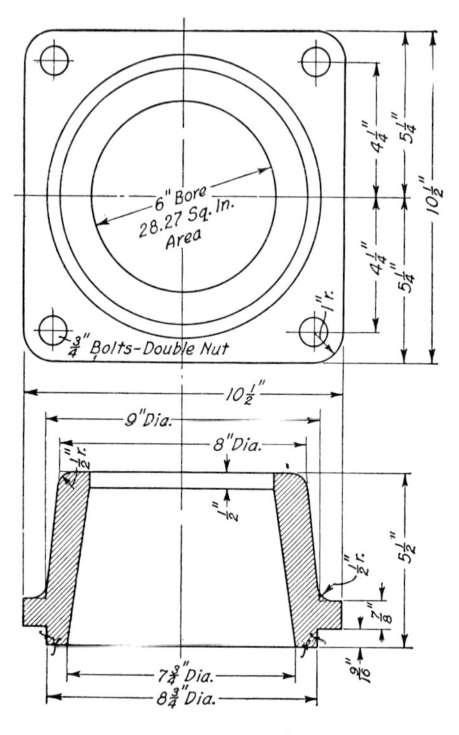

Fig. 16

blower not working, is natural draft. Forced draft, as the name implies, means that the draft is induced by other means. The flow of air through the grates, firebox, and tubes into the smokebox and out of the stack when the locomotive is working is an example of forced draft.

29. Action of Exhaust in Producing Draft.—The action of the exhaust steam in producing a draft is as follows: The exhaust steam passes from the cylinders through the exhaust pipe and the stack to the atmosphere and, in doing so, carries out by induced action some of the smokebox gases and causes a partial vacuum in the smokebox. The movement of the air at atmospheric pressure through the grates and the fire to fill the area of lower pressure in the smokebox causes a draft on the fire. The greater the force of the exhaust, the greater will be the degree of vacuum in the smokebox and the stronger will be the draft through the fire. As the partial vacuum in the smokebox follows each discharge of the exhaust steam, the draft is pulsating rather than steady, this action being more noticeable at low than at high speed. The term, partial vacuum, as here used, means that the pressure of the gases in the smokebox after an exhaust is less than the pressure of the atmosphere, which is about 15 pounds to the square inch. The jet of exhaust steam should not fill the stack, whether it is large or small, until at a point very near the top, as shown in Fig. 17.

30. Blower.—The action of the exhaust steam in passing through the exhaust nozzle produces a draft through the fire only when the engine

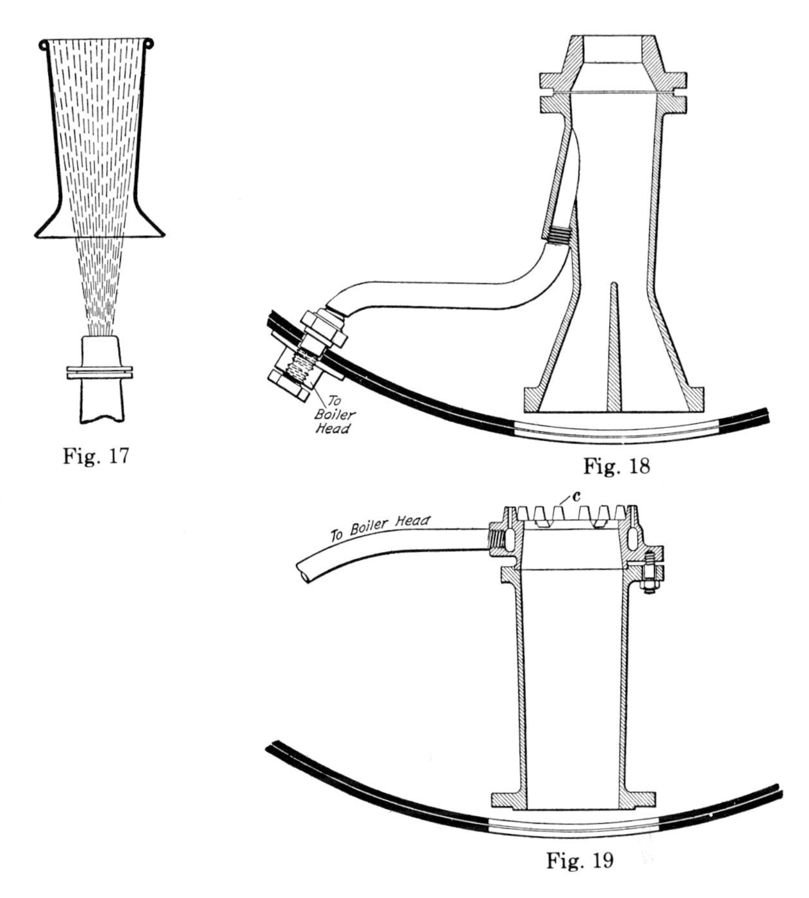

Fig. 17

Fig. 18

Fig. 19

is working steam. The blower provides a means whereby a draft is produced when the engine is standing or is running with steam shut off. The blower consists of a pipe that conveys steam from a valve in the cab to the exhaust nozzle, or at a point near the base of the stack whichever may be found the more convenient.

In Figs. 18 and 19, two different arrangements of the blower pipe at the smokebox are shown. As shown in Fig. 18, the pipe from the boiler head connects outside the smokebox to a pipe that enters the exhaust pipe or the exhaust nozzle. A plug is provided at the junction of the pipes so that a steam hose can be connected and a draft created when the boiler is fired up in the roundhouse. The blower pipe may be connected to the side of the exhaust pipe, as shown in Fig. 18, or it may be screwed into the exhaust nozzle, as shown in Fig. 19, the outlet for the steam being through the small holes in the vertical projections *c*. The simplest arrangement for a blower is to bend a pipe in the smokebox until the end is near the top of the nozzle and directly in line with the center of the stack. When the blower is in operation, a steady flow of steam passes through the blower pipe and the exhaust nozzle. The draft is created by the same action as that of the exhaust jet, that is, by the friction between the steam and the gases in the smokebox. The gases are carried out of the smokebox and are replaced by air through the grates.

FIREBOX

DESCRIPTION

31. Parts of Firebox.—The firebox is a compartment, or box, inserted in the interior of the back end of the boiler shell and so spaced with respect to the shell as to be surrounded by water on all sides. It is in the firebox that the fuel is burned and its heat transmitted to the water. An exterior view of a firebox ready to be applied to a boiler is shown in Fig. 20, and the shell of the boiler, cut away to show a front view of the

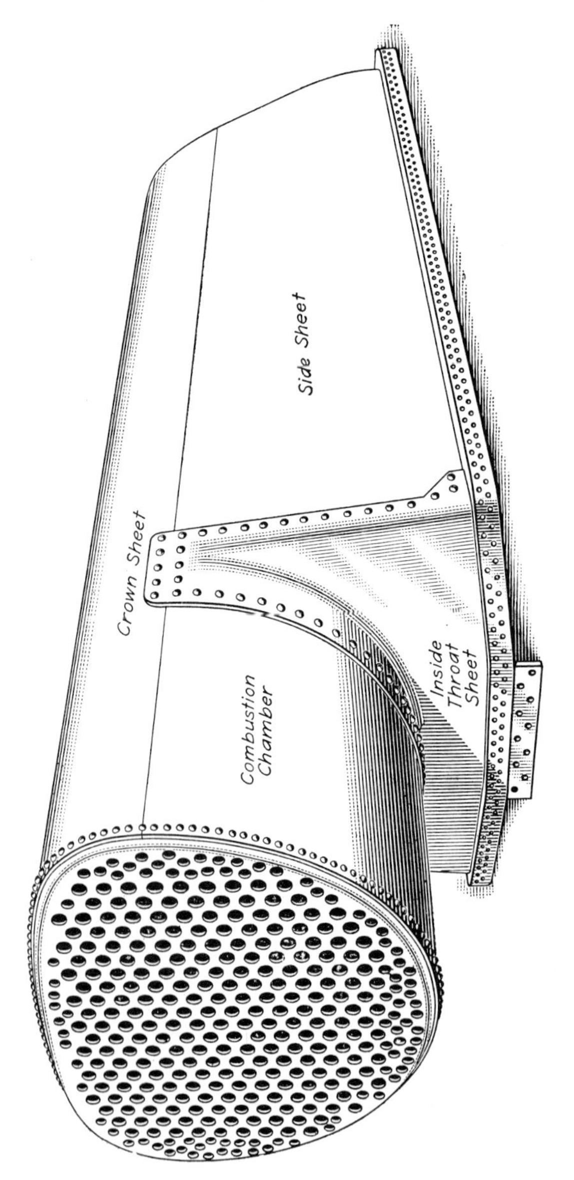

Fig. 20

firebox in the interior, is shown in Fig. 21. The firebox may be considered as being made up of two parts, the rectangular-shaped part, in which the solid part of the fuel is burned on the grates, and the barrel-shaped part, known as the combustion chamber, which serves to promote the combustion of the gases that escape from the coal and thus to secure additional heat from them before they enter the flues.

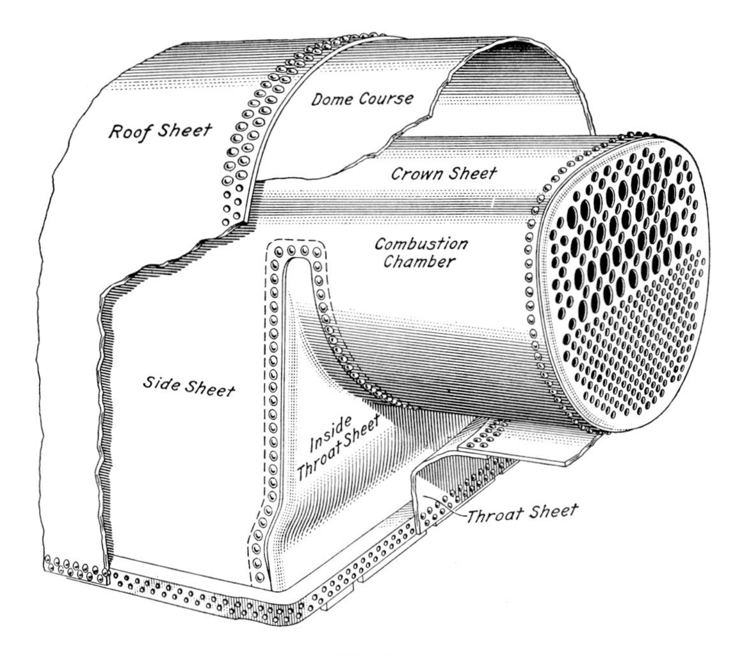

Fig. 21

32. Firebox Sheets.—The firebox comprises certain sheets, generally 3/8 inch thick, which are joined together by either welding and riveting or by welding alone. The sheets that make up the firebox are the crown sheet, which forms the top or roof of the firebox and the combustion chamber, the door sheet, the two side sheets, the back-flue sheet, the combustion-chamber sheet, or the part of the combustion chamber not included in the crown sheet, and the inside throat sheet, which joins the cylindrical combustion chamber to the rectangular part of the firebox.

The crown sheet is welded to the two side sheets and the combustion-chamber sheet, and the inside throat sheet is riveted to the side sheets and the combustion-chamber sheet by a single row of rivets. The backflue sheet, which closes the end of the combustion chamber, is headed as shown, and is secured to it by a single row of rivets, which give the joint ample strength, the flue sheet being further stayed by the flues. The door sheet which forms the back of the firebox, is flanged and is placed within the adjacent sheets, the joint being made by a single row of rivets. The strength of this joint is supplemented by the staybolts used to stay the door sheet to the back head of the boiler.

The heads of the rivets, particularly at the door sheet, are countersunk on the fire side, as shown in Fig. 22, so as to expose less metal to the action of the fire and reduce overheating of the rivet heads with consequent leakage.

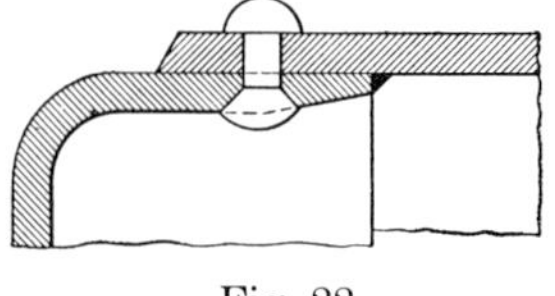

Fig. 22

33. Grates.—The bottom of the firebox is closed by the grates, on which the fuel is retained until it is burned. The grate area, which is found by multiplying the length of the firebox proper by its width, varies between 80 and 100 square feet, depending on the size of the boiler.

The relationship between the grate area in square feet and the firebox volume in cubic feet must be such as to give the carbon and the gases of the fuel time to mix intimately with the oxygen and to burn before entering the flues. A firebox volume of between 5.25 and 6.50 cubic feet for each square foot of grate area gives good results. About the same relationship should exist between the grate area and the firebox heating surface, that is, the surface of the firebox exposed to the heat of the fire on one side and to the water on the other.

STAYING FIREBOX TO BOILER SHELL

34. Necessity for Staying.—The firebox is subjected not only to the pressure of the steam but also to the weight of the water, so the tendency would be to force the crown sheet downwards, and the side sheets, the back-tube sheet, and the door sheet inwards. The firebox sheets are comparatively thin, about 3/8 inch, and would collapse under the load to which they are subjected unless supported by the thicker and more rigid sheets of the boiler shell. These sheets vary in thickness, depending on the boiler pressure, and may be 3/4, 13/16, 7/8, 31/32, or even 1 inch in thickness.

The thickness of the firebox sheets must be kept within certain fixed limits, because, if they are too thick, the transfer of the heat through them to the water would be too slow and they would become overheated. However, the boiler shell may be made of thicker sheets than those of the firebox. The firebox is prevented from collapsing by having all of its surfaces, with the exception of the tube sheet, stayed to the surrounding sheets of the boiler by staybolts spaced about 4½ inches apart. The staybolts vary in length, depending on the distance between the firebox and the boiler, this distance being greatest at the crown sheet of the firebox and least at the mud-ring.

The long staybolts, such as are used at the crown sheet, are called crown stays. The shorter ones referred to as staybolts, and the ones used where the side sheets curve to meet the crown sheet are called radial stays.

35. Staybolts.—The staybolts used to stay the firebox to the adjacent sheets of the boiler are of either the rigid or the flexible type. A rigid staybolt is one in which each end is threaded and screwed into the boiler shell and the firebox and the projecting ends riveted over. The ends of this type of staybolt are then rigid with respect to the sheets and are subject to stresses when the sheets expand and contract with temperature changes. A flexible staybolt is one that is designed to permit of the unequal expansion and contraction of the firebox and the boiler shell without imposing undue strains on either the sheets or the bolt. Flexibility of movement is obtained by making the outer end of the bolt ball-shaped where it contacts the boiler shell.

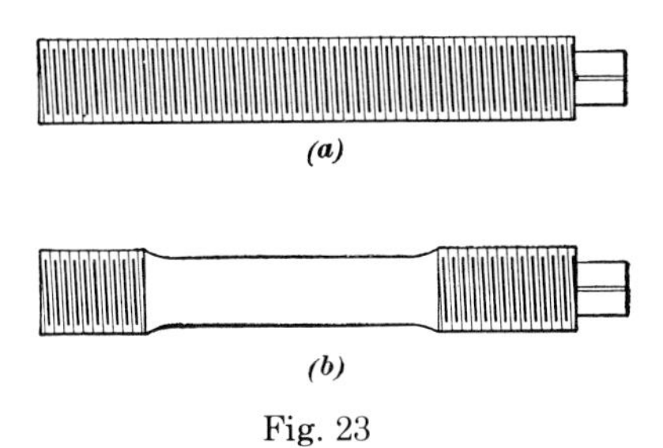

Fig. 23

A rigid staybolt with a straight body is shown in Fig. 23 (*a*), and one with a reduced body to make the application easier is shown in view (*b*). The tell-tale holes that are always drilled in these bolts are not shown. Square ends are forged on the staybolts to permit them to be threaded during manufacture as well as to be screwed into the sheets. The square end as well as any other surplus stock is cut off after application, only enough stock remaining to permit riveting over. These types of staybolts comprise the greater number used to stay the side sheets and the door sheet of the firebox to the adjacent sheets of the boiler.

36. The rigid crown staybolts used to stay the crown sheet to the roof sheet differ from the shorter rigid staybolts in that the firebox ends of the bolts have either a threaded taper head, as in Fig. 24 (*a*), or a button head, as in view (*b*). As the crown sheet will be the first to overheat with low water, it will not pull away from a bolt with the end threaded on a taper or with a button head, as readily as from a straight-threaded bolt. The part of the head of a button-head crown staybolt that seats against the crown sheet is slightly undercut or dished out. Then, when the head is tightened against the sheet, the outer edge of the head contacts the sheet first and makes a tight joint. Most of the staybolts used to stay the crown sheet are of the rigid type.

In Fig. 25 (*a*) is shown on example of a flexible staybolt applied to the side sheet and comprising a two-piece assembly, and in view (*b*) is shown a three-piece assembly. With the two-piece assembly, the seat in the boiler shell is shaped to fit the ball-shaped end of the bolt, whereas in the three-piece assembly the head of the bolt seats in the sleeve *b*, which has a removable cap nut *c*. In both cases, the cap and the sleeve are welded to the boiler shell, as shown at b_1. The head of the bolt will turn slightly on its ball-shaped seat as the firebox expands and contracts; thus, the stress on the staybolt and the sheet will be less than if the outer end were rigid.

37. The practice on some railroads is to use flexible staybolts entirely to stay the firebox to the surrounding sheets of the boiler. On other roads, flexible staybolts are installed only in what is known as the breaking zone, which is that part of the area of the firebox where the movement of

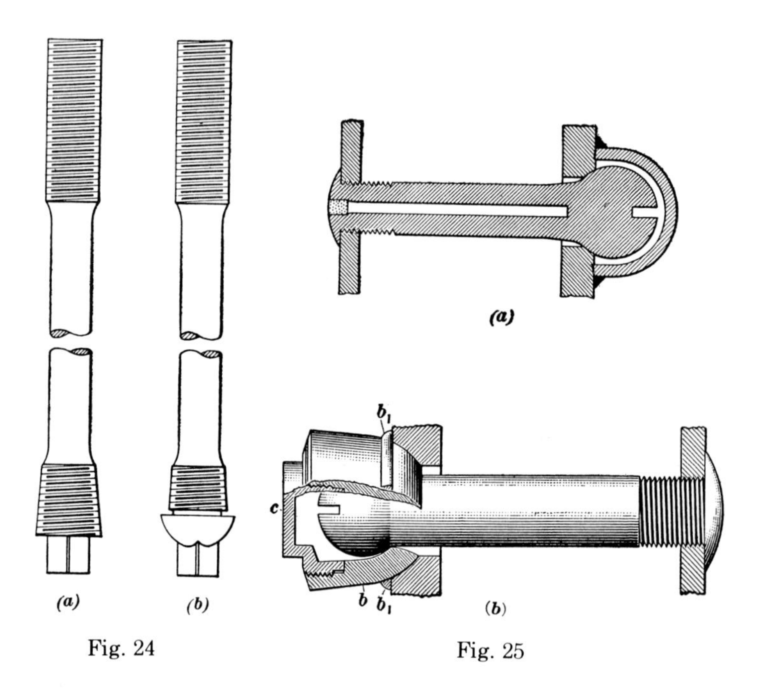

Fig. 24

Fig. 25

the sheets, and hence the tendency to breakage, is the greatest, and staybolts of the rigid type are applied elsewhere. The flexibility of a rigid staybolt increases with its length; hence where the longest staybolts are used, as at the top of the crown sheet, the rigid ones are assumed to be sufficiently flexible for the purpose.

38. Application of Staybolts.—When rigid staybolts are applied, the holes in the sheets, which have been previously drilled or punched, are reamed and tapped in the same operation with a combined tapered reamer and tap, thereby bringing the threads of one sheet in alignment with the threads of the other. The staybolts are then screwed into place and the ends cut off far enough from the sheet to leave sufficient metal for heading over and riveting. The ends are riveted over by backing up one end at a time and riveting over the other end. In Fig. 26 (*a*) and (*b*) is shown the application of rigid staybolts *a* to the side sheets *b* of the firebox and the boiler shell *c*. The application of a button-head crown staybolt is shown in Fig. 27 (*a*), and a taper-head staybolt in view (*b*). The ends of the bolts, before being riveted over, are indicated by *a*. The square ends of the button-head type of staybolts are cut off before the boiler is placed in service. As staybolts break frequently, tell-tale holes, as shown at *p*, Fig. 26, are drilled in the outer ends of the bolts so that they may be detected, when the break, by the discharge of water. The bolts invariably break next to the sheet at their outer ends, because it is here that the leverage induced by the movement of the firebox sheets is the greatest.

When applying the flexible staybolt shown in Fig. 25 (*a*), the holes in the outer sheet and the firebox are reamed and brought into line, and the seat is machined for the ball end of the bolt. The bolt is then screwed into place and the end is riveted over, the ball end being backed up by a tool shaped to fit the head so as to prevent it from being distorted and enable

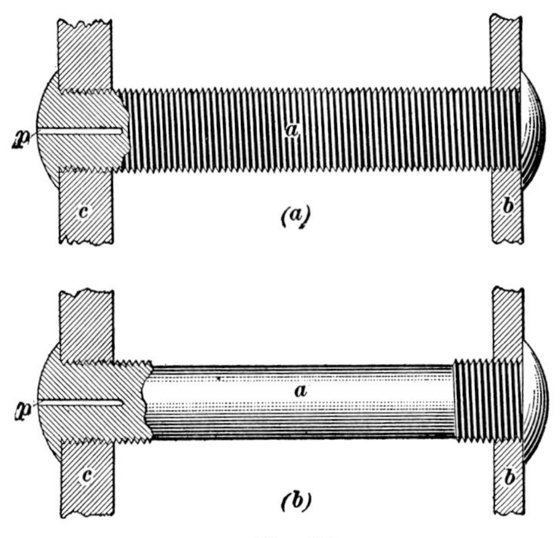

Fig. 26

it to make a true seat; finally, the cap is welded in place. With the type shown in view (*b*), the sleeve is welded in place first, and the bolt is backed up during riveting over by placing a riveting plug over the end of the bolt. After riveting over, the riveting plug is removed and the cap nut replaced.

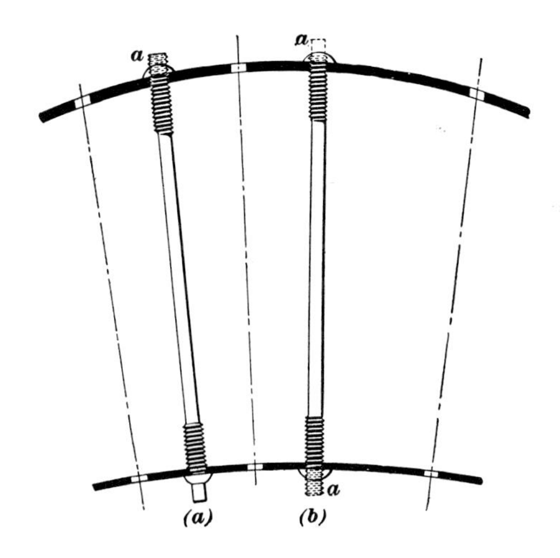

Fig. 27

The tell-tale holes in flexible staybolts are used to discharge water or steam when the bolts are broken and so indicate a defective bolt; also, they permit the making of a test to detect bolts that are not cracked enough to cause a perceptible discharge. If a test made by an electrical device shows that the hole is clear, the bolt is assumed to be intact. But if the test shows that the hole is obstructed, it is assumed that the bolt is cracked and caused a leak so slight as to be unnoticed at the time but large enough for sediment to fill the hole later.

39. Manner of Renewal.—The manner in which a locomotive firebox is renewed depends to a large extent on the practice followed in the shop. Some shops will attempt to carry out as many boiler repairs as possible without removing the boiler from the locomotive proper, whereas, in other shops, the boiler may be taken off and sent to the boiler shop. Needless to say, the application of a new firebox as a unit will be impossible unless the boiler has been removed from its bed.

40. Order of Operations.—When renewing a firebox in its entirety, the staybolts are burned out and various parts of the firebox are removed in sections or as a unit after the mud-ring is taken out. The door sheet, the inside throat sheet, and the back tube-sheet are laid out, flanged, and drilled small in the order given. The laying out can be done from information on the drawings, or templates made from some of the old sheets may be used to aid in locating all of the holes to be drilled. In the case of the flanged sheets, the drilling is done after the flanging has been completed.

The side sheets and the crown sheet are laid out from the drawings or templates and are punched or drilled before being rolled to shape. If the firebox is renewed as a unit, the sections are placed together on the floor and tack-welded at various points. Many fireboxes are entirely buttwelded, while others have the over-lapping flanges riveted. If the firebox is renewed in sections, the back tube-sheet is placed in first, then the side and crown sheet, and then the door sheet, the back end of the boiler being raised just enough to permit each to be placed in proper position. The mud-ring is then lined up and the rivet holes reamed and the rivets applied. The staybolt holes in the roof the side sheets are lined up by means of a long tapered reamer and tap, which, when passed through, not only lines up the holes but also cuts the staybolt thread. The staybolts are then applied and cut off about 3/16 inch from each sheet. Next, they are backed up on the inside and riveted over on the outside, and then backed up on the outside and riveted over on the inside. Flexible staybolts, if of the three-piece assembly, are backed up on the outside by placing a little punching under the cap and then turning the cap up tight. After the staybolts are riveted over on the inside, the punching is removed and the welding of the parts is completed. At the line of welding, the sheets are made corrugated or are pulled slightly out of line so that, after the weld cools, the contraction will pull the weld back into normal position, thereby relieving the sheets of the stress that would result if no precautions were taken.

EXPANSION OF FIREBOX SHEETS

41. Difference in Expansion.—The temperature in the firebox of a properly drafted locomotive approximates 2,000° F. Although the firebox sheets are exposed to a temperature of 2,000°F, the heat is absorbed so rapidly by the water, which, with a pressure of 250 pounds, is at a temperature of 388°F., that the temperature of the sheets, if clean, will not exceed 700°F. But, as the firebox sheets always have a coating of scale of varying thickness, the temperature of the sheets will always be higher than stated. The less scale, the more rapid will be the transfer of heat.

The outside firebox sheets receive all their heat from the water, hence they cannot become heated more than 388°F. Rather, the sheets will be cooler on account of their losing heat by radiation, so there will be a difference of about 300°F. in the temperature of the two sheets. This difference causes the firebox sheets to expand and contract more than the outside sheets and imposes strains on the sheets and the staybolts. The expansion of the firebox sheets is accompanied by a slight bending of the sheets between the staybolts and by a movement of the staybolts into a more or less diagonal position with respect to the sheets.

42. Effect of Expansion.—Firebox steel, like other metals, has a definite expansion per inch for each degree increase in temperature. Although the expansion of 1 inch in length would be very small, yet, when the length of the part is considerable, the minute fraction of expansion in each inch will cause an appreciable increase in the length of the whole part. Thus, a locomotive boiler, when fired up, may show a movement at the furnace bearers of more than ½ inch.

In the case of a box-shaped structure like a firebox, expansion does not affect its length or width to any extent because one sheet resists the movement of the adjoining one, with the result that the accumulated expansion of the sheets meets at the corners and causes a bending of the flanges. For example, the expansion of the side sheets imposes an endwise thrust on the staybolts of the door sheet and the throat sheet, or in a direction in which the staybolts are inflexible, so any movement is restricted to the flanges. However, the staybolts in the side sheets will be bent out of their true position during the expansion of the sheets because the direction of the movement is now at right angles to the staybolts.

Similarly, the effect of any expansion sideways of the door sheet and the inside throat sheet on the side sheets will be opposed by the staybolts in these sheets, and the increase in width will be taken care of at the flanges. The bending action at the flanges during expansion and contraction accounts for the cracks that occur frequently at these points in the firebox.

43. Breaking Zone.—The term, breaking zone, is applied to the area of the firebox where the movement of the sheets during expansion and contraction is the greatest. This area comprises the ends, sides, and corners of the firebox sheets because it is here that the accumulated expansion of the sheets terminates. The breaking zone is stayed to the outside firebox sheets by flexible staybolts, so the location of this zone can be identified by the arrangement of the staybolts of the flexible type in the adjacent area of the outside sheets.

The expansion of a rectangular sheet of steel, free to expand in all directions, begins at the center and extends in all directions. The expansion is a certain amount for each inch in length, so the sheet will increase in length more than in width, the greatest lengthening being at the four corners because these points are farthest removed from the centers.

At the bottom, the sheets of the firebox, with the exception of the combustion chamber, are riveted to the mud-ring so that the sheets cannot expand downwards. Instead, beginning near the center, the side sheets will expand front, back, and upward, the maximum movement being at the corners because of the accumulated expansion here. The crown sheet will expand similarly, the greatest movement being at the ends and the least at the sides, where it curves to meet the side sheets. The extreme movement of the side sheets at the ends is taken care of by two vertical rows of flexible staybolts, with a more numerous arrangement at the two upper corners. The upward expansion of the side sheets and the downward expansion of the crown sheet meet where these join; hence about five rows of staybolts in this area are of the flexible type. The progressive expansion of the crown sheet will accumulate at the ends, but the crown staybolts have, on account of their length, sufficient flexibility to permit of this movement without strains. However, the part of the combustion chamber not included in the crown sheet is nearer to the boiler shell than the crown sheet proper, and shorter stays are necessary; hence the staybolts in this area are all of the flexible type.

DRAFTING OF STEAM LOCOMOTIVES

The following is the report of the Committee on Locomotive Construction, Mechanical Division, Association of American Railroads, which outlines the new standard method of drafting steam locomotives:

Master Mechanics Locomotive Front End Arrangement

Recommended Practice
ADOPTED, 1906; REVISED, 1936

FOREWORD

The following is submitted as a discussion and explanation of the proposed new standard method of drafting steam locomotives, based on a proper proportioning to each other of the gas areas over the brick arch and throughout the smokebox as indicated on the enclosed proposed recommended practice, Sheets Nos. 1 and 2. Employing the same general arrangement of the smokebox details and adhering basically to the design known as the "Master Mechanics" Front End, as described in the 1906 "Proceedings of the American Railway Master Mechanics Association," the proposed method has been developed from an analysis of data secured from standing and road test results while redrafting various classes of bituminous coal burning locomotives of conventional design in a wide variety of service and using all the common kinds and mixtures of bituminous coal.

From a study of the gas areas of properly drafted locomotives and observations made while redrafting, it was discovered that there is a definite and necesary relation of these areas to each other and that this relation is practically identical on all of the locomotives redrafted. By virtue of the foregoing it has been considered logical to use one of these areas, namely, the minimum net gas area through the tubes and flues, as an index to which the other gas areas, including the minimum area of the smokestack, should be compared and proportioned.

Comparison of the stack diameters determined by the method herein recommended with the diameters of existing stacks or stack diameters determined by other methods in general use discloses in a majority of cases that larger stacks may be used. The use of larger stacks permits the use of larger exhaust nozzles with subsequent reduction in back pressure. Reduction in back pressure, when accompanied by satisfactory steaming qualities and fire conditions, results in a saving of fuel. Other advantages of reduced back pressure are increased drawbar pull and a reduction in the general maintenance costs of the locomotive.

While it may appear that this discussion is devoted particularly to an analysis of conditions and redrafting of existing locomotives, the method outlined is equally applicable to new locomotives, and the designs for the brick arch and smokebox details of a new locomotive may be developed in accordance with same as soon as the minimum net gas area through the tubes and flues is known.

As stated, the details of the smokebox and arrangement of same are in accordance with the "Master Mechanics" front end design and consist of an exhaust stand with round bore exhaust nozzle, smokestack and stack extension bolted together, a diaphragm plate or vertical back deflecting plate, a table plate supported by the exhaust stand and attached to the diaphragm plate and sides of the smokebox, and adjustable draft sheet attached to the table plate, and smokebox netting attached to the table plate and the interior of the smokebox and usually applied in a sloping position.

Inasmuch as the principles and method of drafting recommended herin have been gathered from tests made with locomotives equipped with grates having 20 to 30 per cent effective air opening, it is possible that minor changes in some of the proportions outlined may be necessary in order to obtain satisfactory results when drafting locomotives having grates with net air opening not within the above limits. In such cases it is believed the recommended practice will still serve its primary purpose as a guide. Such changes in the proportions as may be found necessary as a result of thorough tests under the latter conditions should immediately be made a part of the proposed method of drafting. In the event the recommended proportions apply without change to locomotives with grates having other than 20 to 30 per cent effective air opening, this practice should be amended at once to include that fact.

Sheet No. 1 illustrates the recommended arrangement of the smokebox details, with recommended gas areas and other pertinent data. Sheet No. 2 illustrates the recommended brick arch design, together with recommended net area between the back end of the arch and the crown sheet.

The following paragraphs and included diagrams cover the design of the various smokebox details and include a discussion of the analysis of gas areas, gas passages in the smokebox, assembly of smokebox details, study of the drafts and tests to determine locomotive and fuel performance.

I. Analysis of Smokebox and Firebox Design

(a) Calculation and Tabulation of Actual Gas Areas:

When preparing to redraft a locomotive or make some changes to improve the steaming qualities, the first step to be taken is to calculate and tabulate the actual gas areas. For convenience these may be calculated in square inches and are as follows:

1. Net area between the top of the brick arch and the crown sheet at the rear end of the arch.
2. Minimum net gas area through the tubes and flues.
3. Maximum area under table plate.
4. Minimum net area under table plate (opposite exhaust stand and steam pipes).
5. Area under draft sheet.
6. Net area through smokebox netting.
7. Area of smokestack at minimum diameter.

In tabulating the gas areas a simple graphical chart such as Chart No. 1 is recommended. The minimum net gas area through the tubes and flues is used as the base for plotting the other areas and is rated at 100 per cent. A percentage tabulation of the other areas is also given. For convenient reference and comparative purposes each chart should carry any important data such as locomotive classification, size of stack, etc. A typical group of actual gas areas or a U. S. R. A. Mikado type locomotive are shown in Chart No. 1.

(b) Recommended Gas Areas:

The recommended gas areas are illustrated graphically in Chart No. 2 as a percentage of the minimum net gas area through the tubes and flues. The areas shown form the basis of the proposed new method of drafting and have been successfully employed in redrafting several hundreds of locomotives. Many new locomotives drafted to these proportions have been placed in service and operated under varying conditions without any change in the smokebox details.

ACTUAL GAS AREAS
CHART NO. 1

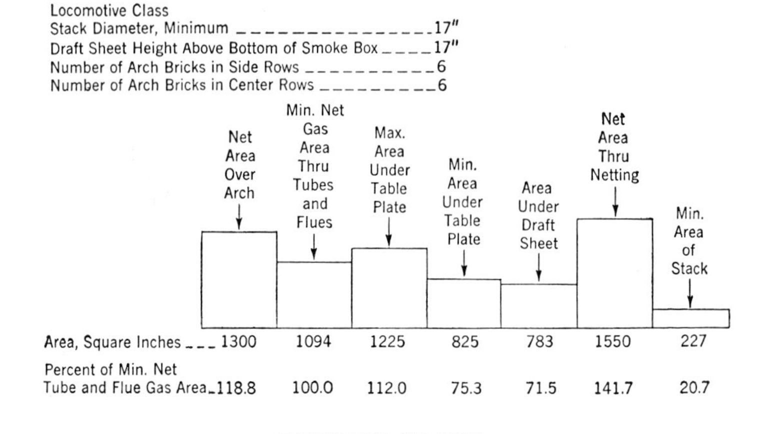

RECOMMENDED GAS AREAS
CHART NO. 2

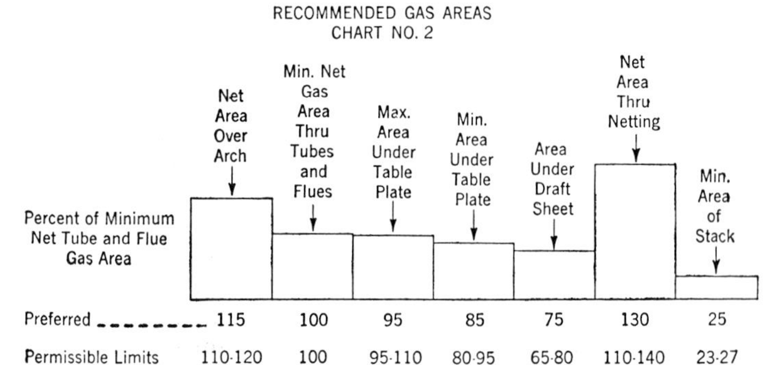

In Chart No. 2 it will be noted that there is a gradual stepping down in the preferred areas from the area over the arch to the area under the draft sheet. While this condition is ideal it will be found necessary in some cases to have the maximum area under the table plate somewhat in excess of the minimum net gas area through the tubes and flues in order that the minimum area under the table plate shall not be less than the area under the draft sheet. Where the latter condition exists it has been found that the draft sheet loses the most of its value as a regulator of the drafts.

Application of the recommended proportions to the locomotive for which the actual gas areas are tabulated in Chart No. 1 permitted an increase in the stack diameter from 17 in. to 20 in. This, in turn, made it possible to increase the exhaust nozzle diameter from 7 in. to 8 in. with entirely satisfactory results and with a substantial saving in fuel.

(c) Important Gas Passages in Smokebox:

1. Space Between Front Flue Sheet and Diaphragm Plate:

Not infrequently on some of the older designs of locomotives it is found that the diaphragm plate is less than 30 in. ahead of the front flue sheet. This condition is usually responsible for and accompanied by excessive heat at the firedoor. In exaggerated cases the flames in the firebox have a tendency to roll and not move freely over the brick arch.

In all cases where the diaphragm plate is less than 30 in. ahead of the front flue sheet it is recommended that the diaphragm be sloped forward at the bottom from a point in line with the bottom of the superheater header. The total amount of slope will usually be determined by the distance between the exhaust stand and the diaphragm and should not exceed 15 in. In cases where the flare of the smokestack interferes with obtaining the desired slope in the diaphragm a small portion may be cut off the stack flare without harmful effects. The portion of table plate projecting backward beyond the new location of the diaphragm plate should be cut off. A diagram illustrating a sloped diaphragm plate is shown in Figure 1.

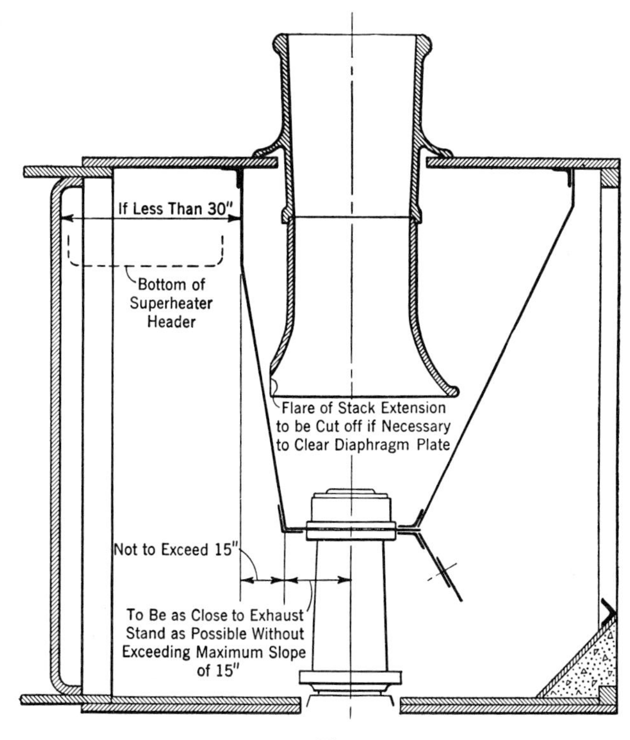

Fig. 1

On some of the more modern designs of locomotives the reverse of the foregoing condition is encountered, and the diaphragm plate is an excessive distance from the front flue sheet. With some of the smaller locomotives this condition m ay be responsible for the difficulty in obtaining sufficient draft on the fire. Where such an effect is recognized it has been found helpful to install an additional diaphragm plate in back of the existing plate and approximately 36 in. ahead of the front flue sheet. The table plate should be extended over to the extra diaphragm and the compartment thus formed made perfectly air tight. The application of the additional diaphragm is illustrated in Figure 2.

The superheater damper, on locomotives equipped with same, is located in the passage between the front flue sheet and diaphragm plate. Some railroads have removed all superheater dampers while others have retained them. It is not intended in this discussion to approve or criticize either practice. However, it has been noted that drafting of the locomotives by the method herein described has been much more successful when the dampers have been removed. All accompanying diagrams have been prepared on that basis.

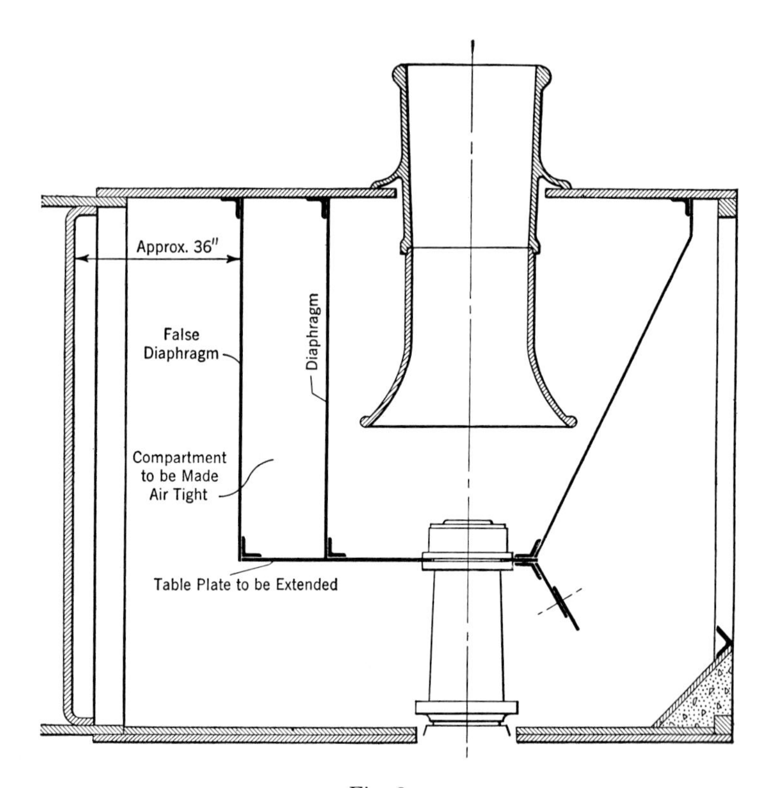

Fig. 2

2. Space Between Diaphragm Plate and Back of Stack:

The space in back of the stack may also be excessive, although this is not as likely to be responsible for poor steaming qualities as excessive space behind the diaphragm. On smaller power, however, if the existing diaphragm is more than 12 in. behind the back side of the stack some improvement may be noted if an additional diaphragm plate is applied as close to the smokestack as possible. The compartment thus formed should be made perfectly air tight. A diagram illustrating the additional diaphragm is shown in Figure 3.

3. Space Between Front Edge of Draft Sheet and Smokebox Front:

The space between the front edge of the draft sheet and the smokebox front is very important and in no case should the area between the draft sheet and the smokebox front be less than the area under the draft sheet. It is preferred to have the table plate extend forward from the center of the exhaust stand as little as possible, providing only sufficient plate to attach the smokebox netting and the draft sheet.

4. Space Below Table Plate:

Too much attention cannot be given to keeping the space below the table plate free from any obstructions which may hinder the free flow of gases from the firebox. The presence of large pipes, such as the booster exhaust or pipes connecting the exhaust stand or exhaust passages of the cylinder to the feedwater heater, particularly when located on the floor of the smokebox, are bound to set up eddies in the flow of gases and may be responsible for undesirable fire conditions

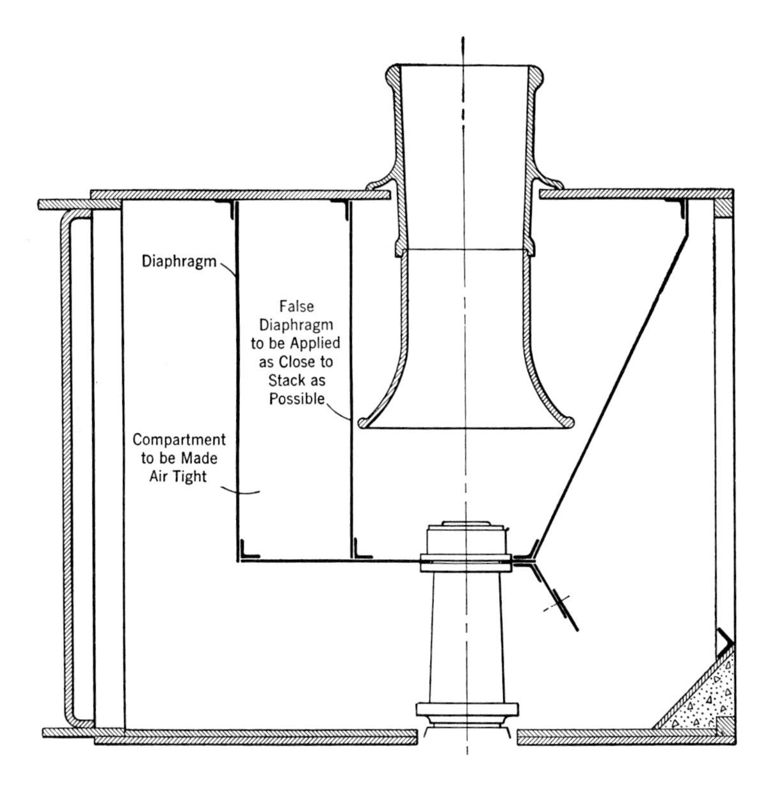

Fig. 3

and poor steaming. The injurious effects of these pipes may exist even though the net area under the table plate, deducting the areas of the pipes, may be well within the recommended limits. Every attempt should be made to remove these pipes from the smokebox entirely. If it is impractical or impossible to do this, the pipes should be applied directly under and close to the table plate or placed in line with the exhaust stand in order to offer as little resistance as possible to the gas flow.

In some cases it will be found that main steam pipe casings within the smokebox are unnecessarily large, making it difficult to obtain the recommended minimum area under the table plate. Extra large steam pipe casings may also have an undesirable effect on the fire by creating eddies in the flow of gases from the firebox. When these conditions exist it is recommended that the size of the steam pipe casings be reduced in order to secure the desired minimum area, rather than raising the table plate for this purpose. However, in no case should the minimum area under the table plate be less than the minimum recommended on Sheet No. 1.

(d) Design of Brick Arch:

The importance of the brick arch construction is emphasized since it plays a most important part in the combustion process. The net area over the arch at the rear end should be within the limits recommended in Chart No. 2 in order to provide ample space for the passage of the gases of combustion and yet confine the stack loss to a minimum. Care should be taken to see that the arch is free from holes of any kind. For best results the arch should be sealed at the throat sheet. The use of "Toe" brick at the throat sheet is usually accompanied by and accountable for excessive stack loss, smoke and unequal draft distribution.

II. Recommended Design of Smokebox Details

(a) Smokestack:

The diameter of the smokestack will be obtained from Chart No. 2, using that dimension which, in even inches, provides an area closest to that recommended. It should be understood that this will be the minimum diameter or the diameter of the stack at the choke.

A two-piece smokestack, consisting of the stack proper and stack extension, is recommended. The stack proper should have a tapered bore throughout its length, the taper being 1 in. in diameter in 15 in. of length. While this taper is preferred, satisfactory results may be obtained with stacks having a taper of 1 in. in diameter in 12 in. of length. However, it is recommended that the stack taper be kept within these limits, namely, 1 in. in diameter in 12 in. to 15 in. of length. Where the locomotive design permits, it is recommended that the entire length of the stack proper be made 30 in.

The stack extension should have a parallel bore equal to the minimum bore of the stack and end in a flare 28 in. to 32 in. in diameter, depending on the size of the stack. The flare should be approximately 15 in. in length and be designed with a long sweeping curved surface. The length of the stack extension will be determined by other conditions and should be such as to provide a space 15 in. to 16 in. in height between the top of the exhaust nozzle and the bottom of the stack extension.

Figure 4 illustrates the recommended design of stack and stack extension.

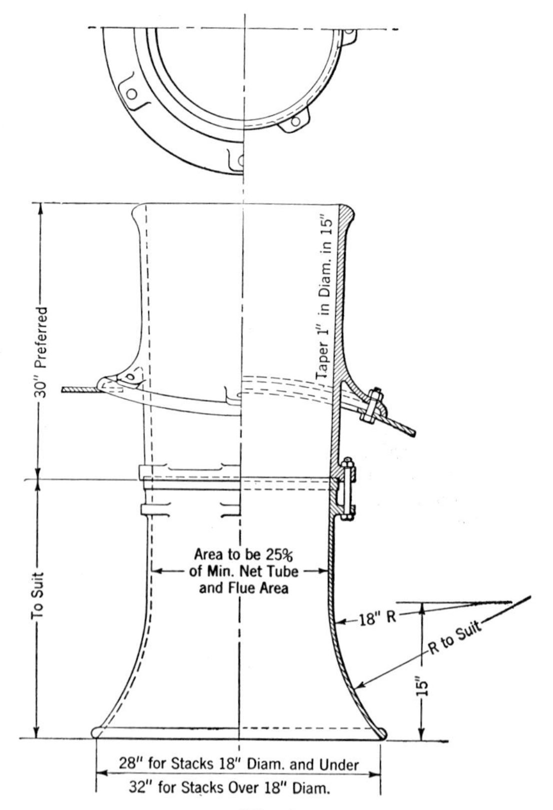

Fig. 4

(b) Exhaust Stand:

Figure 5 illustrates the recommended design of exhaust stand. It will be noted that no provision is made for expansion of the exhaust steam within the exhaust stand as has been done in some designs. The barrel of the exhaust stand should taper directly from the rectangular shape at the bottom to the cylindrical at the top. The parting rib in the bottom of the stand should be 8 in. to 9 in. in height. The design illustrated is applicable only to two-cylinder locomotives.

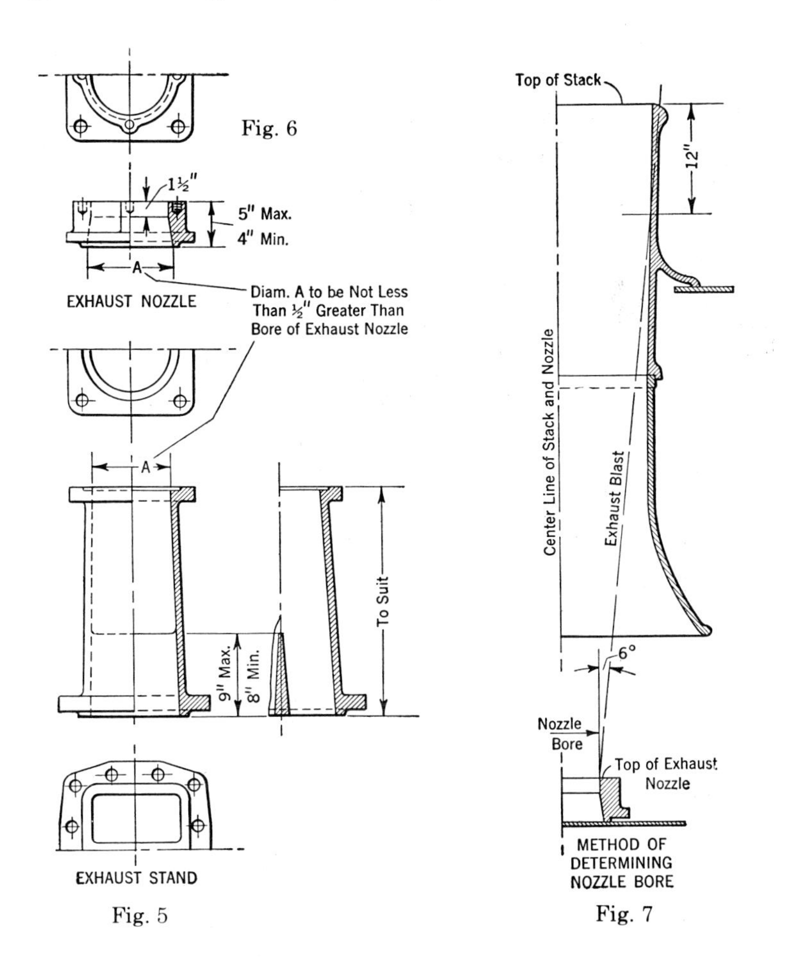

Fig. 5

Fig. 7

Attention is directed to the note on Figure 5 stating that the bore of the exhaust stand at the top should never be less than ½ in. greater than the bore of the maximum diameter exhaust nozzle used.

In some cases the supply of exhaust steam for the feedwater heater is taken from the exhaust stand. This practice is not recommended as it affords an additional source of steam leaks within the smokebox. Furthermore, the pipes applied for conveying the exhaust steam to the heater, when applied below the table plate, very often offer serious restriction to the free flow of gases. It is much preferred that these pipes be connected to the exhaust passages of the cylinders and connected with the feedwater heater either entirely outside of the smokebox or in depressions built into and sealed from the smokebox.

(c) Exhaust Nozzle:

Experiments have been in progress for many years to develop the "perfect" exhaust nozzle. As a result there are several exhaust nozzles of radically different design in use at the present time. The most recent experiments along this line were those conducted at the University of Illinois by Professor Young. In his tests practically all designs of exhaust nozzles now in use were tried and the efficiency of each determined by its ability to provide a steam jet which would entrain the greatest volume of air. It was determined by Professor Young that the ordinary round bore nozzle, when provided with some sort of a spreader or bridge to roughen the steam jet, is, for all practical purposes, the equal of any other type of nozzle.

Because of its simplicity of construction and widespread use, as well as its proved efficiency, the round bore exhaust nozzle if recommended. Figure 6 illustrates the recommended design. No provision for the blower is made in the nozzle. The bore of the nozzle should be parallel for approximately 1½ in., and the total height of the nozzle, 4 in. to 5 in. The bore of the nozzle at the junction with the exhaust stand should never be less than ½ in. greater than the bore of the largest diameter exhaust nozzle used.

In determining the correct bore of the exhaust nozzle the theoretical shape of the exhaust steam blast and the point on the stack bore at which it is desired to have the exhaust blast make its "seal" must be taken into consideration. It has been found by tests with round bore exhaust nozzles equipped with square bar across spreaders that the exhaust steam leaves the nozzle at an angle of approximately six degrees when exhausted at normal working back pressures of 8 to 10 pounds. It has also been observed that best results are obtained when the theoretical "seal" of the exhaust steam jet with the bore of the stack is at a point approximately 12 in. below the top of the stack.

From the foregoing the bore of the exhaust nozzle is determined as follows and is illustrated diagrammatically in Figure 7.

Make a layout showing the inside surface of the stack. In its correct relation to the top of the stack draw a line representing the top of the exhaust nozzle. From a line parallel to the top of the stack and intersecting the stack bore at a distance of

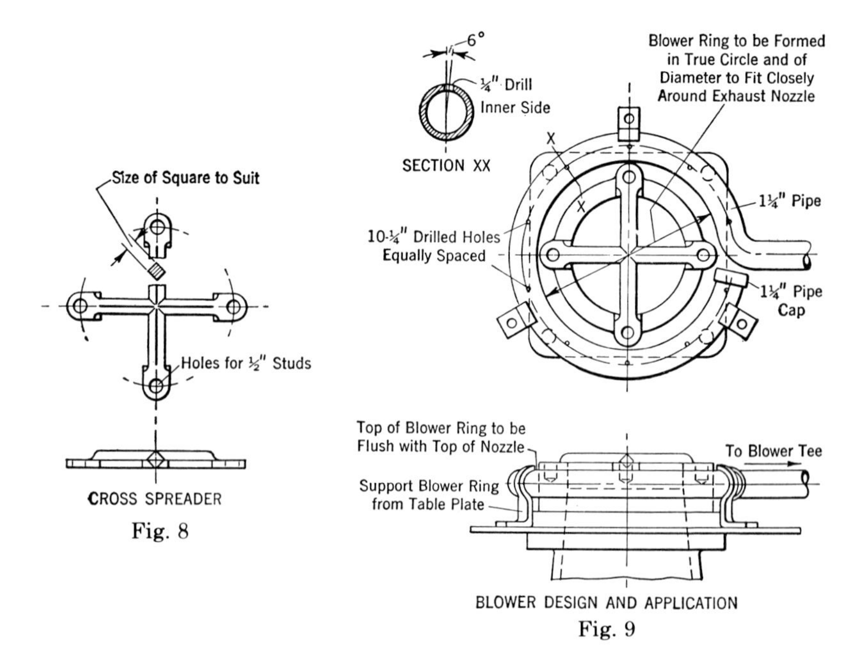

Fig. 8

Fig. 9

12 in. from the top of the stack project two lines, each at an angle of six degrees from the vertical, to intersect the line representing the top of the exhaust nozzle. The distance between these lines, measured on the top of the nozzle, will be the recommended bore of the nozzle. For practical reasons the nozzle should be bored to the nearest even dimension in quarter inches.

When making changes in the bore of the exhaust nozzle in order to imporve steaming qualities, it is suggested that increases or decreases in the bore be made in increments of one quarter inch with nozzles of 8 in. bore and over. For nozzles under 8 in. bore the changes should be made in increments of one-eighth inch.

(d) Exhaust Nozzle Spreader:

In the course of the tests made while redrafting locomotives, various types of exhaust nozzle spreaders or bridges were tried. These included the square bar cross spreader, the basket bridge, the single bar spreader, and the Goodfellow prongs. Tests were also made with an open nozzle, but without notable success except on yard engines in comparatively light service.

By far the most satisfactory results were obtained with the square bar cross spreader, and this type is recommended. In making the square bar spreader the diagonals of the cross section of the bar are perpendicular and horizontal. The recommended design is shown in Figure 8.

The size of the bar to be used for the spreader depends largely on the size of the nozzle, although there is no fixed rule on this. Based on the nozzle bore, the suggested sizes of the bar for cross spreader are as follows:

Nozzle Bore	Size of Square Bar for Cross Spreader
5″ to 6⅞″	⅜″
7″ to 7⅞″	7/16″
8″ to 8⅞″	½″
9″ and above	⅝″

Where satisfactory steaming qualities and fire conditions can be obtained by so doing, it is recommended that the cross spreader rest on top of the nozzle. However, in the course of drafting certain locomotives it may be found that improvement in the fire conditions can be made by setting the bottom edge of the cross spreader ⅛ in. or ¼ in. below and into the top of the nozzle. Likewise, in some cases it may be found that a change in the size of the bar in the spreader will prove a great benefit.

(e) The Blower:

In many instances too little attention has been given to the blower design, although the blower is used innumerable times and for indefinite periods during each day's service of the ordinary steam locomotive. An inefficient blower is wasteful of fuel as well as being unsatisfactory as a draft producing device.

Because of its effectiveness in filling the stack and creating draft, and because of the simplicity of construction, the "ring" type blower, made of ordinary 1¼ in. pipe, is recommended. Figure 9 illustrates the details of the design and the recommended application of the blower.

(f) Design and Application of Draft Sheet:

The draft sheet should be securely bolted to an angle or plate attached to the front end of the table plate and should fit neatly against the sides of the smokebox. While it is recommended that this sheet be applied at an angle of 30 degrees from the vertical, better results are secured in some cases when it is set at a greater or lesser angle than 30 degrees. The bottom edge of the draft sheet should be perfectly straight and perpendicular to the vertical center line of the boiler. A typical application is illustrated on Sheet No. 1.

(g) Deflecting Plate in Bottom of Smokebox:

A deflecting plate applied at an angle of 45 degrees in the bottom of the smokebox, as illustrated on Sheet No. 1, is recommended because of its protective value to the smokebox front and because it serves to prevent cinder accumulation at this point. Application of an angle iron across the top edge of this plate as shown has very successfully reduced cinder cutting of the smokebox door, door flange and bolts.

(h) Construction of Stack, Exhaust Stand and Nozzle for Test Purposes:

In order to provide the details necessary to redraft a locomotive for test purposes without the necessity of having patterns constructed and castings purchased, a smoke stack and exhaust stand constructed of steel plate may be utilized. Very satisfactory results have been obtained in this manner. A typical plate stack and exhaust stand are illustrated in Figure 10. It will be observed that a removable bushing, held in place by the cross spreader, is used for the exhaust nozzle. This makes it possible to determine the final nozzle size to be used at a minimum of cost for labor and material.

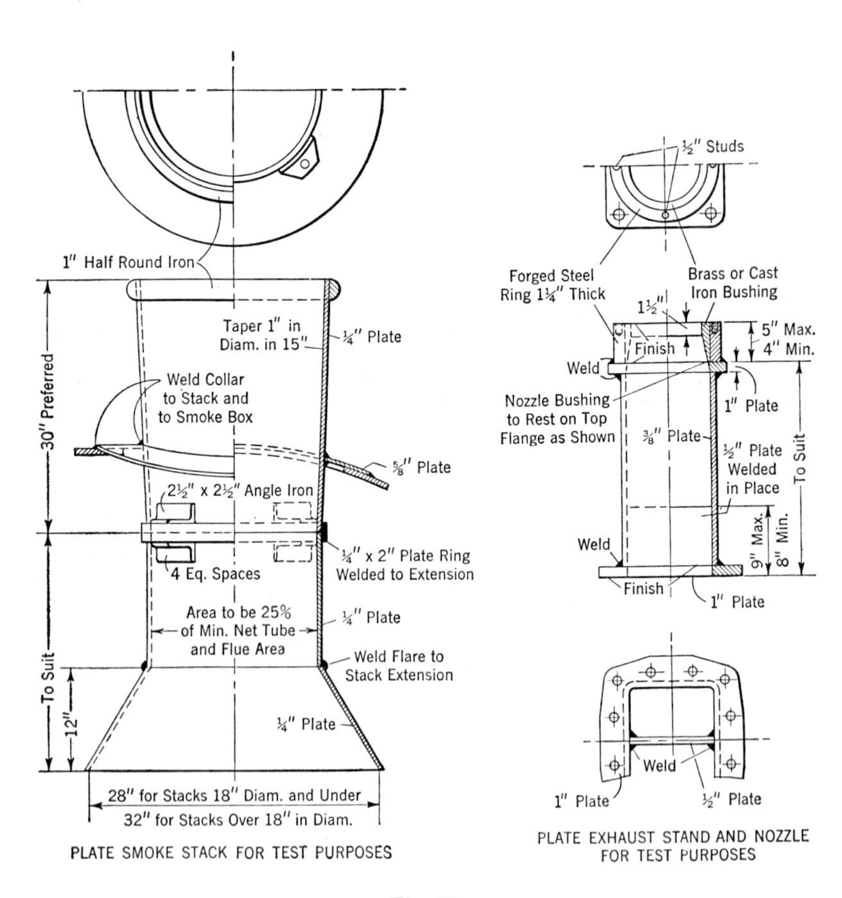

Fig. 10

III. Assembly of Smokebox Details

(a) Typical Recommended Arrangement:

Sheet No. 1 illustrates a smokebox and Sheet No. 2 an arch brick arrangement prepared in accordance with the recommendations outlined herein. For convenient reference the recommended gas areas are also shown, together with other pertinent data mentioned elsewhere in this discussion.

(b) Points to Be Observed in Assembly of Smokebox Details:

Too much care cannot be taken in assembling the various smokebox details if the utmost efficiency is to be realized. It is particularly essential that there be perfect alignment of the stack and exhaust nozzle. All plates should be applied exactly in accordance with the drawings. The diaphragm plate, table plate and draft sheet should be tight and free from holes.

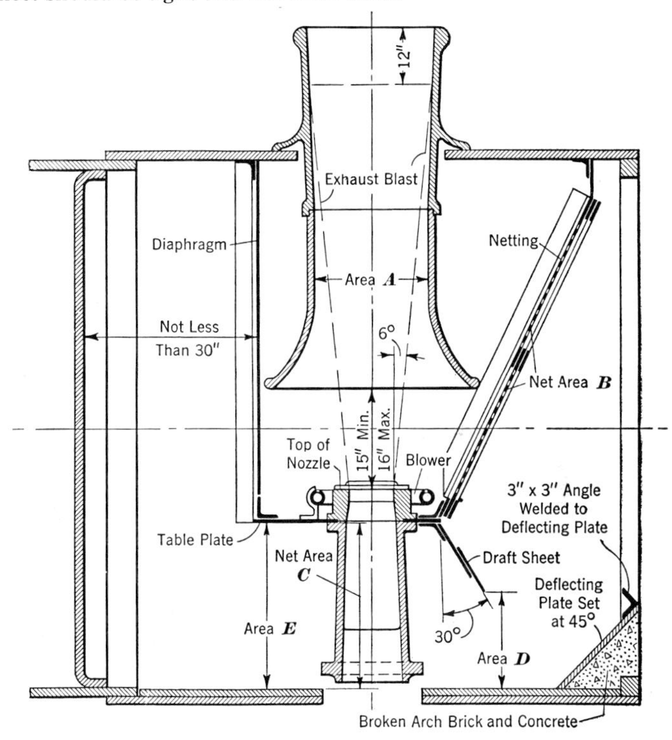

GAS AREAS—PERCENT OF MINIMUM NET AREA THRU TUBES AND FLUES		
Area	Percent of Minimum Net Gas Area Thru Tubes and Flues.	
	Preferred	Permissible Limits
A	25	23 to 27
B	130	110 to 140
C	85	00 to 95
D	75	65 to 80
E	95	95 to 110

Sheet 1

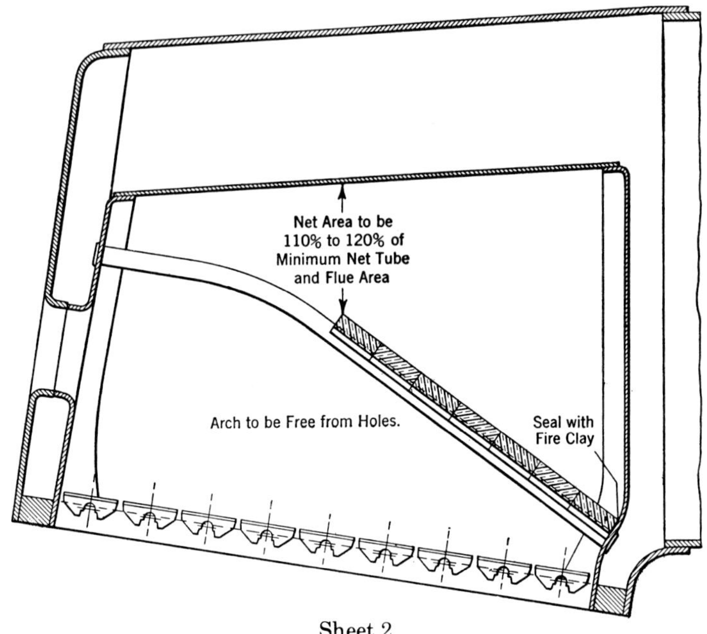

Sheet 2

(c) Test for Steam Leaks:

After applying the exhaust stand a hydrostatic test should be applied. The joints between the exhaust stand and cylinder, and between the exhaust nozzle and stand should be made perfectly tight during this test. Superheater units should be observed for leaks and tightened if necessary. All pipe joints in the smokebox must be made absolutely tight. Steam leaks in the smokebox can offset the most capable efforts to make a locomotive steam properly and lead to incorrect analysis of the fire conditions.

(d) Test for Air Leaks:

Air leaks, like steam leaks, are responsible for much of the difficulty encountered in obtaining and maintaining good steaming qualities and economical fuel performance.

A simple test for disclosing air leaks in the front end is known as the "smoke" test and is conducted as follows: Place a cover over the entire top of the stack and then throw a quantity of coal on the fire. All air leaks of consequence will be indicated by the escaping smoke.

IV. Discussion of the Drafts

(a) Drafts of Most Significance:

While it is not necessary to know and record the actual drafts obtained in the combustion area and smokebox in order to satisfactorily draft or redraft a locomotive, this information forms a valuable record, especially where an extensive program of redrafting is undertaken.

Due to the difficulty of securing accurate readings of firebox and ashpan drafts on the road, these drafts are given no further consideration in this discussion. If it is desired to obtain a record of these drafts it is recommended that standing tests be made.

Smokebox drafts can be readily obtained in road service and furnish all the data necessary for comparing the effects on the drafts brought about by redrafting. The smokebox drafts of most significance are those taken at the following location in the smokebox: Above and below the table plate at a point just in back of the junction between the smokebox netting and the table plate, and in back of the diaphragm at a point approximately on the horizontal center line of the smokebox. Draft gage pipes applied at the above positions should extend in to the vertical center line of the smokebox with the inner end of each pipe capped and provided with six staggered ⅛ in. drilled holes within a space of four inches from the capped end. The draft gage pipes may be extended to a draft gage panel in the cab thus providing safe, convenient reading of the drafts. One-quarter inch pipe is satisfactory for draft gage pipe.

(b) Plotting of Draft Curves:

Draft readings should be plotted as illustrated in Figure 11, plotting draft in inches of water against back pressure in pounds. It has been found helpful when plotting comparative draft curves to illustrate the effect of various smokebox changes to plot the draft at only one position in the smokebox on one sheet. A composite draft sheet such as shown in Figure 11, illustrating the drafts at all three positions in the smokebox, should be prepared for record after changes to develop satisfactory steaming qualities and fire conditions have been completed. Draft curves for the original smokebox arrangement should be plotted in order to determine and compare the exact effect of the modified smokebox arrangement on the drafts.

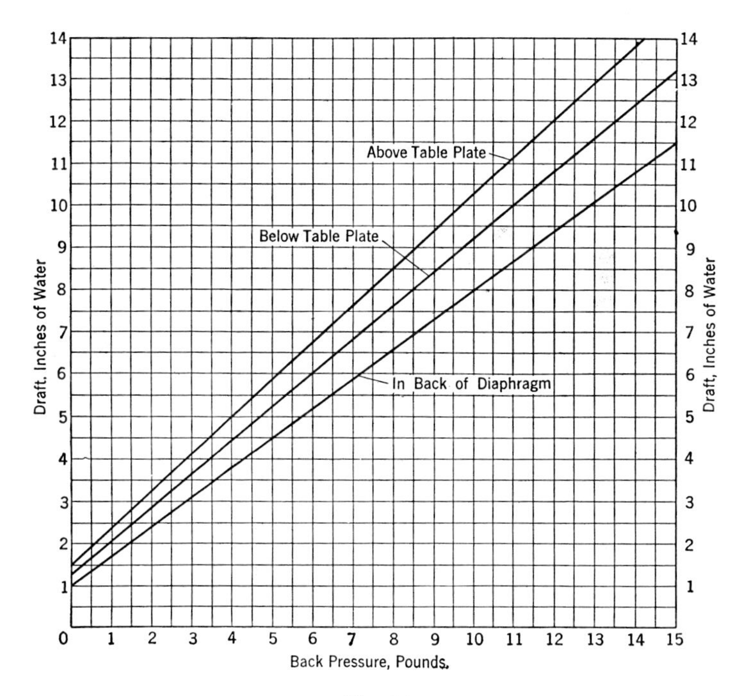

Fig. 11

(c) Analysis of the Draft Curves:

It will be noted that the draft curves illustrated in Figure 11 are straight lines with practically the same "falling off" in the drafts from the draft above the table plate to the draft in back of the diaphragm. While the latter condition is ideal it cannot always be obtained due to limitations of design and the impossibility of obtaining the preferred proportional gas areas. It should be possible, however, to obtain draft curves represented by straight lines from any locomotive which is properly drafted. Draft curves which fall off in the upper back pressure ranges indicate improper seal of the exhaust blast in the stack. This fact may account for smoke or other undesirable fire conditions when working the locomotive at its maximum output.

A brief explanation of the draft curves plotted in Figure 11 is offered at this point. It will be observed on these curves that some draft is indicated at zero pounds of back pressure. This condition will be found to exist in all cases where locomotives are redrafted in accordance with these principles. Due to the increase in the nozzle bore, back pressure will not be indicated on the gage until the steam chest pressure is from 25 to 50 pounds or higher, depending on the nozzle bore. This is probably due to the lack of sensitiveness of the ordinary back pressure gage. Draft is indicated on the draft gages, however, whenever steam is exhausted from the cylinders, regardless of the amount of steam chest pressure. The drafts illustrated in Figure 11 at zero pounds back pressure were taken at the point back pressure was about to register on the gage.

(d) Necessary Amount of Draft:

While no attempt should be made to state definitely how much draft is necessary to produce satisfactory steaming qualities, with good fire and stack conditions, it has been observed on a large number of locomotives, redrafted in accordance with these principles, that the best performance has been obtained when the draft in back of the diaphragm in inches of water is approximately eight-tenths of the back pressure in pounds in the normal working range of the locomotive, considered at eight to ten pounds back pressure. This figure is empirical and is offered for its possible value as a guide.

In summary it is considered advisable to state that each class of locomotives should only have sufficient draft to burn the fire satisfactorily under all operating conditions with free steaming qualities and without smoke. Excessive drafts are to be avoided as they are largely responsible for excessive stack loss and cinder cutting of staybolts, flue sheets and various smokebox details.

V. Locomotive and Fuel Performance Tests to Determine Advantages Due to Redrafting

(a) Standing Tests:

Extensive use of the standing test was resorted to during the early experiments in redrafting and assembling the data upon which this practice is based. The standing tests made it possible to quickly make the changes needed to produce satisfactory steaming qualities and provided information of inestimable value in arriving at the proportions recommended herein.

These proportions and principles of front end design have proved so reliable that a great number of locomotives, redrafted in accordance with them, have been placed in revenue service without preliminary trial. Only minor changes have been required to produce altogether satisfactory steaming qualities and stack conditions. These changes have been made without loss of time to the locomotive in any instance. It is also worthy of note that in no case has a steam failure occurred while redrafting a locomotive. In view of this, standing tests for the purpose of redrafting are not considered necessary and are not recommended.

(b) Dynamometer Car Tests:

Where a Dynamometer Car is available, accurate information on the improvement made in a locomotive by redrafting may be obtained by making a series of comparative tests before and after redrafting. In making such tests a division should be chosen which will provide the most consistent operation from the standpoint of tonnage and speed, and with a minimum of drifting distance. In many cases it is preferable to make the tests over only that portion of a division providing the desired conditions, thus eliminating many of the variables which affect the locomotive performance. It is always preferable to make the test with the standard or original smokebox arrangement first, making sufficient tests to obtain accurate average results.

Tests with the redrafted engine to secure comparative data should not be started until it is reasonably certain that the steaming qualities and stack conditions are the best that can be obtained.

On all such tests made with the Dynamometer Car the coal should be weighed and water measured. The locomotive should be equipped with a back pressure gage, steam chest pressure gage, steam pyrometer and draft gages. The reverse gear should be calibrated. Gage readings may be taken at mileposts or at specified intervals of time. The usual dynamometer data should also be recorded. All of this information is essential to determine the actual benefits of redrafting and affords very interesting data for permanent record and study.

In making comparative fuel performance tests it is very essential that the locomotive be worked at the same capacity on all the tests. Maintaining an equal average drawbar horsepower on tests with the locomotive before and after redrafting assures results which can safely be compared, providing this equal drawbar horsepower is obtained with fairly equal average speeds in each case. The coal per drawbar horsepower hour should be used to measure the locomotive fuel performance.

Dynamometer tests may also be conducted to determine the comparative ability of the original and redrafted locomotive to handle trains. In such tests the increased tonnage hauled by the redrafted locomotive or the reduction in running time over the division with equal tonnage will afford comparative data. On tests of this nature in which the average drawbar horsepower of the redrafted locomotive will be higher than that of the locomotive before redrafting, it may also happen that the coal per drawbar horsepower hour of the redrafted locomotive will equal or even exceed that for the locomotive before redrafting. This will be governed very largely by the actual improvement in the fire conditions and the amount of reduction in back pressure brought about by redrafting and the amount of the increase in speed or tonnage, or a combination of both, of the redrafted over the original locomotive.

(c) Road Tests Without Dynamometer Car (Observation Tests):

Road tests to determine comparative fuel performance of a redrafted locomotive, when made without a Dynamometer Car and where coal per 1000 gross ton miles is used as a basis for comparison, are of no particular value, and may often be very misleading, even though the coal may be weighed on such tests. While tonnage and average speed may be kept comparable there are other uncontrollable factors entering in, which may affect the coal consumption and the locomotive performance generally.

Increase in tonnage or speed for the redrafted locomotive may be determined without the use of a Dynamometer Car. Tests or trial runs for this purpose should certainly be made in order that the advantages in this respect, brought about by redrafting, may be utilized. Whenever tests of this nature are conducted without a Dynamometer Car it is recommended that cab gage readings, including the draft readings and cut-off, be taken as on dynamometer tests. The data secured will prove of considerable value.

VI. Conclusion

Although in some cases immediate improvements in the fuel and general performance of a locomotive may be obtained by making partial changes in line with these recommendations, the greatest success from an application of the foregoing principles of drafting can be realized only when the procedure indicated is carried out in its entirety as outlined.

LOCOMOTIVE BOILERS
(PART 1)

EXAMINATION QUESTIONS

Notice to Students.—*Study the Instruction Paper thoroughly before you attempt to answer these questions.* **Read each question carefully and be sure you understand it;** *then write the best answer you can. When your answers are completed, examine them closely, correct all the errors you can find, and* **see that every question is answered.**

(1) Explain how the exhaust steam produces a draft.

(2) What is meant by the minimum gas area of a flue?

(3) Why is welding not permissible on the cylindrical part of the boiler?

(4) What sheet closes the end of the boiler course next to the smokebox?

(5) What is the purpose of the diaphragm, the table plate, and the damper?

(6) What relation should the net internal gas area bear to the grate area?

(7) What is the purpose of the foundation ring?

(8) What is the real function of the diaphragm apron or damper?

(9) Name the smokebox details in the Master Mechanics' front-end design.

(10) What is the effect (*a*) of raising the damper and *(b)* of lowering the damper?

(11) What is the purpose of the blower?

(12) Name the sheets that make up the firebox.

(13) What is the purpose of the steam dome?

(14) Name the sheets and courses that comprise the boiler shell.

(15) What prevents the steam pressure from collapsing the firebox?

(16) What are the joints called that connect the various boiler courses?

(17) What is the difference between a rigid and a flexible staybolt?

(18) To what space does the term water-leg refer?

(19) What type of joint is used to connect the ends of a boiler course?

(20) Define a locomotive boiler.

(21) What causes a greater expansion of the firebox sheets than the outside firebox sheets?

(22) To what areas is the term, breaking zone, applied?

(23) Why does not the high firebox temperature overheat the firebox sheets?

(24) What is the purpose of the stack extension?

(25) Define a triple-riveted, double-welt butt joint.

(26) How is the firebox connected to the boiler shell?

Locomotive Boilers
Part 2

By
J. W. Harding

1967B Printed in U.S.A. Edition 2

1945 Edition

LOCOMOTIVE BOILERS
(PART 2)

CONSTRUCTION, DETAILS, AND MAINTENANCE
(Continued)

FIREBOX APPLIANCES

PURPOSE

1. The purpose of the different appliances that are installed in the fireboxes of modern locomotive boilers is to secure either a more complete burning of the fuel, or a more rapid circulation of water around the firebox and thereby a quicker generation of steam. These appliances comprise firebrick arches, combustion chambers, arch tubes, and thermic syphons.

ARCH TUBES

2. An arch tube is an iron or steel tube, 3 to 3½ inches in diameter, that is used to promote the circulation of the water around the firebox and to support the brick arch.

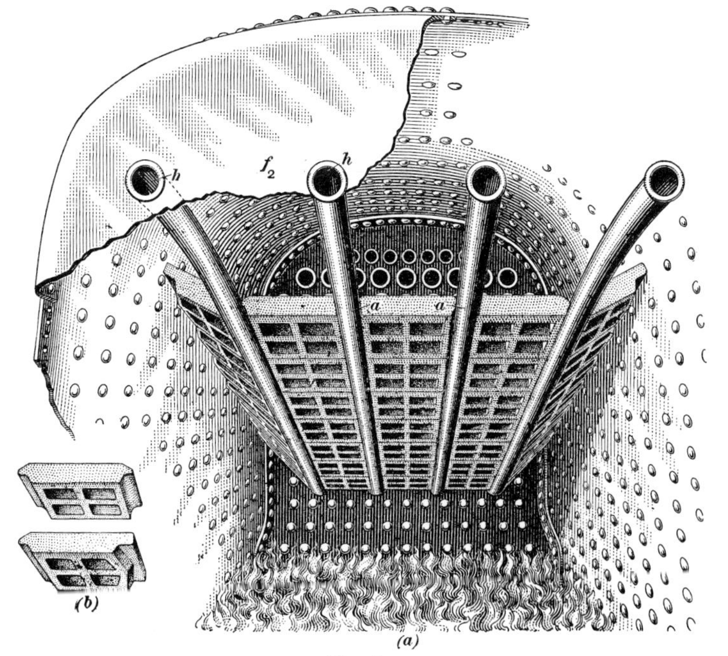

Fig. 1

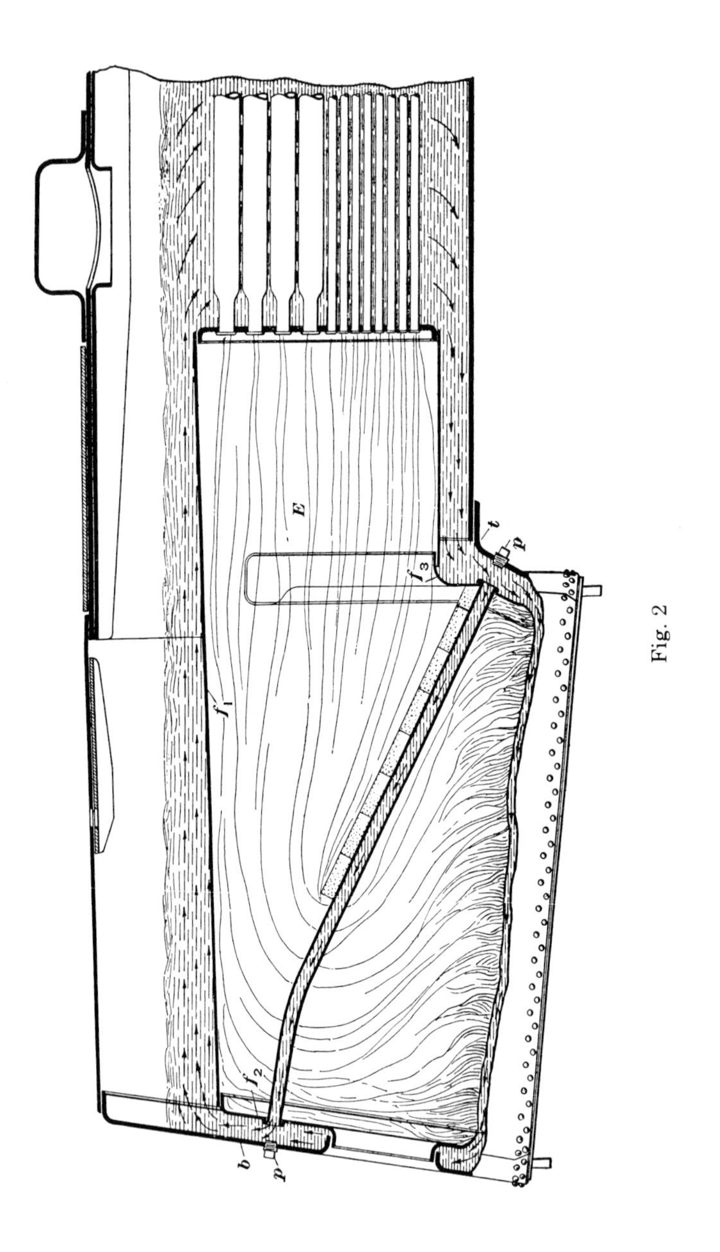

Fig. 2

The arch tubes h are shown in the firebox in Fig. 1, in which the back head of the boiler, the roof sheet, and the outside firebox sheets are omitted, the sheets shown being the two inside firebox sheets, the tube sheet, and a part of the door sheet. One end of the tube h is inserted in the door sheet f_2, and the other end, when a combustion chamber is used, is connected to the inside, or firebox, throat sheet. When a combustion chamber is not used, the front ends of the arch tubes are inserted in the tube-sheet below the tubes.

Arch tubes promote the circulation of the water, because the front ends of the tubes are in such a position that they receive the water from a point at the bottom of the boiler shell where it is comparatively cool. The water in the arch tubes evaporates very rapidly and is replaced by the cooler water which enters the tubes at their front ends, the result being a strong flow of water, as indicated by the arrows in Fig. 2. The rapid flow of the water tends to make the other heating surfaces more effective. The number of arch tubes that are used varies from two to six, depending on the width of the firebox, and they are so placed that the firebrick will fit snugly on the top of them.

3. Application.—The method of applying the arch tubes to the firebox sheets is shown in Fig. 2, in which f_1 indicates the crown-sheet, f_2 the door sheet, f_3 the inside throat sheet, *b* the back head, and *t* the throat sheet. The holes in the firebox sheets are drilled 1/32 inch larger in diameter than the arch tube, so that it can be easily applied. The tubes are then bent to the required shape and cut to length. From 3/8 to 1/2 inch of the tube is allowed to project into the water space so as to provide material for beading over as shown in Fig. 3, which is an enlarged view of the back end of an arch tube, and an arch-tube plug.

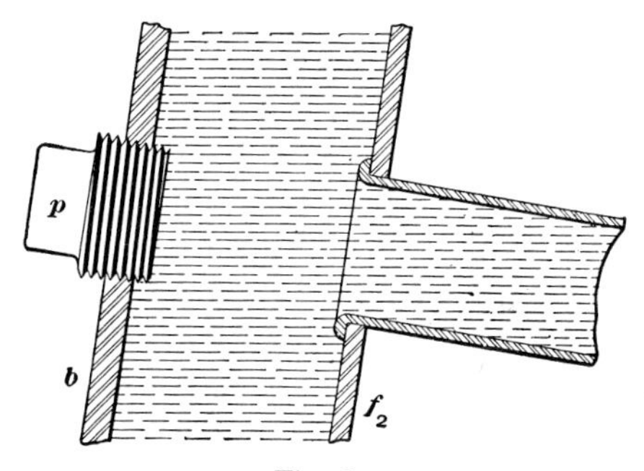

Fig. 3

Arch-tube plugs *p*, Fig. 2, are inserted in the back head *b* and throat sheet *t* so as to provide a means whereby the tubes can be rolled and beaded and also to give access for cleaning. The plugs are of standard taper, 3/4 inch in 12 inches, and twelve threads per inch.

FIREBRICK ARCH

4. Description.—As the name indicates, the firebrick arch consists of an arch of firebrick, built up in sections and set in the firebox in an inclined position.

The firebrick arch is applied for two principal reasons, first to aid combustion and so save fuel, and second to increase the life of the flues by reducing flue leakage.

In Fig. 1 (*a*) is shown a firebrick arch installed in a wide firebox, as seen from the firebox door. The ends of the bricks at *a* are shaped to fit the point where they rest on the arch tubes *h*. Next to the side sheets, the bricks rest against the sheets. The surfaces of the bricks exposed to the fire are dished out as shown. This construction retards the passage of the gases along the arch and so causes them to mix more completely with the air. An opening is left at the back end of the arch below the crown-sheet to allow the gases to pass to the tubes. View (*b*) shows two of the firebricks removed from the arch. The lower brick is one that is placed along the side of the firebox, and the upper one is a middle brick.

The position of the brick arch divides the firebox into two sections; one below the arch, in which the fuel is consumed, and the other above the arch, which acts as a combustion chamber in case no other one is provided, and as an extension of the combustion chamber when there is one. The arch is extended back until the opening between the top of the arch and the crown-sheet is equivalent to about 115 per cent of the total flue area.

5. Position of Arch at Tube-Sheet.—The ideal method of setting brick arches is to place them tight against the tube-sheet. In this position, the tubes and flues receive the most protection from the currents of cold air that may enter through the grates at the front end of the firebox.

The claim is sometimes made that with a closed arch, or one set tight against the flue sheet, the fine coal would lodge on top of the arch and tend to plug the lower flues, whereas with an opening between the arch and the sheet the draft through the opening would carry the coal through the flues without stopping them up. This in a measure is correct, but an open arch would have serious drawbacks. Tests have shown that the draft with an open arch is twice as great immediately under the opening than at any other part of the firebox. The effect is that the finer particles of coal at that point are lifted up and carried out of the stack unburned. Also, with a dead fire under the opening, the cold air is drawn through directly into the bottom flues and causes them to leak.

With a closed arch, regardless of the position of the damper in the smokebox, the draft is weakest at the front of the firebox and strengthens progressively toward the rear. This is caused by the fact that the arch has a gradual slope upwards from the front toward the rear of the firebox, the opening between the arch and the door sheet accounting for the greater draft on the back section of grates.

To equalize the draft, installations have been made, in some cases, of grates with a greater percentage of air openings in the front ones than in the rear.

6. Advantages of Firebrick Arch.—The advantages of the firebrick arch are: (1) It lengthens the path of the hot gases to the tubes and assures the more intimate mixing of the gases by surface friction and consequent turbulence; (2) it constitutes a reservoir from which heat is radiated to the firebed when the latter is cooled by the addition of fuel, that is, the brick arch serves as a radiation screen, the temperature of the lower surface of the brick being maintained sufficiently high for this purpose by the heat transferred thereto by radiation from the incandescent fuel below it. The greater part of the heat from the firebed therefore is returned to the fuel by radiation, thus maintaining the temperature of combustion.

COMBUSTION CHAMBER

7. Description.—The combustion chamber *E*, Fig. 2, is a compartment or space in a locomotive boiler, between the firebox and the back tube-sheet, and is really an extension of the firebox forward of the grates; it may vary in length from 6 inches to 6 or 7 feet, depending on the size and capacity of the boiler.

The combustion chamber is formed by lengthening the firebox sheets and extending them beyond the mud-ring; this portion of the firebox then becomes more or less circular, as shown in Fig. 4. In order to connect the rectangular portion of the firebox, which is riveted to the sides and back of the mud-ring, to the circular portion that extends beyond it, an inside throat sheet of the shape shown must be employed. At the front, this sheet is riveted at the bottom to the inside of the mud-ring, the sides are riveted to the firebox proper, and the top flange is riveted to the part of the firebox that forms the combustion chamber.

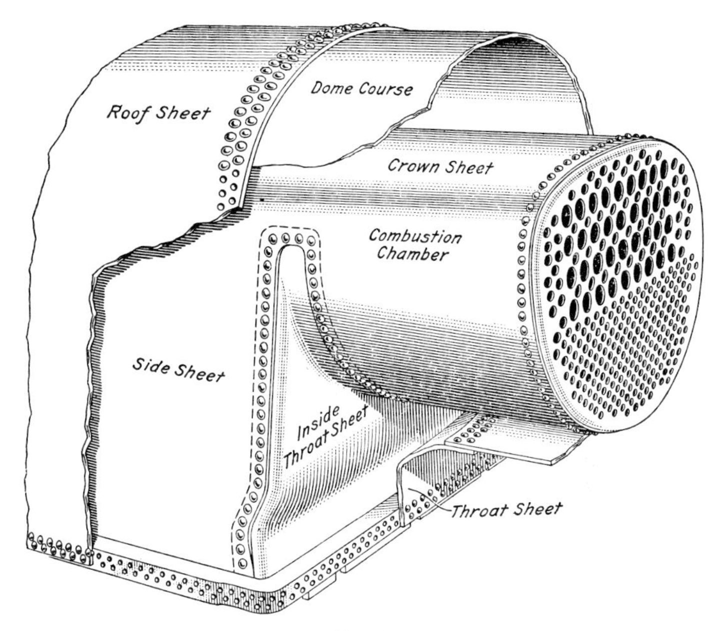

Fig. 4

The type of combustion chamber shown is known, on account of its appearance, as the barrel, or built-in, combustion chamber. When the tube-sheet is directly in front of the grates, the space above the brick arch is sometimes referred to as the combustion chamber. The use of combustion chambers requires a large number of additional staybolts and therefore more attention must be given them. The installation of flexible staybolts properly applied will prevent much of the staybolt trouble in combustion chambers.

8. Advantages.—The surface of the firebox evaporates about five times as much water per square foot as does the surface of the tubes. Therefore, the advantage of the combustion chamber is that it gives more firebox heating surface and a greater evaporation with a specified grate area. The use of a combustion chamber in large boilers also permits the use of shorter tubes. Tubes of an excessive length cause the draft to lag at their front ends. With a combustion chamber the gases have to travel farther before they enter the tubes and they therefore have a better chance of being thoroughly burned than if a combustion chamber is not used.

THERMIC SYPHONS

9. Purpose.—A thermic syphon may be considered as a special type of arch tube. Two or three may be installed in the firebox; Fig. 5 shows a two-syphon application, and one is often placed in the combustion chamber. One end of the syphon *s* is inserted in a diaphragm *h* welded to the inside throat sheet *i*, and the other end connects to the crown-sheet in which an opening is cut to correspond to the opening s_2 in the syphon.

The function of the syphons is to stimulate and accelerate the circulation of the water in the boiler and hence increase the rate at which the water is being evaporated into steam. An increase in the evaporation rate results in the development of a greater drawbar pull at speed; that is, the boiler horsepower is increased.

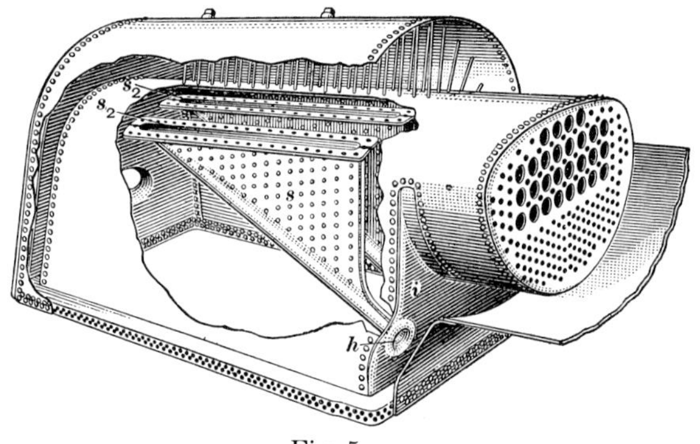

Fig. 5

The syphons also not only provide additional and highly effective heating surface, but also prevent excessive damage to the crown-sheet in cases of low water. The upward circulation through the syphons deposits water on the crown-sheet after the general water level has receded below it, and this limits the rupture to a smaller area than otherwise.

10. Construction.—An exterior view of a thermic syphon is shown in Fig. 6. It is made from an approximately square plate of ⅜-inch firebox steel folded over a 6¼-inch mandrel along a diagonal in such a manner as to form a triangular water leg. The sides that are drawn together sufficiently to provide a water space 3 inches wide, are supported by staybolts in the ordinary manner. The bottom or channel portion s_3 of the syphon, being circular, is self-supporting and is extended to form a cylindrical intake neck 6¼ inches in diameter. The vertical edges of the sheets are flanged inwardly and are joined by autogenous welding. The welded seam extends from the top of the syphon down to the end of the neck and is supported throughout its length by staybolts. The upper edges of the sheets s_4 are flanged out to an over-all width of about 13 inches and to a contour to correspond with the crown-sheet of the firebox, and are drilled to agree with the location of the radial stays.

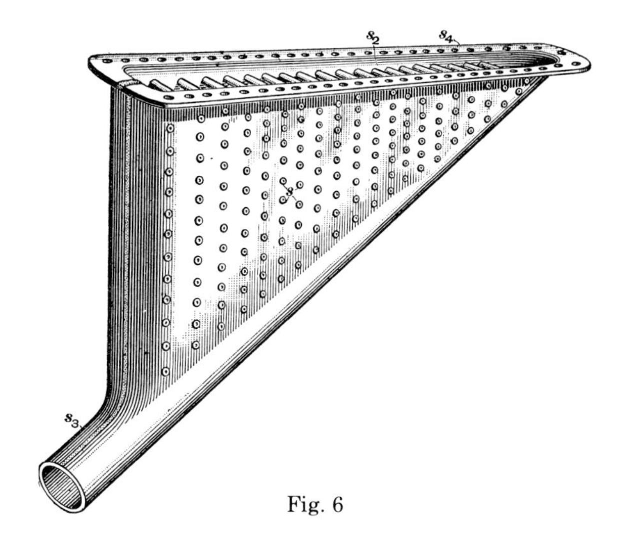

Fig. 6

11. Application.—The manner in which thermic syphons are applied to the crown-sheet is shown in Fig. 7. A space is cut out of the crown-sheet to correspond to the area of the top of the syphon, then its edges are set flush with the crown-sheet and welded. The radial stays are so arranged that a row comes between the welds and the syphon, thereby relieving them of stresses. It will be noted that the circular-shaped channels of the syphons serve to form a support for the brick arch.

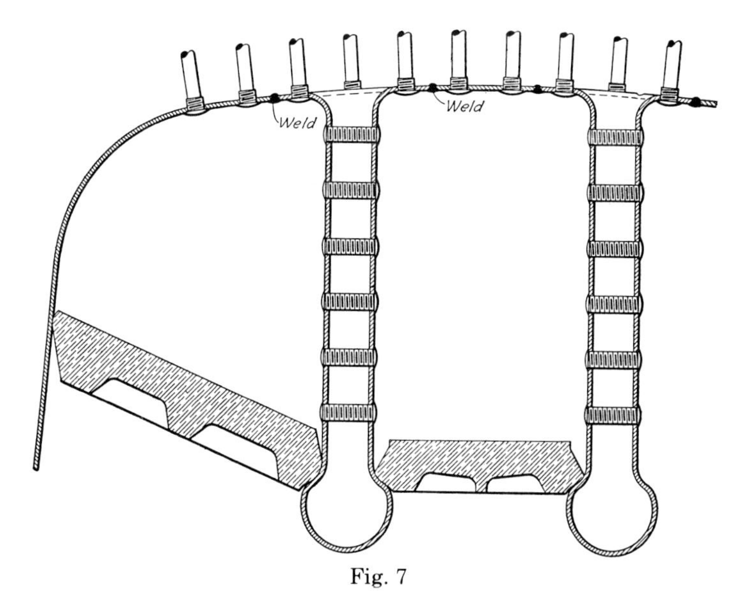

Fig. 7

At the front, the intake neck of the syphon is inserted into and welded to a flexible diaphragm plate *h*, Fig. 8. This diaphragm, or breathing plate, is ½ inch thick and is corrugated to permit the body of the syphon to come and go with the crown-sheet. The diaphragm plate is applied by cutting a hole in the flue sheet or in the inside throat sheet if a combustion chamber is used to suit the plate, which is then welded in place and further supported by flexible stays. If desired, the diaphragm corrugations may be pressed directly into the flue sheet or throat sheet.

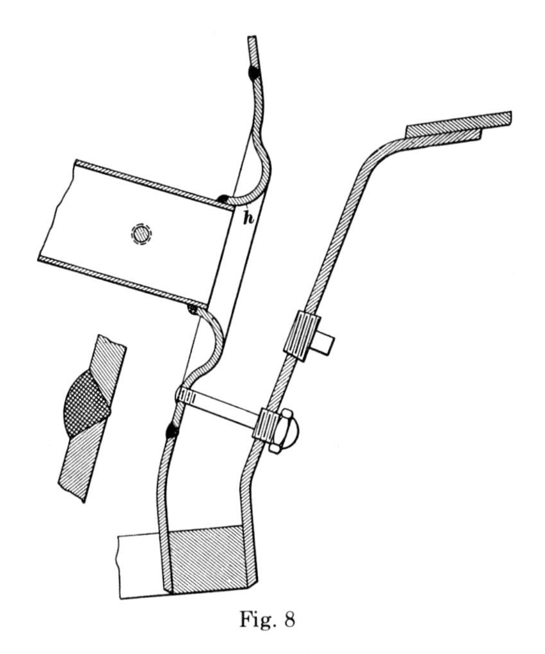

Fig. 8

OIL-BURNING FIREBOX

12. A sectional view of a firebox for burning oil is shown in Fig. 9. The firebox differs from that of a coal-burning locomotive in that the grates and ash-pan are not used. The grates are replaced by brick, and a hopper provided with a door *a* is used instead of an ash-pan. The sides of the firebox are lined with firebrick to a point above the direct path of the oil jet, so as to protect the sheets against the intense heat of the jet. The oil, which is vaporized in the burner *b*, is thrown against the flash wall at the back of the firebox. The air is supplied through the air holes *c* and also through the hopper door when it is opened and through a specially shaped firedoor.

BELPAIRE FIREBOX

13. Description.—In Fig. 10 is shown a cross-sectional view of a Belpaire firebox. The difference between this and other types of fireboxes lies in the shape of the roof sheet and the crown-sheet, the Belpaire firebox being designed so that the crown-sheet *a* and the roof sheet *b* are comparatively flat and parallel with each other. With this arrangement the radial, or crown, stays *c* are at right angles to both of the sheets, and this is a very desirable feature in staying flat surfaces under pressure. The upper section of the roof sheet is still further stayed by transverse

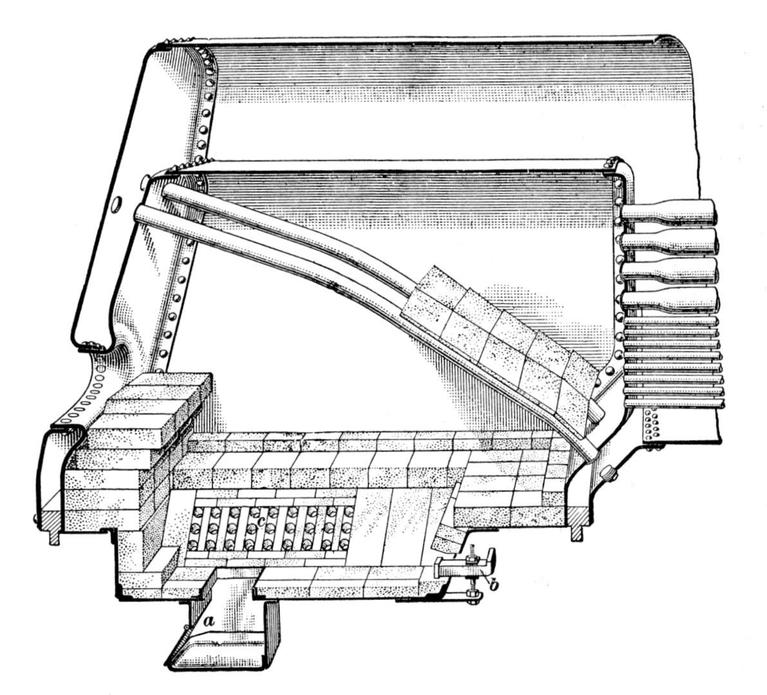

Fig. 9

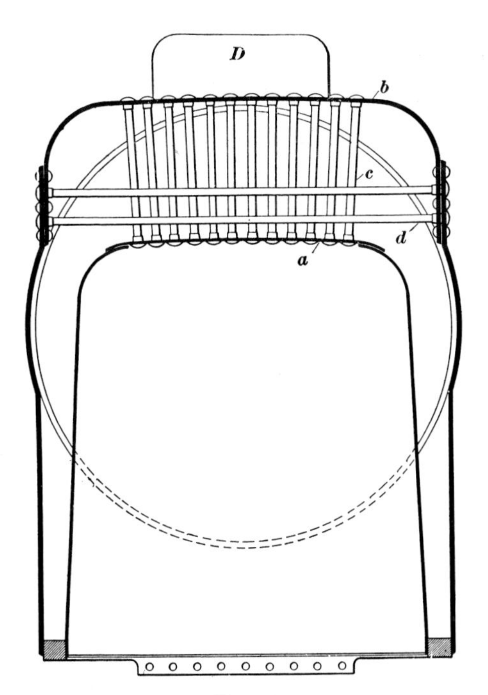

Fig. 10

rods *d*, which are placed at intervals in the steam space and are securely fastened at the ends. The Belpaire firebox gives an increased steam space owing to enlarging the upper part of the boiler but its principal advantage is the right-angle position of the crown stays with respect to the crown-sheet.

The cylindrical part of the boiler in front of the firebox does not differ from any other type, and may be straight-top extended-wagon, or conical. However, the steam dome *D* is always placed on one of the courses in front of the firebox, because, on account of the flat surface, it cannot be very well placed over the firebox.

WATER-TUBE FIREBOX

14. Design for Withstanding Pressure.—When it is required to confine liquids and gases under pressure, the ideal shape of the container is spherical. When it is impracticable to use a spherical container, the next in order of strength is one of cylindrical shape with convex ends. With the same thickness of material, a cylinder with convex ends will withstand a greater internal pressure than one with flat ends. A locomotive boiler has a cylindrical shell, but it is not possible to have the ends made convex. Strength is therefore obtained by having both ends, or the front tube-sheet and the back head, securely stayed to the shell, in which case the boiler nearly approaches the ideal for resisting pressure. The firebox, however, with its flat surfaces, is poorly designed to withstand pressure, and for this reason it requires careful staying.

The flat crown-sheet of the ordinary type of firebox cannot be safely stayed for pressures in excess of about 300 pounds, hence the boiler pressure is limited to the pressure for which the crown-sheet can be stayed with safety. When higher boiler pressures are desired, the watertube firebox must be employed, but as high boiler pressures involve difficulties from scale as well as complications such as three cylinders in order to secure the proper expansion of the high-pressure steam, the water-tube firebox is in very limited use, and is here described merely to show the principle of construction.

15. The water-tube firebox is practically ideal from the standpoint of withstanding internal pressure, as all of the parts that contain water are cylindrical so that no staying is necessary. The principal objection to its use, unless the feedwater is pure, is the difficulty experienced and the time consumed in keeping the water tubes clean and free from scale. With the ordinary form of locomotive boiler, scale is not a serious problem or is at least one that can be controlled because the storage space in which it can collect is large.

However, with a water-tube firebox the tubes are comparatively small and on this account can be easily partly obstructed or blocked by an accumulation of scale, thereby causing a flue failure. Water-tube boilers can be used in stationary practice because the steam after being used in the turbines is converted into water in the condensers. This water, which is practically free of scale-forming matter, is again sent to the boiler and evaporated and used over and over again so that there is only a limited amount of scale to be dealt with.

16. An exterior view of a water-tube firebox is shown in Fig. 11 (*a*). It comprises a drum *a*, 40 inches in diameter, that is in communication with

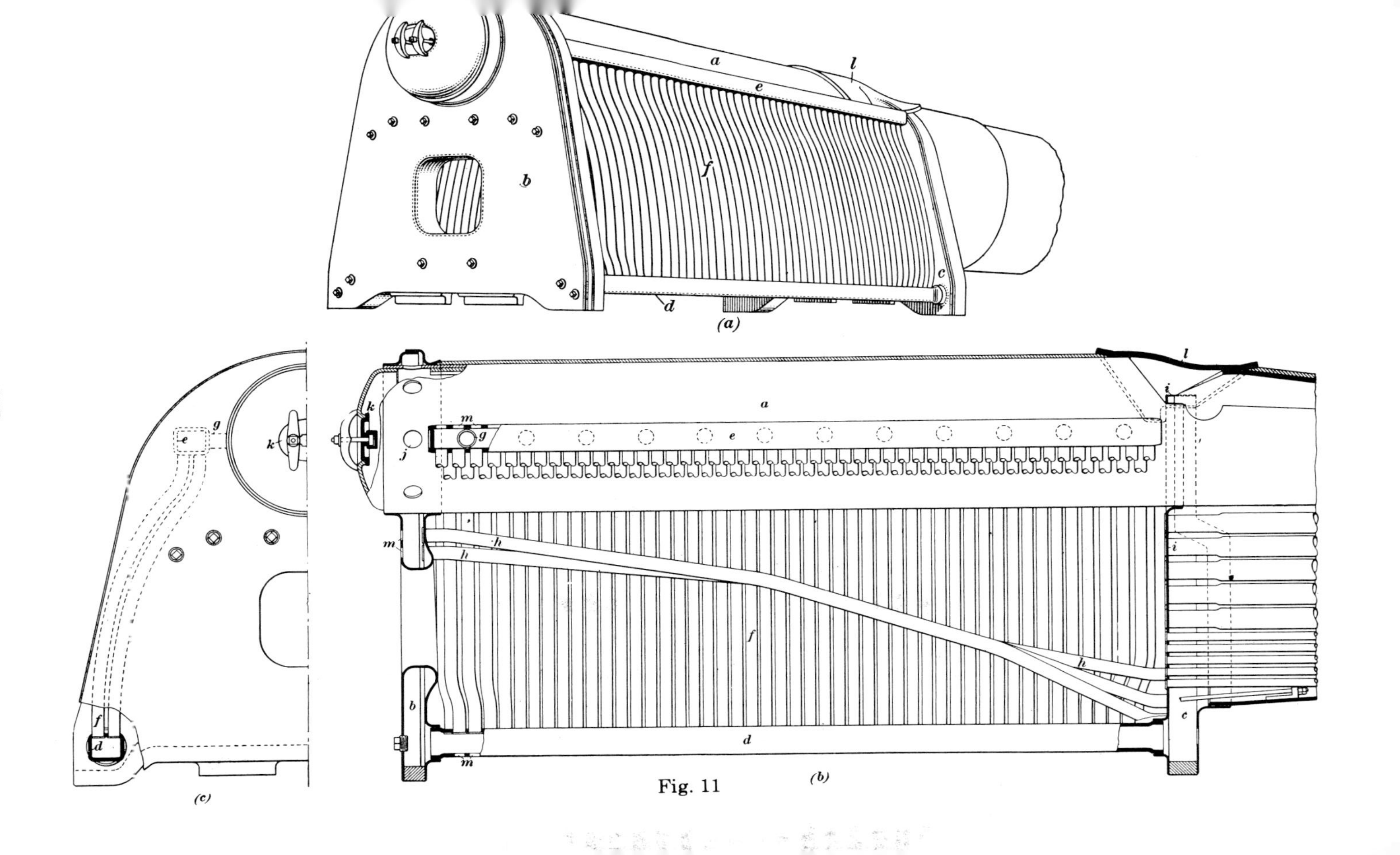

Fig. 11

the hollow back head *b*, and a water leg *c* formed by the throat sheet and back tube-sheet, a bottom side header *d* on each side, a top side header *e*, one on each side, and numerous tubes *f*, 2½ inches in diameter, on each side that make up the sides of the firebox. These tubes are expanded into the header at the top and into the foundation ring at the bottom. The back head and the door sheet are flanged and riveted together so as to form the customary water leg found on the ordinary boiler. The front and rear water legs are supported by staybolts, not shown.

A sectional view taken lengthwise is shown in view (*b*) and an end view is given in (*c*). In view (*b*) is shown a full view of the drum *a* broken away at the rear, the back water leg *b* and the front water leg *c* are shown in section, and the bottom side header *d* is broken away at the ends to show its method of connection to the water legs *b* and *c*. All of the tubes on the front side of the firebox are shown broken away, so the tubes shown in full are on the far side. As shown in view (*c*) the bottom side header is rectangular except at the ends, where it enters the water legs; here it is cylindrical in order to make a steam-tight connection. The header *e* is connected to the drum *a* by twelve short pipes *g*, 3½ inches in diameter. The tubes *f* are staggered and arranged in two rows, and six arch tubes *h*, three shown, connect the two water legs and also serve as supports for the brick arch.

17. The drum *a*, Fig. 11 (*b*), extends through the back water leg *b* and the back tube-sheet *i*. The sheets affected are flanged and are riveted all around to the drum by rivets, not shown. The circulation between the drum and the back water leg occurs through eight holes *j*, equally spaced; the front end of the drum is not closed by a head but opens into the boiler proper. Access to the drum at the rear is made through the head *k*. The junction between the front end of the drum and the shell of the boiler is made by a heavy plate *l*, views (*a*) and (*b*).

The connecting pipes *g* and the arch tubes as well as the tubes that make up the sides of the firebox are rolled and expanded and in some instances beaded, all of which is done through holes *m*, at the end of every tube, normally closed by plugs. These plugs are removed when necessary to clean out the tubes.

The firebox is insulated to prevent the loss of heat and is airtight so that all air must enter through the grates. The firebox measures 11 feet in length and is 8 feet wide, but the distance between *b* and *c* is 16 feet, owing to an allowance being made for a combustion chamber 5 feet long formed by a firebrick wall across the firebox, and joined at the top to the brick arch.

BOILER DETAILS

TUBES AND FLUES

18. The tubes and flues connect the back and the front tube-sheets, and are used to convey the products of combustion from the firebox to the smokebox. They also break the hot gases into small columns so the heat is more readily transmitted to the water which surrounds them, thereby increasing the heating surface. The term *flues* is applied to those tubes which contain the superheater units, and the term *tubes* to all the others. The tubes and flues are fitted into holes that are drilled in the front and the back tube-sheets, and they also act as stays for these sheets. Tubes and flues are either of lap-welded or of seamless construction. Lapwelded tubes are made from strips of highly refined charcoal iron, or steel bent to shape and lap-welded at the joint. Seamless tubes are drawn from a block of refined iron or steel by special machinery which produces a tube without seams.

The tubes used in locomotive boilers are generally either 2 inches or 2¼ inches outside diameter. Experience and experiments have proved that these sizes are the most desirable. If smaller sizes are used, the tendency is for the tubes to become stopped up, while increasing the size decreases the total tube heating surface and reduces the capacity of the boiler to generate steam. Small tubes have a greater heating surface in proportion to the volume of gases passing through them than do large ones, because the small tubes break the hot gases into smaller columns and thus utilize more of the heat, and the hot gases escape to the smokebox at a lower temperature.

19. Application of Tubes.—The tube holes in the tube-sheets are drilled to size before the sheets are put in place. The tubes are always applied and removed through the front tube-sheet and, to facilitate the application and the removal of the tubes, the tube holes in the front tube-sheet are drilled 1/16 inch larger in diameter than the holes in the rear tube-sheet to permit the tubes to be slipped through freely. In some cases a single hole in the front tube-sheet is made ¼ larger in diameter than the others, and all of the tubes are applied and removed through this one hole when making a complete installation of new tubes. This method has the disadvantage of making it difficult to remove a single tube, as the heavy coating of scale that always accumulates on a tube makes it hard to drive the tube out, but the method has the advantage of doing away with the necessity of expanding the tubes so much at the front tube-sheet.

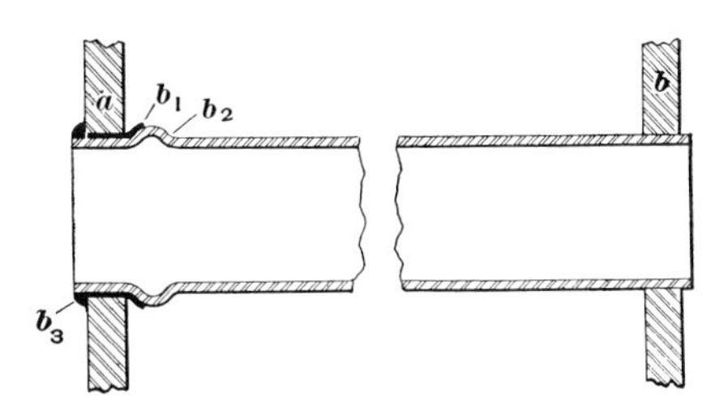

Fig. 12

The holes in the back tube-sheet *a*, Fig. 12, are slightly larger than the tubes so as to permit a copper ferrule b_1 to be inserted in the hole between the sheet and the tube.

The ferrule is a piece of copper pipe 1/16 inch in thickness and ¾ inch long and is rolled lightly in the hole before the tube is inserted. The tube extends about 3/16 to ¼ inch beyond the face of the sheet into the firebox and is rolled tight against the ferrule and the sheet by means of roller expanders. The tube is then expanded by sectional expanders to bell it out on the inside as shown at b_2. The end which extends beyond the sheet is then turned over and beaded down firmly, as shown at b_3, by use of a beading tool of special form. Finally, the tube is electrically welded all the way around as a positive means of preventing leaks.

20. By this method, the tube, in addition to being securely rolled in place, is clinched and welded against the sheet on both sides, and therefore makes a strong joint. The ferrule b_1 is placed flush with the face of the sheet on the fire side and serves to make a tight joint between the tube and the sheet, because when the tube is expanded the ferrule, being soft, will fill completely any opening between the tube and the tube hole. The ferrule also prevents the tube from leaking, because it keeps the space between the expanding and contracting surfaces of the tube-sheet and the tube always filled.

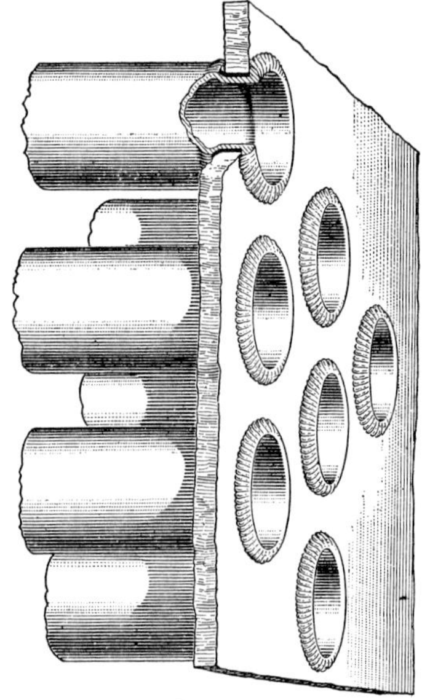

Fig. 13

The section of the tube-sheet in Fig. 13 shows how the tubes appear when they are beaded over and welded.

The tubes are rolled at the front tube-sheet b, Fig. 14, in the same manner as at the back tube-sheet, but the sectional expander is not used, and the tubes, according to local conditions and practice, may or may not be beaded as shown at b_3 on the other end, but some are always beaded for staying purposes.

21. Superheater Flues.—The superheater flues necessarily have to be of a larger diameter than the tubes to make provision for the superheater units.

To permit the installation of the units and still have sufficient space for the hot gases to pass through and circulate around the units, the tubes, or flues, as they are generally called, are made 5⅜ or 5½ inches outside diameter with the firebox end swaged down to 4⅜ to 4⅝ inches, as shown in Fig. 14. The effective area through which the gases can pass when the superheater units are in place is of course greatly reduced so that there is no reason why the firebox end of a tube cannot be made smaller and still be of sufficient size to provide space for the flow of the gases.

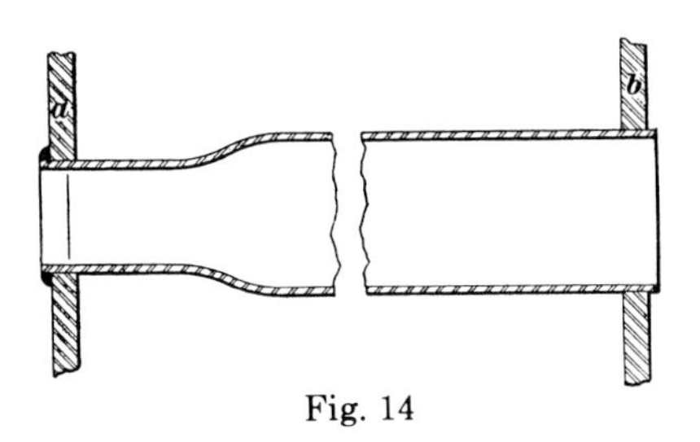

Fig. 14

The flues, Fig. 14, are beaded at the firebox end so that they may be more securely held in position. In some cases, ferrules are used, the same as with the small tubes, while in other cases they are omitted; in this event the flue is welded to the sheet, as shown. At the front end, the superheater flues are merely expanded.

When removing a tube or flue, the beading and welding at the back tube-sheet is chipped off and the front end of the tube is burned off just back of the ring made by the expander. The piece that now remains in the front tube-sheet is chipped out and the tube or flue is driven forwards and outwards.

THE STEAM DOME

22. Description.—The steam dome is a cylindrical receptacle on the top of the boiler, and is used for collecting and holding dry steam. It is usually placed on the boiler course in front of the firebox, although with the smaller and older types of boilers the steam dome was placed directly over the crown-sheet. As is shown in the sectional view, Fig. 15, the steam

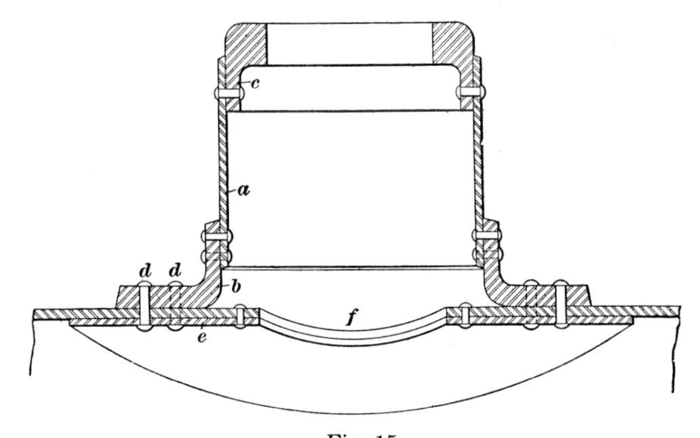

Fig. 15

dome is made up of a cylindrical barrel *a*, which has a double-riveted butt joint on one side, not shown, a heavy flanged collar *b* to which the barrel is riveted, and a cast-steel ring *c* to which the cover, not shown, is secured by bolts. The dome may also be made in one piece, as shown in Fig. 16, by pressing it to shape in a flanging press. This method eliminates the riveting in the butt joint and also in the collar and ring. The dome, whether built up or in one piece, is riveted to the boiler shell by a double row of rivets *d* which extend through the dome base, the boiler shell, and the dome liner *e*. The liner is necessary to strengthen the boiler shell, which is weakened by the cutting of the large hole *f* in the plate. This hole permits dry steam to enter the dome and also provides an opening for the throttle pipe f_1, which connects to the dry pipe f_2. The throttle valve f_3 controls the passage of steam from the dome to the throttle pipe and dry pipe.

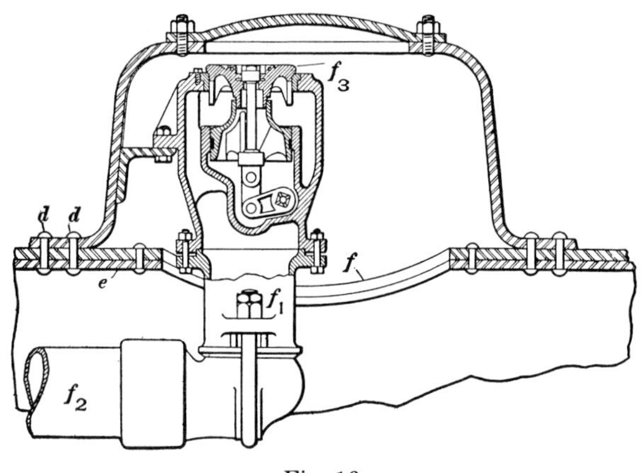

Fig. 16

23. The hole *f* also provides an opening through which a man may enter the boiler to inspect the interior or to make necessary repairs, and it is the only opening in the boiler that is large enough for this purpose. The opening *f* should not be made larger than is absolutely necessary, because its size seriously affects the strength of the shell; the larger the opening the weaker the shell, other things being equal. The opening is about 33 inches in diameter.

The dome cover, which closes the opening on the top of the dome, should be kept reasonably small, as the only method of staying is by bolting it to the outer edge of the ring or to the inner edge of the dome, if a one-piece dome is used.

Domes are made about 2 feet 6 inches in diameter and from 12 to 36 inches high. The size of the dome depends on the size of the boiler and the clearance of the right of way, the latter preventing the use of a high dome on large boilers. The dome, being the highest point above the surface of the water, is an ideal place in which to put the throttle valve; for to reach it there the steam must travel farther before it enters the valve and this insures dryer steam because the steam is more thoroughly relieved of any water that may be entrained in it.

In boilers of very large capacity, an extra dome, called an auxiliary dome, is sometimes used to which the safety valves and the whistle are attached. In such case the dome that contains the throttle valve can be located nearer the front end. This arrangement is an advantage with long boilers because the length of the dry pipe is reduced.

24. Description.—The grates usually consist of a set of parallel bars called grate bars, made of cast iron, and placed at the bottom of the firebox. The purpose of the grates is to hold the fuel as it is being burned.

Grates are classified as rocking, or shaking, grates; dead, or dump, grates; and corner grates. In Fig. 17, which shows one-half of a section of grates, the rocking grates are marked a, the dump grates a_1, and the corner grates a_2. The shaking grates have trunnions c on the ends of the grate bars which are carried by the grate bearers d and d_1.

Rocking or shaking grates, which are shown in section in Fig. 18, consist of a series of flat, or finger, grate bars. These grate bars can be turned out of a horizontal position by connecting rods b that are pinned to the projections f on the lower side of the grate bars. The rods b are connected by rods b_1 to levers d, which extend through to the cab deck and are attached to brackets on the back head. The grates are turned by applying a bar to the lever d, as shown, or by means of mechanical shakers, if the grates are very large.

25. The grate bars are connected up in sections to the rods b, Fig. 18, and therefore one section of the grates can be moved independently of another section. Care should be taken not to connect too many sections to the same arm, as difficulty will then be experienced in shaking the grate.

A finger grate bar is constructed with projections or fingers on each side, which come between the fingers on the grate bar next to it. When the grates are rocked or turned, the fingers of one grate bar disengage with the fingers of the next and allow the ashes to fall through into the ash-pan. A flat shaking grate has no fingers.

Corner grates a_2, Fig. 17, have square trunnions instead of round. They are set stationary in place and are not connected with the shaking or rocking grates. Drop, or dump, grates a_1 are of large size and are so arranged that they can be dropped independently of the other grates. Their use allows large clinkers as well as ashes to be dumped into the ash-pan. Dump grates are usually placed at the back, and sometimes at the front of the firebox as shown.

In narrow fireboxes, the grate bars extend entirely across the box and are supported by grate bearers bolted to the side sheets. The bearers are so set that the top of the grate bars will at least be level with the top of the foundation ring or slightly above it. In wide fireboxes a center bearer d_1 is located in the center of the firebox and serves the same purpose as the side bearer. The use of a center bearer permits the grate bars to be made shorter and lighter and thus more easily handled.

The grate area is expressed in square feet and is obtained by multiplying the total length of the grate, in feet, by its width, in feet.

26. Grates With Restricted Openings.—Formerly it was thought necessary to have the air openings through the grates in excess of 40 per cent of the total grate area, but it was found with improved front-end design and larger stacks that this amount of opening permitted too much air to enter the firebox for proper efficiency. The loss in boiler efficiency was due to the fact that the excess air had to be heated up from that of the outside temperature to that of the firebox. Good results with improved front ends have been obtained with the openings in the grates reduced to about 15 per cent of the total grate area.

Another advantage of grates with restricted air openings is a reduction in stack losses. Under ordinary operation there is a difference of only about 2 ounces in pressure between that in the ash-pan and that in the firebox, yet this difference is sufficient to set up an air or gas current with a velocity of about 20 miles per hour immediately above the fire and increasing very rapidly in its passage to the flues. A blast of this velocity is capable of lifting and carrying with it large particles of coal and especially where the grate openings are fairly large and so spaced as to

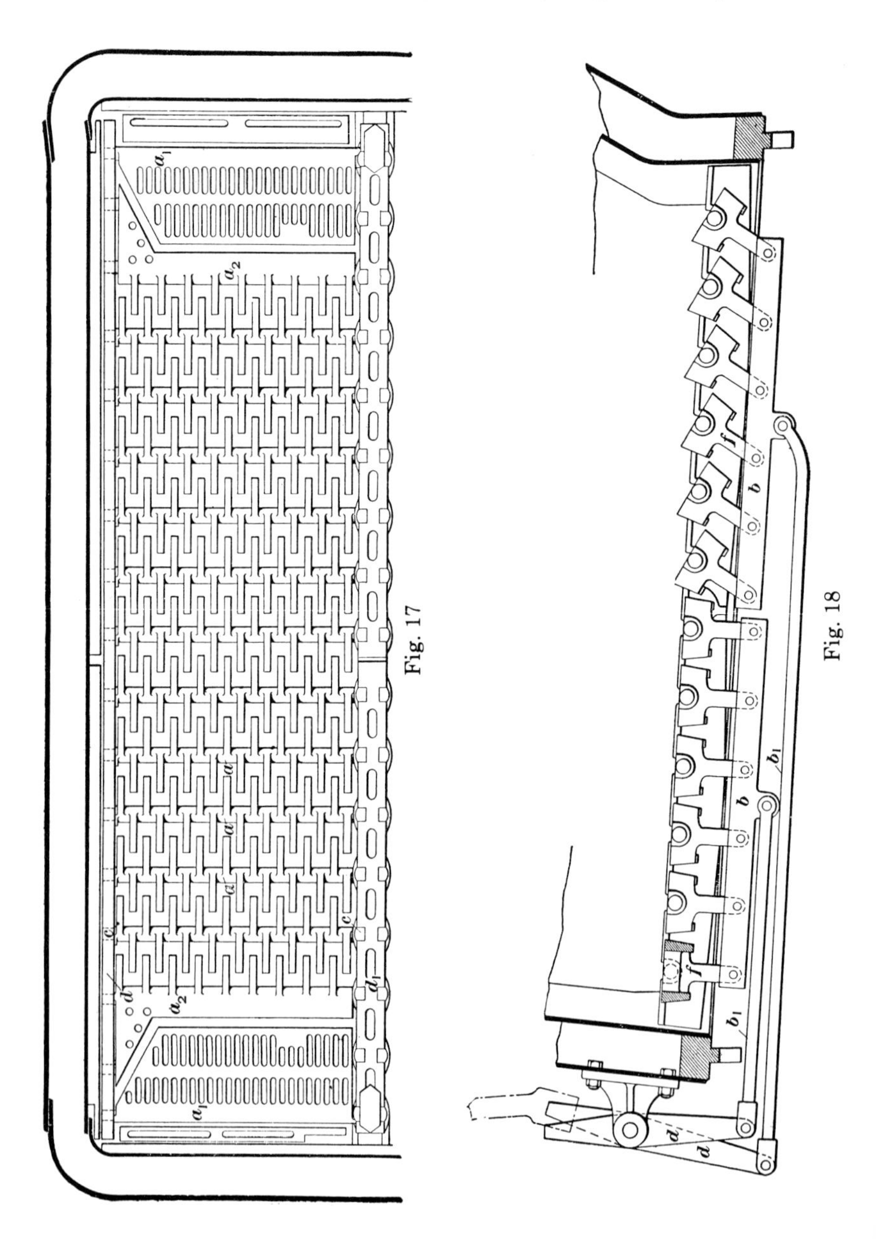

Fig. 17

Fig. 18

permit the formation of holes through the fire bed. Air moving at this velocity can be considered as the direct cause of the stack loss due to the discharge of unburned or partially burned coal commonly called cinders.

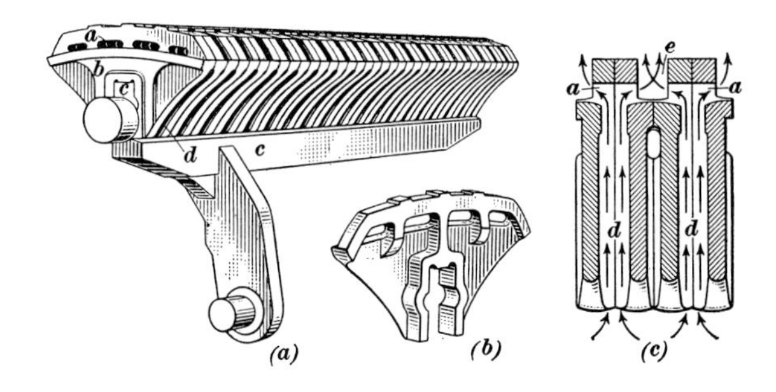

Fig. 19

27. The Hulson tuyere type of grate, which is designed to restrict the air openings to any amount desired, is shown in Fig. 19. This grate takes its name from the fact that the air is admitted through tuyéres *a*, view (*a*), in the castings *b* that are assembled one at a time from the end of the grate bar *c*. These castings, view (*b*), are identical and when placed with the inside faces together, view (*c*), passageways *d* are formed between them, open at the bottom and covered at the top. From these passageways the air enters valleys *e* formed between each pair of fingers by the spacing ribs shown. The incoming streams of air baffle each other in the valleys, thereby reducing their velocity but not their volume. The total air opening is governed by the number and the size of the tuyeres and not by the area of the valleys, and this opening may vary from 37 per cent of the total grate area or less, as may be desired.

ASH-PAN

28. The ash-pan is a receptacle placed under the firebox for the purpose of collecting and holding the ashes which drop or are dumped through the grates. Ash-pans should be constructed so as to prevent, as far as possible, live coals from dropping through and starting fires along the right of way.

In Fig. 20 is shown an exterior view of the side of an ash-pan used with a wide firebox. The principal parts of an ash-pan are the receptacle proper, an arrangement for dumping the ashes, and in some cases dampers to control the admission of air to the fire. Ash-pans are made in a large number of different styles and shapes, depending on the size of the firebox, the wheel arrangement of the locomotive, and the service in which the locomotive is used.

The size of the firebox materially affects the type of ash-pan used, as the pan must extend under the entire grate surface.

The ash-pan shown in Fig. 20 is made up of two hoppers *a*, hence the term *hopper ash-pan*. The hopper arrangement is necessary because the trailing wheels come in about the center of the pan. The bottoms of the hoppers are closed by cast-iron doors a_1, which may be operated by a series of levers from the cab or by a compressed-air apparatus. The hoppers have a width of 24 to 30 inches at the bottom and gradually extend upwards and outwards to the foundation ring *m*. The sides of the pan extend up over the frames to a point level with or slightly above the bottom of the ring. In some cases the pan is attached directly to the ring, and in others a space of from 3 to 5 inches is left between the pan and the ring so as to provide a space through which air can enter to the under side of the grates. When the sides of the pan are bolted directly to the ring, pockets or perforated plates are inserted along the sides and also at the ends of the pan to provide the necessary air inlets. The preferred way is to have the air inlets at the top of the pan away from the ashes.

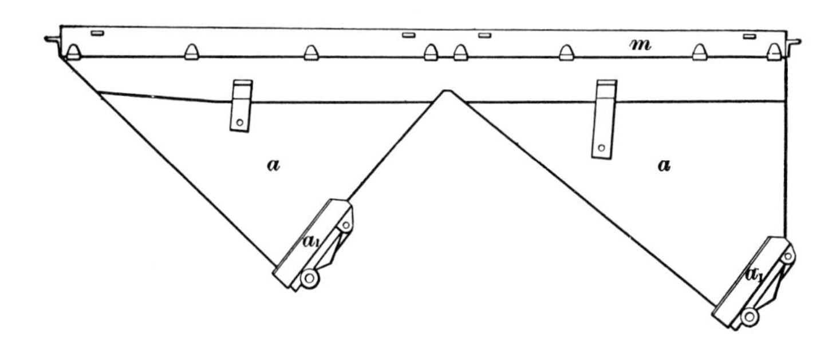

Fig. 20

When very wide fireboxes are used, the ash-pan may have a series of hoppers between the frames and also on the outside of the frames on each side, but the principle of construction is the same as already described.

On account of the numerous classes of locomotives, there is no one style of ash-pan that can be considered as a standard. Nearly all the large railroad companies have adopted styles to suit their own special purpose.

By orders of the Interstate Commerce Commission, all ash-pans must be equipped with some device whereby they can be cleaned or dumped without the necessity of going under the engine to do it. The means provided for this purpose are operated from the cab either by hand, air, or steam, at the convenience of the engineer. The hoppers and the hopper doors are made of cast steel or malleable iron. The plates used in the construction of the rest of the pan are usually 1/4 inch, or 5/16 inch thick. Some pans are made of one piece of cast steel but this practice is not general.

29. Air Inlets in Ash-Pans.—Although the ash-pan is intended primarily as a receptacle for ashes, yet it is one of the most important adjuncts affecting combustion that is placed on the locomotive. The function of the appliances in the front end is to maintain a partial vacuum in the firebox, and the amount of vacuum necessary in the firebox to burn the fuel at a certain rate depends on the difference in pressure above and below the grates. Under ordinary conditions the pressure in the firebox is only about 2 ounces less than that of the surrounding air. If, therefore, the ash-pan is so designed that sufficient air cannot flow under the grates to maintain atmospheric pressure, it follows that to obtain the necessary difference in pressure a higher vacuum must be created in the firebox. This can be obtained only by reducing the size of the nozzle, which in turn results in higher back pressure and affects adversely locomotive operation.

It has been generally admitted that openings in the ash-pan equivalent to about 15 per cent of the total grate area are sufficient under practically all conditions to maintain atmospheric pressure under the grates. Tests have demonstrated, however, that even with this ratio of air opening, a partial vacuum equal to .6 inches of water frequently occurs under the grates under certain conditions of operation, and this partial vacuum is equivalent to reducing the vacuum above the grates by an equal amount. Therefore, the practice is to employ air openings somewhat in excess of 15 per cent of the total grate area.

CIRCULATION IN BOILERS

30. Importance of Circulation.—Circulation in locomotive boilers is the name given to the motion of the water under the influence of heat. The water nearest to the source of heat is heated and expands, and thereby becomes lighter and rises to the top. As this water rises to the surface, other cooler water will take its place, become heated, and also

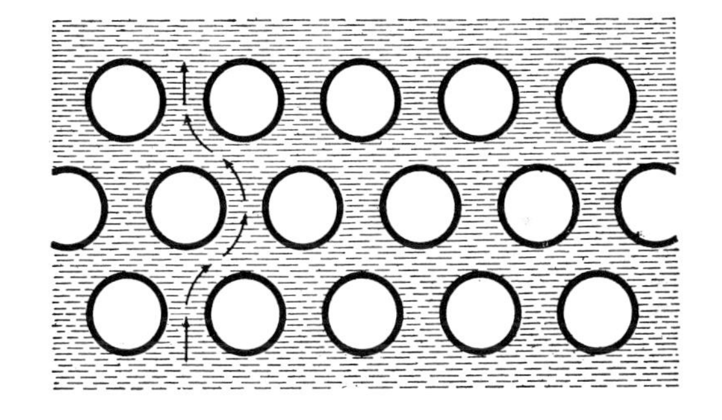

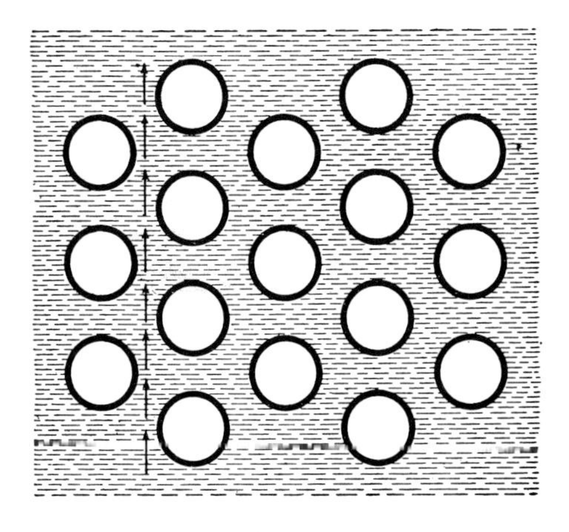

Fig. 21

rise. The circulation of the water keeps the heating surfaces covered with water at all times and thus prevents any part from becoming overheated or burned. Circulation secures regularity and steadiness in the generation of steam, tends to equalize the temperature throughout the boiler, and prevents unequal expansion and consequent straining of the different parts. The circulation of the water tends to maintain the greatest possible difference of temperature between the water on one side of the plates and the hot gases on the other side, thus enabling the heat to be absorbed readily. In addition, circulation tends to keep the heating surfaces as clear of mud and other deposits as possible. The more rapid the circulation, the greater the amount of heat that will be transferred from the firebox and tubes to the water.

31. Advantages of Proper Circulation.—The advantages derived from proper circulation may be summed up as follows: The boiler is kept cleaner, so more water will be evaporated per pound of fuel, which results in a saving of coal. There are fewer strains on the boiler due to unequal expansion and contraction, and there is less corrosion and pitting of the sheets.

The circulation of water in a locomotive boiler is shown in Fig. 2. The heated water from around the firebox rises to the surface and takes steam bubbles with it. The cooler water from the front of the boiler, where nearly all feed-water inlets are located, flows back to take the place of that which has moved upwards. In this way a circulation of water is maintained in the general direction indicated by the arrows.

In locomotive boilers, the large number of tubes and flues makes it important that they be so spaced as to interfere as little as possible with the circulation of the water around them. There are two methods of arranging the tube spacing in boilers. In Fig. 21 (*a*) is shown what is called zig-zag spacing, and view (*b*) shows vertical spacing. The second method is more generally used because, as shown by the arrows, there is less obstruction to the steam bubbles as they rise to the surface than with the arrangement shown in view (*a*).

HEATING SURFACE

32. Definition.—The heating surface of a locomotive boiler is that part of the boiler that is exposed to the direct heat of the fire or the gases of combustion on one side, and which has water available for evaporation on the other side.

33. Kinds of Heating Surfaces.—Heating surface is usually divided into firebox heating surface and tube and flue heating surface. The firebox heating surface consists of the sheets of the firebox above the grate level, and comprises the crown-sheet, the side sheets, and door sheet less the area of the door hole, and the back tube-sheet less the internal area of the tubes and flues. Tube or flue heating surface comprises the total outside surface of the tubes and flues between the front and the back tube-sheets. The area of the front tube-sheet is not taken into consideration in calculating the heating surface. Heating surface is usually expressed in square feet.

34. Efficiency of Heating Surface.—The efficiency of the heating surface depends on the following factors: The location and condition of the heating surfaces, and the arrangements by which the water has rapid and free circulation throughout the boiler. The location of the heating surface has an important influence; for the surface in direct contact with the fire will evaporate water at a much higher rate than an equal amount of heating surface that is not in direct contact with the fire, such as the tube heating surface.

If the tubes and flues become coated with scale on the water side or are partly or entirely plugged with soot on the fire side, these surfaces will conduct less heat to the water, and the efficiency of the surface is lowered because it does not evaporate the amount of water that it should.

The location of the entry of the feed-water may be such as to hinder proper circulation. The tubes and flues may be placed too closely together, in which case the water may be retarded from coming into contact with the proper heating surfaces; thus the rate of evaporation and hence the efficiency of the surface will be lowered.

Of the different heating surfaces of the boiler, the crown-sheet surface is the most effective, then that of the side, tube, and door sheets in the order named, and the tube and flue surfaces last. In calculating the evaporation values, all the sheets of the firebox are taken at the same rate, likewise the rate for the tubes and flues is taken as the same.

35. Direct and Indirect Heating Surfaces.—The heating surface of the firebox sheets and arch tubes is termed the direct heating surface, as these parts are subject to the direct action of the fire. The surfaces of the tubes and flues are called the indirect heating surfaces because these surfaces are only exposed to the heat of the gases that have passed out of the firebox. The sum of the direct and indirect heating surfaces is the total heating surface of the boiler.

BOILER EVAPORATION

36. Calculating Evaporation.—Boiler evaporation refers to the total amount of water evaporated into steam by the heat generated in the firebox and the tubes and flues.

The total evaporation with a modern locomotive boiler is very difficult to determine with accuracy for any given condition. The rate of heat transfer by radiation of firebox heating surface, upon which the greater amount of the evaporation depends, involves so many factors that it forms a very complicated problem. The rate of heat transfer varies with the grate area, the firebox heating surface, the rate of firing, the ratio of firebox heating surface to combustion volume, the distance from the source of heat to the heating surface, the temperature of combustion, the amount of air used per pound of fuel in the process of combustion, as well as other factors.

37. The generally accepted figure was an evaporation of 55 pounds of water per square foot of firebox heating surface per hour on a basis of 12,000 B.t.u.'s per pound of coal, but this was with a firebox without a combustion chamber or syphons. With modern fireboxes, on account of their greater heating surface, the total maximum evaporation will be greater, but the unit rate of evaporation per square foot of heating surface will doubtless be less because of the lower average temperature in the firebox. The evaporating rate for the tubes and flues of earlier types of boilers was taken as about 10 pounds of water for each square foot of outside heating surface per hour for a 2¼-inch tube 18 feet long, but this figure will not apply to modern boilers.

The total evaporation of the combined heating surfaces of the firebox and the tubes and flues can be found from road tests in which a check is made of the weight of water supplied to the boiler per hour. Such a test would show an evaporation of about 7 or 8 pounds of water per hour for each square foot of heating surface of the firebox, tubes, and flues.

38. Steam Required per Cylinder Horsepower Hour. Tests have shown that the development of one cylinder horsepower for one hour, which is the equivalent of raising 1,980,000 pounds one foot in one hour, requires 20.8 pounds of steam for a locomotive using steam of a moderate degree of superheat, as obtained with a type A superheater. With steam pressures of from 225 to 275 pounds and 300 degrees of superheat, as with a type E superheater, and with feedwater heaters and large grates, one cylinder horsepower for one hour can be developed from 17½ pounds of steam.

Therefore, the steam required per hour is equal to the horsepower of the locomotive multiplied by either 20.8 or 17.5.

EXAMPLE.—How much steam is required per hour for a steam locomotive of 2,000 horsepower equipped with a type A superheater?

SOLUTION.—Each horsepower developed requires 20.8 lb. of steam, hence the total amount of steam for 1 hour is $2{,}000 \times 20.8 = 41{,}600$ lb. Ans.

As it requires one pound of water to generate one pound of steam, the weight of the water used will be 41,600 pounds, or almost 5,000 gallons per hour.

39. Coal Required per Cylinder Horsepower Hour.—The quantity of coal required for the development of one cylinder horsepower for one hour will average about 3.25 pounds for locomotives with moderate boiler pressures, equipped with type A superheaters. With high pressures and high superheat, a fair average would be 2.5 pounds.

40. Boiler Horsepower.—The boiler horsepower can be found from the following rule:

Rule.—*To find the boiler horsepower, divide the total evaporation per hour by the pounds of steam required per cylinder horsepower hour. This is 20.8 pounds for moderate superheat and 17.5 pounds for high pressure and high superheat.*

EXAMPLE.—The total steam evaporated with a locomotive equipped with a type A superheater is 39,600 pounds per hour. What is the boiler horsepower?

SOLUTION.—The boiler horsepower is equal to $\frac{39{,}600}{20.8} = 1904$. Ans.

As already pointed out, the application of the foregoing rule is complicated by the difficulty in calculating the total evaporation with modern locomotive boilers.

It is desirable to proportion the heating surface so as to make the boiler horsepower equal to at least the cylinder horsepower. It has been shown that a boiler horsepower in excess of the cylinder horsepower results in considerable economy in fuel and water.

CALCULATING THE GRATE AREA

41. Rate of Combustion.—The rate of combustion is stated in terms of number of pounds of coal burned per square foot of grate area per hour. If the grate area is 60 square feet, and if 3 tons of coal is burned in 1 hour, the rate of combustion is 100 pounds of coal per square foot of grate area per hour.

42. Grate Area Required.—For high-grade bituminous coal, containing about 14,000 British thermal units per pound, the maximum rate of combustion for economical evaporation has been found to be about 100 pounds per square foot of grate area per hour. Larger amounts than this have been burned, but careful tests show that such practice is wasteful of fuel and results in low evaporation per pound of coal.

When the horsepower to be developed is known, the amount of coal that must be burned to produce it can be determined as explained in Art. 39, where it was stated that horsepower x 3.25 or x 2.5 = pounds of coal required per hour.

The grate area required to burn the amount of coal required can then be found by dividing the total coal that must be economically burned per hour by the number of pounds that can be burned per square foot of grate area per hour. The rule, then, for finding the grate area is as follows:

Rule.—*To obtain the grate area, in square feet, for burning bituminous coal, divide the total coal to be burned, in pounds per hour, by 100.*

EXAMPLE.—Find the grate area required for a locomotive of 2,542 horsepower burning bituminous coal.

SOLUTION.—The coal required to be burned per hour is 2,542 x 3.25 = 8,261.50 lb. The grate area is therefore $\frac{8,261.50}{100}$ = 82.6 sq. ft. Ans.

LOCOMOTIVE BOILER DESIGN

43. Principles of Design.—The design of the locomotive boiler and firebox and their appurtenances should be such as will permit of the greatest possible evaporation from a given amount of fuel. It is an advantage if the boiler capacity is equal to or in excess of the cylinder requirements. When this is the case, the maximum hauling capacity of the locomotive can be obtained even when working under somewhat adverse conditions.

In modern locomotive practice, all things are made subordinate to the boiler and the cylinders. Of these two, the boiler is first in importance, as the performance of the locomotive depends on the ability of the boiler to generate steam. If the boiler fails to supply the required amount of steam, the result will be a locomotive that will not haul the train without losing time, except with a reduced tonnage.

Again, a boiler may provide sufficient steam, but the grate area may be incorrectly designed and proportioned. In this event the locomotive will be wasteful of fuel.

In order that modern locomotives shall do the work required of them, it is necessary for the boiler to be properly constructed, for the firebox to be of sufficient size to produce the best possible combustion of the fuel, and for the grate area to be such that the fuel will be burned economically.

Reliable tests have shown that a square foot of firebox heating surface will evaporate about five times as much water per hour as a square foot of tube heating surface. It is therefore an advantage to have a relatively large firebox heating surface.

The demand for heavy power has necessitated increasing the size of the boiler to such an extent that locomotive boiler design is a problem that requires careful study.

Of all steam-generating devices, the locomotive boiler is undoubtedly the most severely taxed as a structure. The reasons are as follows: The boiler cannot be perfectly insulated from cold; it must stand the effects of severe strains due to rapid and uneven heating up and cooling off while in service; it is subjected more or less to sudden cooling at the ash-pit and on the way from the ash-pit to the engine house; and, when required for service, steam is very often raised quickly, thereby causing very rapid expansion and the setting up of stresses that ultimately will cause rupture of some part.

CONSTRUCTION OF BOILERS

44. Principles of Construction.—Two of the fundamental principles in boiler construction are, the boiler must be constructed so that it will withstand a pressure considerably in excess of the working pressure, and the construction should be such as to prevent disastrous results if the boiler should explode through neglect or carelessness.

The art of boiler making has been developed to such an extent that it is possible to build boilers that will withstand the high pressures now required with less danger of explosion than existed with the lower pressures that were used with the older types.

45. Order of Construction.—The construction of a locomotive boiler may be divided into a series of operations which briefly are as follows: (1) Procuring and testing materials; (2) laying out or marking off the sheets or plates; (3) punching the plates; (4) rolling or bending the plates to shape and flanging the plates that require it; (5) machining the plates, such as drilling the tube and flue holes, etc.; (6) assembling the several pieces; (7) riveting; (8) applying staybolts and stays; (9) calking all joints; (10) testing.

46. Materials.—The materials used in the construction of locomotive boilers are wrought iron, mild steel, and cast steel. Mild steel and alloy steel are used almost exclusively for the shell, heads, and firebox sheets. Tubes and flues are made from mild steel or highly refined iron, and staybolts and stays from refined iron or mild steel. The foundation ring is made of cast steel or wrought iron.

The steel used in boiler construction must be of the best quality obtainable, and should be designated and marked either as *flange steel* or *firebox steel.* Plate stamped by the manufacturer with either of these designations and also with the tensile strength of the piece is usually accepted as being made in accordance with the requirements of Federal and state laws governing the manufacture of plate. The plates should have a tensile strength, that is, a resistance to the pulling apart of the particles, of not less than 55,000 pounds per square inch, and not more than 65,000 pounds per square inch.

Wrought iron, mild steel, cast steel, and refined iron used for the details of the boiler should be in accordance with standard practice and the laws governing it. All iron and steel should be free from seams and mechanical defects and should show a uniformly fibrous structure throughout.

47. Laying Out Plates.—The first step in the construction of a boiler after the plates are received is to lay out or mark off the different parts that go to make up the boiler. These parts consist of one or more cylindrical courses; the taper, or conical, course if there is one; the front and back tube-sheets; the door sheet; the throat sheet; the side and crown sheets; the roof sheet; the back head, and the smokebox. Great care must be taken in the laying-out of the plates so that all rivet holes and staybolt holes are in proper alinement.

48. Punching, Flanging, and Rolling.—The sheets after they are laid out are taken to the punching machine where all holes for the rivets, staybolts, etc., are punched. The plates are then trimmed to the required size and shape either by shearing, or burning off the surplus material with the oxyacetylene torch. The sheets that require flanging are then taken to the flange shop where they are flanged by a flanging press. The sheets are heated, and then the flanging press by the use of specially formed dies, forms the flange by hydraulic pressure in one or two operations depending on the shape of the sheet. The sheets that require flanging are the front and the back tube-sheets, door sheet, throat sheet, and back head. After the front and the back tube-sheets are flanged, they are taken to the drilling machine where the tube and flue holes are drilled in them.

The cylindrical courses and the taper and conical courses, after being punched and trimmed to size and shape, are taken to the rolls, where they are rolled to a circular shape. It is very important that these sheets be rolled to a perfect circle, as otherwise the action of the pressure when the boiler is under steam will tend to force the sheet into a circle. This action is liable to start in the seams and joints and may result in a leaky boiler.

This is especially true of the construction of the longitudinal butt joints, where, unless care is exercised, a flattened surface will result by reason of the several thicknesses of plate being placed together. The inside firebox sheet and the roof sheet are finally rolled to conform to the shape of the firebox.

49. Assembling.—As all the sheets are now shaped, they are set up or bolted in their respective places. The barrel, or shell, is usually put together first. After being securely bolted together, all holes are reamed out to the required size of the rivets. The barrel is then taken to the riveting machine, where by hydraulic pressure the various sheets are riveted together under a pressure of from 75 to 150 tons. The firebox sheets are then assembled and riveted together in much the same manner, and the firebox is then placed in the shell in its proper place in relation to the rest of the boiler. At this time, the holes for the staybolts that support the flat surfaces of the firebox are tapped out and the staybolts inserted and riveted over at the outer ends, if stays of the type requiring riveting are used. The stayrods that support the back head and the front tube-sheet are also fitted and riveted to the outer shell. The various sheets being now assembled and all rivets and staybolts in place, the seams or joints are made steam-tight, or *calked*, as it is called, by means of a calking tool held in a pneumatic hammer. The tubes and flues are inserted in place, and rolled, expanded, and beaded or welded. On the completion of this, the boiler is ready to be tested.

CARE OF BOILERS

50. Low Water.—The efficiency of a locomotive boiler depends largely on the care it receives. It is of first importance that those in charge see that there is at all times a sufficient quantity of water in the boiler to cover the crown-sheet and thereby make it impossible for the sheet to become overheated.

The firebox sheets and the tubes are in contact with the fire, and they would become heated to the temperature of the fire if it were not for the water on the other side. The temperature of the water depends on the boiler pressure, and rarely exceeds 400° F. Therefore, while the sheets are in contact with the water they cannot greatly exceed this temperature, although the temperature in the firebox may exceed 2,500° F., which is about the fusing point of firebox steel. The heat in the firebox is conducted through the plate to the water and is absorbed, thereby preventing the sheet from heating to the temperature of the fire.

If, however, the transmission of the heat to the water is obstructed by scale or grease, or if the water, owing to being light and foamy, fails to absorb the heat, the plates will retain the heat and may become red-hot; or if, from any cause, the sheets are unprotected by water, they may become overheated. Metal loses its strength when heated, and, if heated to a high temperature, has comparatively little strength to resist the pressure within the boiler; as a result, the sheets are forced off the stays and failure occurs. It is a well-recognized fact that scale or grease may be the direct cause of an explosion.

The crown-sheet is the highest sheet in the firebox and it will be the first to become dry in the event of low water. An overheated crown-sheet invariably results in an engine failure or a boiler explosion. An important factor in the prevention of low water is the proper maintenance of the water registering devices.

51. Examples of Overheated Crown-Sheets.—In Figs. 22 and 23 are shown examples of crown-sheets damaged by low water, but not to such an extent as to cause a boiler explosion. In Fig. 22, the portion of the crown-sheet that was overheated is indicated by the staybolts shown in black. The two staybolts within the dotted enclosure and indicated by wings, pulled through the sheet and the sheet itself was bagged 3/8 inch. All the other staybolts in black were loose, and at the point indicated by heavy marks the sheet was bagged from 1/16 to 1/8 inch.

The area of the crown-sheet within the large dotted triangle would have been burned were it not for the flow of water induced by the syphons, thus probably preventing a disastrous boiler explosion.

In Fig. 23 the area within the dotted line *abc* was overheated. The four staybolts within the area *d* were pulled out and this area, which was about 10 inches square, was bagged to a depth of 1 1/4 inches. The two rows of staybolts within the individual circles were loose, while those with the wings were partly pulled out. The narrow area of the sheet between the two rows of loose staybolts was bagged from 1/8 to 3/8 inch. The entire area within the dotted lines *ef* would have been damaged were it not for the syphons.

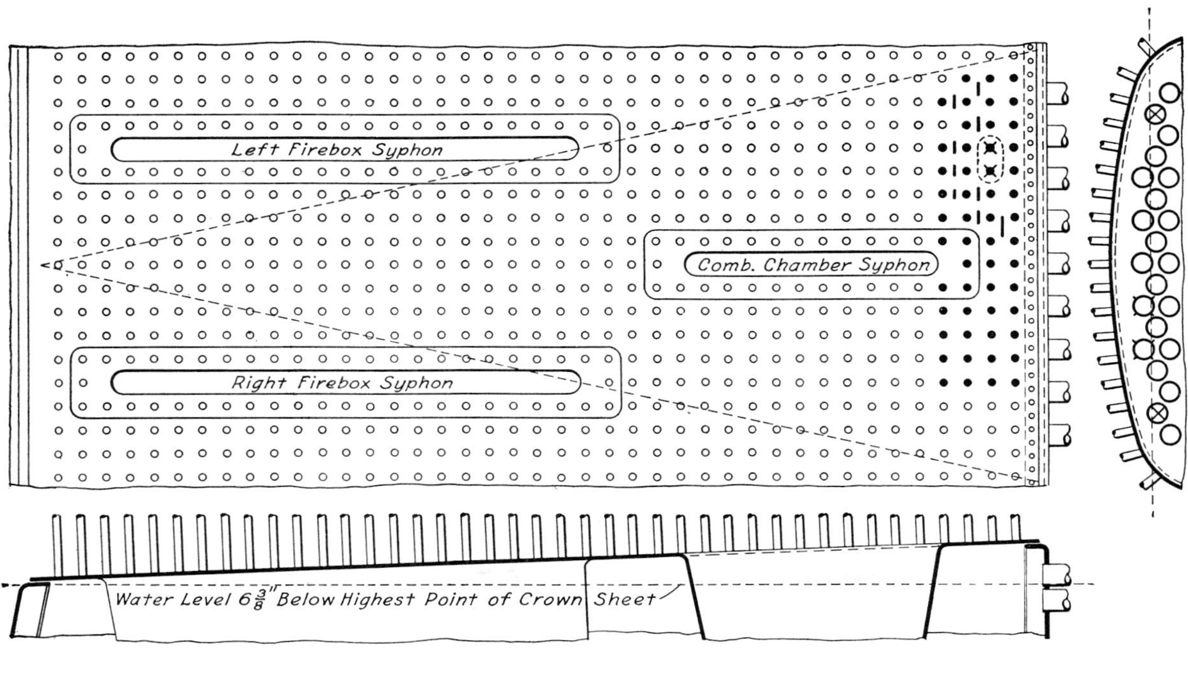
Left Firebox Syphon
Comb. Chamber Syphon
Right Firebox Syphon
Water Level 6 3/8" Below Highest Point of Crown Sheet

Fig. 22

52. Loose Crown Stays.—The most striking feature of an overheated crown-sheet is the number of crown stays that are loose; in fact, the first indication of a leaky crown-sheet is leaky crown stays. The reason is that, in the application of the crown stays, the stays are screwed in tightly, thus putting the metal adjacent to the holes under a strain, and causing the metal to grip the stay and keep it tight. That is, the particles of metal immediately around a crown stay will be crowded closer together than those farther out.

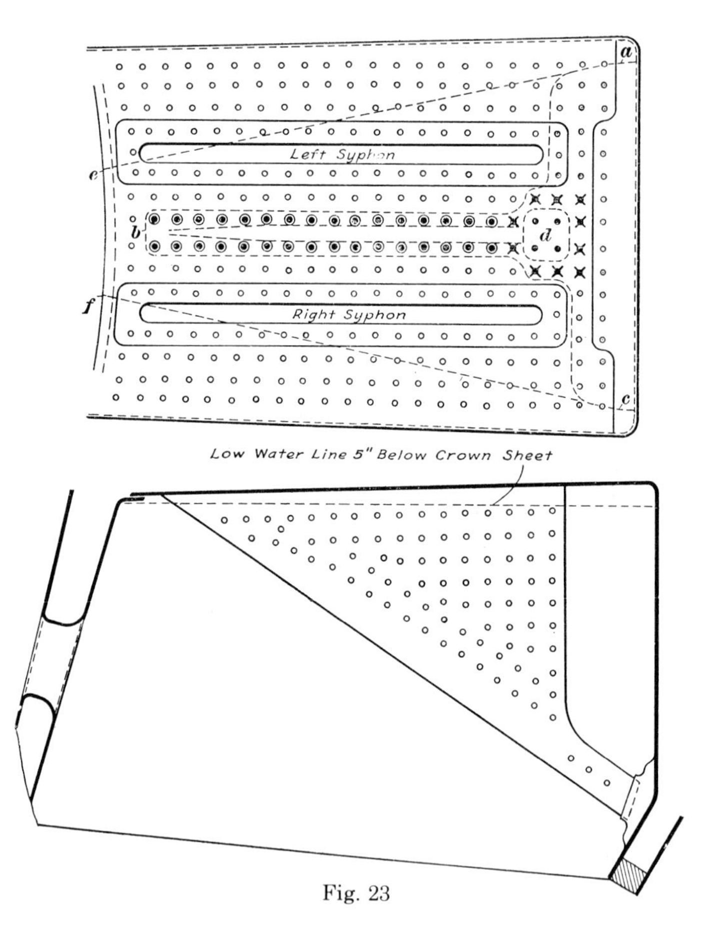

Fig. 23

Now, it is a well-known fact that heating a metal to the proper temperature, as is done when annealing main rods, side rods, etc., relieves it of any internal strains because of the physical change that takes place in the metal owing to the heat. As soon as the absence of water causes the crown-sheet to begin to heat up, an annealing action immediately starts in the metal, and its elasticity around the crown stay is removed or the grip of the sheet on the stay relaxes, so that the particles of the metal that heretofore were crowded closely around a stay will begin to move away from it toward an area where they are not in such close contact.

As soon as this action, which is actually a slight shrinkage of the sheet around the stay, starts, the stay loosens, and steam and water will begin to leak through. A continuation of the heating action softens the plate, so that it will be bulged downwards by the internal pressure in the boiler. The bulging enlarges the hole around the stay and if continued long enough the sheet will be pulled off the end of the stay, this action being assisted by the crown stay itself becoming soft.

53. The action that follows the overheating of a crown-sheet is therefore progressive. First, the sheet anneals and then begins to leak at the crown stays; next, the sheet begins to bag, or bulge, around the stays and finally it pulls away from them. After the first stay fails, those adjacent are liable to go immediately, as they are subjected to more than their share of the boiler pressure. If a large area of the sheet fails, the result will be a violent explosion, as the force of the explosion is in proportion to the size and suddenness of the initial rupture, and the volume and heat of the water in the boiler.

The portion of the crown-sheet that has been overheated is easily identified by the water side being black, all of the scale having been burned off. The blackened portion of the sheet will be surrounded by the rusty, scale-covered area that was covered by water, so that the area of the overheated portion as well as the high-water line will be plainly shown.

54. Reason for Violence of Explosion.—The temperature at which water will begin to turn into steam depends on the pressure at which the water is subjected. At atmospheric pressure, about 15 pounds per square inch, water will begin to turn into steam if it has a temperature of 212°F. If the pressure was suddenly and considerably reduced on water at this temperature the whole body of water would turn into steam. At a pressure of 200 pounds to the square inch, the water in the boiler will boil when its temperature reaches 388°F. Now if the crown-sheet ruptures suddenly over a wide area, the water in the boiler at, say, 388°F. is so far above its boiling temperature at atmospheric pressure that every particle of water instantaneously flashes into steam and increases in volume about eighteen hundred times. The steam is unable to escape fast enough from the boiler so that the result is a tremendous increase in pressure far above the pressure the boiler is designed to withstand. It is this high pressure that produces the disastrous results that often follow boiler explosions.

55. Clean Heating Surfaces.—Another important item in the care of a boiler is to see that all heating surfaces are clean on both the inside and the outside. The interior of the boiler can be kept reasonably clean by frequently washing it out. All the exterior heating surfaces should be kept clear of soot, etc., because such things obstruct the transmission of the heat. Tubes and flues that are plugged, or stopped up, will materially affect the steaming of the locomotive, because they not only reduce the effective heating surface but they also retard to a great extent the draft on the fire.

Tests have proven that with one half of the superheater flues plugged, the performance of the locomotive was reduced to that of a saturated-steam locomotive, with an increase in fuel consumption of 25 per cent. It has also been shown that with one hundred small tubes plugged the increase in fuel consumption is nearly 50 per cent.

This shows that the cleaning of the tubes and the flues as well as the other heating surfaces should receive careful attention. To insure economical operation, the tubes and flues, especially the superheater flues, should be blown out at the end of each trip. In bad-water districts if the boiler is not washed out as frequently as it should be, scale will form a coating over the surfaces and is liable to cause serious trouble. If the scale should become very heavy, the heat will not be carried away by the water fast enough to prevent a rise in the temperature of the plate.

The increase in temperature beyond a certain limit will cause the plate to become more or less seriously distorted. Unless cared for and remedied promptly, the plate will be affected to such an extent as to make a rupture very likely.

56. Blowing Out.—In no case should a boiler be blown out with a full pressure of steam, because the sudden reduction of pressure is almost sure to cause excessive strains on parts of the boiler as well as cause scale to form on the heating surfaces. To blow out the water, the pressure should be reduced to about 20 pounds per square inch before the blow-off cock is opened. A boiler should not be refilled with cold water until it has cooled off, because the sudden contraction of the several parts is sure to result in leaky seams or tubes.

57. Use of Blower.—Care should be exercised in the use of the blower, especially in locomotives without a brick arch. The improper use of the blower draws an excessive amount of cold air into the firebox and directly against the tubes and the flues. When the fire is low, or when there are holes in it, the draft caused by the blower will cool off the firebox sheets and tubes. The contraction that results will usually start the tubes leaking to an extent that will in some cases cause an engine failure.

The improper use of the blower will also cause leaks at the staybolts. It is at the ash-pit, however, that the most damage is done by the abuse of the blower. Usually those in charge of the dumping and the cleaning of the fire think more of their own comfort than the care of the boiler, and use the blower strong enough to take all smoke, gases, and fine ashes away from the firebox. This causes a strong current of cold air to pass through the firebox and into the tubes, and the sudden contraction of the various parts invariably causes leaky tubes.

58. Honeycomb.—Care should be exercised in the handling of a locomotive boiler that small masses of clinkers, called honeycomb, are not allowed to form on the firebox sheets, especially on the heads of the crown stays and on the ends of the tubes and flues. Honeycomb is caused by the foreign matter in the coal adhering to those parts of the firebox surfaces which are exposed to the direct action of the flames and which are not properly covered with water on the other side. The clinker should be knocked off with a long rod through the fire-door, as when on the tube-sheet it obstructs the flow of hot gases through the tubes and reduces their effectiveness the same as if they were plugged.

59. Even Firebox Temperature.—When firing, it is important that the fireman maintain a firebox temperature as nearly uniform as possible. This avoids the alternate rise and fall in temperature, as the fire is first forced and then allowed to burn down. A high firebox temperature causes the tubes to expand rapidly, and when the cold currents of air strike them, they suddenly contract. The alternate expansion and contraction soon causes loose tubes and flues and, very often, engine failures.

With oil-burning locomotives, on account of the high firebox temperature when the engine is working, care should be taken to see that the dampers are immediately closed when the fire is reduced. This prevents cold air from striking the hot surfaces of the firebox. The possibilities of damage to the firebox are much greater with an oilburning locomotive than in the case of a coal burner.

60. Leaks.—All leaks in the firebox, especially around the foundation ring, should be repaired as soon as possible, because the leakage represents a considerable loss of fuel. The steam also has a tendency to unite chemically with the sulphur in the coal and attack the metal in the sheets. All steam leaks into the front end from the steam pipes and the superheater should be attended to promptly. The leakage of steam causes a cutting action that soon wears away the parts it comes in contact with, and also reduces the vacuum in the smokebox. The smokebox door and the steam-pipe connections through the smokebox should be as nearly air-tight as possible so as to prevent the air from entering the smokebox and destroying the vacuum created by the action of the exhaust.

ACCIDENT INVESTIGATION REPORT

61. Copy of Form.—A copy of the form used by the inspectors of the Bureau of Locomotive Inspection of the Interstate Commerce Commission in reporting accidents resulting from boiler failures is herewith given. The form has been partly filled in to show how it is to be compiled in reporting an accident, which in this case is assumed to be a failure of a crown-sheet due to low water.

Form No. 25. **File No.**__________

INTERSTATE COMMERCE COMMISSION

BUREAU OF LOCOMOTIVE INSPECTION

Accident Investigation Report

Attention is directed to the following extract from the law under which this investigation was made:

"Neither said report nor any report of said investigation nor any part thereof shall be admitted as evidence or used for any purpose in any suit or action for damages growing out of any matter mentioned in said report or investigation."

Investigation made at ____________________,

____________________, 19____

Name of railroad A B C.

Initials of locomotive A. B. C. Number 503

Type 2-8-2 Steam pressure 185 lbs.

Class of service ____Freight____ Speed ____35 m. p. h.____

Operated by ____A. B. C. Railroad Company.____

Nature of accident ____Crown sheet failure.____

Place ________________ Date ________

Time ________ Engineer ________________

Residence ________________

Fireman ________________

Residence ________________

Conductor ________________

Residence ________________

CASUALTIES TO PERSONS

Name and Address	Employed as	Nature and Extent of injuries	Probable disability
None injured.			

(Boiler Failure.)

Builder's number of boiler ________________

Where built ________________

Type ____Straight top____ Type of fire box ____Radial stay____

Were fire-box sheets overheated? If so, to what extent ________

Crown sheet overheated at both front and back ends.

If water was low, give distance from crown to line of low water ________

Approximately 10 inches.

Thickness of scale on sheet ______________________

____

Description of accident, including condition of various parts or appurtenances affected, and all other material details:

Freight train Extra North, consisting of 64 loads, 4,160 tons, drawn by A.B.C. locomotive 503, in charge of Engineer............., Fireman............................ and Conductor................, left...................... at 10:30 A.M., May 14, 19--, and when passing through ...., at 11:32 A.M. a failure occurred in the fire box which at the time was assumed by the crew to be a burst flue.

The locomotive was towed to............. where the railroad company's inspector made an investigation of the boiler and all appurtenances on May 15. Our investigation of this accident was made at............. on May 30--, after the appurtenances had been removed and examined by the company's inspector.

EXAMINATION OF FIREBOX:

The firebox was of three-piece construction, and was equipped with two (2) Nicholson thermic syphons. The crown sheet was supported by 26 transverse rows and 16 longitudinal rows of crown stays, the ten center rows were 1-1/8 inch buttonhead radial stays, except the first row on each side of each syphon, which were hammered-head radial stays and the three outer rows on each side, which were flexible stays with telltale holes. Below these stays, on each side, were four rows of flexible staybolts having telltale holes. Flexible staybolts with telltale holes were also applied in the breaking zones at all four corners of the firebox, all other staybolts were 1-inch rigid bolts. All radial stays and staybolts were spaced approximately 4 inches by 4 inches at the largest spacing.

(If above space is insufficient for description of accident, insert additional sheets.)

The crown sheet was found to have been overheated in spots. At the front end, the heat showed back to between the fourth and fifth crown stays from the flue sheet at center of crown sheet and from this point tapered toward the flue sheet at both sides and downwards to the first bolt from the flue sheet and the third bolt from the center, then extended downwards to the tenth bolt on each side of the center. At the back end of the firebox the heat showed between the two center rows of radial stays from the thirteenth transverse row from the door sheet back to the second row from the door sheet, where it widened out to a distance of about 20 inches on both sides of the center.

The line of demarcation at the flue sheet showed about 10 inches below the highest part of the crown sheet and tapered back as per the taper in the crown sheet, which was 4 inches.

The door sheet pulled from four staybolts in the center at the top and the door sheet flange came down about 5-inches at the lowest point.

The riveted seam at the top of the flue sheet was sprung and the welded beads on the top row of the superheater flues, the top row of fire tubes, and the tubes between the top and second rows of superheater tubes, were broken loose.

The irregular demarcation by the heat was evidently due to upward circulation through the syphons depositing water on the crown sheet after the general water level had receded below the crown sheet.

WATER GLASS AND GAUGE COCKS:

The bottom water-glass mounting and the gauge cocks were applied in boiler back head.

The gasket had been removed from the nut on the bottom of the tubular water glass and with this nut screwed down on the water-glass mounting the lowest reading of the water glass was 3½ inches above the highest part of the crown sheet.

The bottom gauge cock was 3½ inches above the highest part of the crown sheet.

SAFETY VALVES:

The safety valves were removed and placed on locomotive 504, and were left there after being tested.

INJECTORS:

The injectors were removed and placed on locomotive 505 where they were left after being tested.

FIRE DOOR:

The boiler was equipped with a mechanically-operated fire door.

INSPECTION AND REPAIRS:

Locomotive 503 was last given general repairs in August,19 at company's..........shop, at which time this firebox was applied, and since that time to April 1, 19-- the locomotive had made 99,423 miles. The last annual inspection was made at .... August 1, 19--. The last monthly inspection was made at ....enginehouse April 15,19--. This report had been removed from the cab case and was in the roundhouse foreman's office. All firebox sheets, syphons, and flues were reported in good condition.

Contributory defects found:

There were no contributory defects found during this investigation; however, all the appurtenances had been removed from the boiler and examined by the railroad inspectors before this inspection was made.

General condition of locomotive:

Good.

Cause of accident:

Crown sheet failure caused by overheating due to low water.

______________________________, *Inspector*

______________________________, *Inspector.*

Remarks:

WASHING OUT BOILERS

62. The washing out of a locomotive boiler is a most important item as successful maintenance and boiler efficiency depend largely on the care that is taken in washing out. Nearly all water contains more or less impurities, which settle to the bottom when the water is boiled. In the case of a boiler, the impurities settle on the tubes and the firebox sheets or at the bottom of the water legs, and, unless removed promptly, soon cause the formation of a hard scale. The scale reduces the efficiency of the heating surface to a considerable extent, because the heat must penetrate the scale before it reaches the water. Tests have shown that a deposit of scale 1/8 inch thick results in a loss of 15 per cent of the fuel burned. Another result of scale, especially on the firebox sheets, is the liability of the sheet to become overheated and collapse.

Washout plugs are provided to facilitate the washing out of a boiler and are placed in the most advantageous places. A certain number are placed in the back head above the crown-sheet, and also at about the same level in the wrapper sheet. Other plugs are located in the boiler shell on each side slightly above the top of the flues and in the front tube-sheet, as well as at each outside corner of the firebox and in the water legs. Washout holes closed by washout covers are also provided in the belly of the boiler, one being often placed in each course.

When washing out a boiler, all washout plugs and covers are first removed. A suitable nozzle is inserted in each of the various washout holes in the back head and in the wrapper sheet, and the crown-sheet is then washed off, warm water under a pressure of at least 50 pounds being used. Next, the flues are washed down by inserting the nozzle in the holes in the boiler shell. By this time, the sediment has been washed into the water legs and the belly of the boiler, and any that has not already escaped through the openings at the latter point can be washed out by inserting the nozzle in the holes in the front tube-sheet. The water legs can be cleaned out by forcing water into the holes provided there.

63. It will be noted that the system used in washing out a boiler is to start at the higher washout plugs and wash the sediment into the lower parts. Scrapers are used when sediment cannot be removed by washing. A large mass of sediment or scale on the crown-sheet or the water legs, which will cause the sheet to be burned and which cannot be dislodged by the ordinary methods, must be removed at all costs, even to the extent of taking out a number of crown stays or staybolts to do so.

TESTING OF BOILERS

64. Method of Test.—The Interstate Commerce Commission prescribes that every boiler before being put into service and at least every 12 months thereafter shall be subjected to hydrostatic pressure 25 per cent above the working steam pressure.

The test is made by filling the boiler with warm water and then raising the pressure of the water by means of a force pump or an injector. When the boiler is being filled with water, all air is expelled by removing a plug or by opening a valve in the highest part of the boiler; the pressure can be applied more quickly and easily when there is no air present. The safety valves are either removed and plugs inserted in their places or else they are closed and clamped. An accurate test gauge is applied to the boiler and the pressure must be watched closely to see that the prescribed testing pressure is not exceeded, as it is very easy to strain, unduly, some part of the boiler.

Cold water is not satisfactory for testing, because the boiler plates are cold and contracted to a minimum and leaks will appear. A boiler that is in good condition and tight under steam will usually show numerous leaks when full of cold water under pressure, so that the use of cold water makes the test unnecessarily severe.

After all defects, such as leaky seams and broken staybolts and crown stays have been taken care of, the boiler is fired up and the pressure raised to not less than the working pressure so as to determine the permanency of the repairs made.

Broken staybolts are more easily detected when the boiler is under pressure, as when making a hydraulic test, than when it is not. The reason is that the broken ends are now separated slightly, whereas when the boiler is not under pressure the broken ends are in contact, thus giving a sound like a good staybolt.

POOR-STEAMING LOCOMOTIVES

65. Causes of Poor Steaming.—A poor-steaming locomotive is one in which the boiler will not generate steam fast enough to meet the requirements of service. With such a locomotive it is very difficult to maintain full boiler pressure except under very favorable conditions of operation.

The modern locomotive is designed and proportioned in accordance with experience and observation, and therefore when an engine steams badly, the trouble should not be too hastily charged to faulty designing. The fault may occasionally be in the design, such as cylinders too large for the amount of heating surface, insufficient water spaces, or the diaphragm too close to the tube-sheet, but it is far more likely to be due to other causes over which the engineer has control. Aside from faulty design, there are three general causes that will make a locomotive steam poorly: (1) Improper or insufficient draft; (2) the heat of the fire not being fully utilized; and (3) poor management on the part of those in the cab.

66. Insufficient Draft.—Insufficient draft may be due either to some obstruction in the firebox, tubes, smokebox, or stack, or to insufficient vacuum being formed in the smokebox by the exhaust steam, this meaning that the difference between the atmospheric pressure and the smokebox pressure is not enough to create the proper draft, or to the fact that, while a sufficient vacuum may be formed by the exhaust steam, its effect on the fire is partly destroyed through some defect in the draft apparatus. The obstruction to the draft may perhaps be in the grates, which may be fitted so close as to prevent the admission of a sufficient quantity of air through the fire.

Some of the tubes may be stopped up, owing to careless firing or to poor regulating of the draft appliances. If the draft is weak, ashes and cinders will gradually accumulate in the tubes until finally they become stopped up.

If the draft is too strong, large cinders will be drawn into the tubes and flues and produce the same effect as when the draft is too weak. Plugged tubes and flues reduce the heating surface of the boiler and hence its steam-generating capacity. If the tubes and flues are obstructed at the firebox end by honeycomb, the effect will be the same as if the tubes were plugged.

If the firebox is reasonably clean and there are no ashes or cinders in the tubes and flues, the cause of the poor draft may be in the smokebox. The diaphragm or deflector plates may be too close to the tube-sheet, the diaphragm apron or damper may be too low, the netting may be clogged or the smokebox may be filled up with cinders. The foregoing shows that as little obstruction as possible should be imposed on the flow of the gases to the atmosphere, other than what is necessary to produce a uniformity of draft over the whole heating surface.

67. Measurement of Draft.—The draft is measured by a draft gauge, a view of which is shown in Fig. 24. This device consists of a U-shaped tube containing water, and having a sliding scale graduated in inches, the whole being suitably mounted on a part *b*. The upper end of the leg *a* of the tube is connected to and opens into the smokebox so that the water within the tube is subject to smokebox pressure, and the end of the other leg *c* is open and exposed to atmospheric pressure. The water in both legs will stand at the same level when the pressures in the legs are equal, as when the locomotive is not working. With the locomotive working, the draft reduces the pressure in the smokebox and on the water in leg *a*, and the greater pressure of the atmosphere acting on the water in leg *c* raises the water in leg *a*.

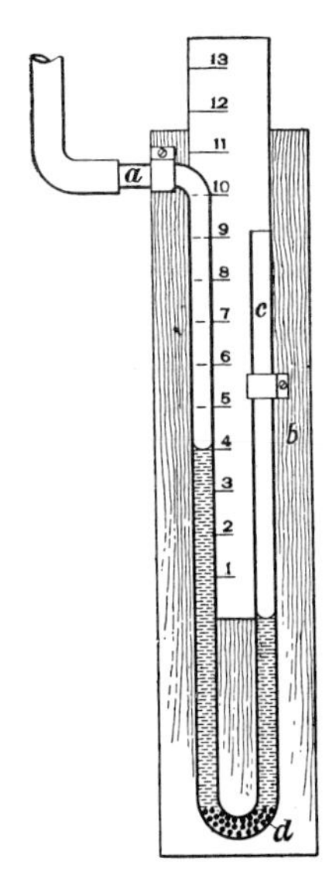

Fig. 24

The draft is measured in inches of water as follows: The sliding scale is moved so that the zero reading is in line with the lower water level, as shown, and the figure which then comes opposite to the higher water level is noted, which in Fig. 24 is 4. With this reading, the draft would be said to be equal to 4 inches of water, this meaning that the difference between the pressure of the atmosphere and the pressure in the smokebox is sufficient to maintain a column of water 4 inches high. A pressure equal to 1 inch of water in a draft gauge is equal to about ½ ounce, so that 4 inches would be equal to about 2 ounces of pressure per square inch.

In locomotive practice the draft varies between 4 and 10 inches of water. The shot *d* in the bottom of the tube prevent a too rapid movement of the water due to change in pressure.

68. Insufficient Vacuum.—The firebox and the tubes may be clean and the smokebox appliances may be properly adjusted, but still the engine may not steam. The cause may then be due to insufficient vacuum being formed in the smokebox. Insufficient vacuum may be due to any of the following causes: The nozzle may be so large that the exhaust steam will pass through the smokebox at a very low velocity; the exhaust pipe may be set out of line with the center of the stack, thus throwing the jet more on one side of the stack than the other, or part of the jet may strike the inside of the smokebox proper, or the stack extension may be so set as to have the same effect on the exhaust jet as if the exhaust pipe were out of line.

The principle causes, however, of an insufficient vacuum in the smokebox are air leaks in the smokebox or steam leaks from the steam pipes and superheater units into the smokebox. If the door of the smokebox is improperly set and air can enter through the joints, the action of the exhaust jet will draw air through the openings and less will come through the fire.

69. Poor Circulation.—Another cause of a poor-steaming engine is a poor circulation of water, or a dirty boiler with the heating surfaces covered with mud and scale. To promote circulation, the feedwater should enter the boiler near the front tube-sheet so that it can work along to the tubes and the flues to the hot surfaces and rise with the steam.

70. Poor Management.—Another cause of a poor-steaming engine is poor management and lack of cooperation on the part of those in the cab. To obtain the best results, the firing and feeding should be carried on according to approved methods. The use of the injector has a very important effect on the steaming of a locomotive. The best fireman cannot maintain a good head of steam if the engineer insists on flooding the boiler. Neither can the required amount of steam be produced if the engineer works the engine at a long cut-off when a shorter one obtains the same if not better results.

CALCULATING STRENGTH OF BOILER JOINTS

DEFINITIONS RELATING TO BOILER PLATE

71. Tensile Strength of Boiler Plate.—The tensile strength of boiler plate is determined by actual test in a testing machine, and the unit by which it is measured is the force, in pounds, necessary to pull apart a piece of plate of the cross-sectional area of 1 square inch, as indicated in Fig. 25. The square inch may, however, be in any form as long as it contains a square inch of metal; that is, it may be 1 inch each way, or it may be wider and thinner, as 2 inches wide and ½ inch thick. Increasing the cross-sectional area of a piece of plate to 2 square inches doubles the amount of the metal, and naturally doubles the tensile strength, so that tensile strength depends on crosssectional area. The dimensions of a standard test piece are shown in Fig. 26.

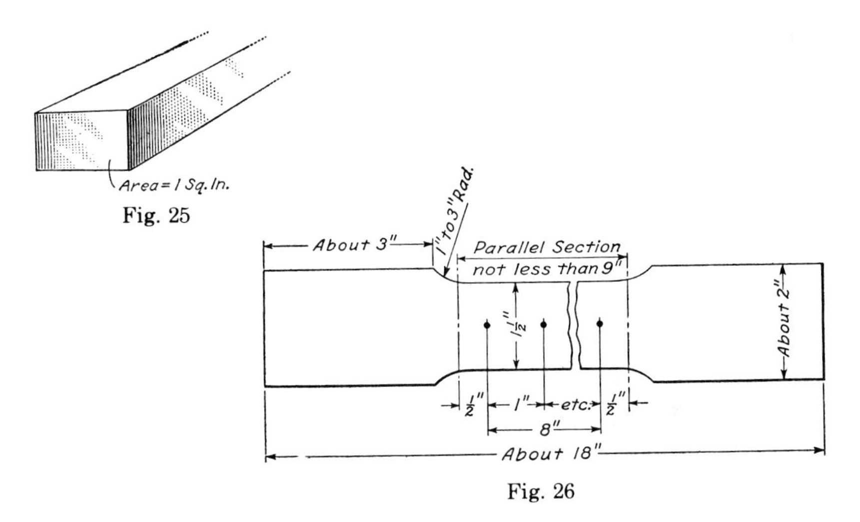

Fig. 25

Fig. 26

The recommended strength for boiler plate used in the boiler shell is a tensile strength of 60,000 pounds per square inch of cross-sectional area but a variation of 5,000 pounds either way is permissible. When the tensile strength is unknown, a value of 50,000 pounds is to be used. The recommended tensile strength for firebox steel is 57,000 pounds, and the same variation is permitted as with boiler plate.

As yet the A.A.R. has no specifications for other than carbon-steel boiler plate. Nickel-steel boiler plate as well as other alloy-steel boiler plate have a much higher tensile strength. For example, a nickel-steel boiler shell has a minimum tensile strength of 75,000 pounds per square inch and the firebox plate a tensile strength of 65,000 pounds per square inch. According to the manufacturer's tests the shearing strength of nickel-steel rivets, hand driven, in most cases exceeds 65,000 pounds per square inch. This is permissible according to Rule 6 of the Federal regulations.

Nickel-steel boiler shells permit of higher pressures with the same thickness of plate as compared with carbon-steel shells.

72. Elastic Limit.—The elastic limit is the point at which the test piece of boiler plate will not return to its original length when relieved of tension during the test. All metals will stretch when subjected to a pull, and if the pull is not too great the metal will return it its original length when the pull is removed. However, any pull in excess of a certain amount will stretch the metal beyond the limit of recovery and it will then not return to its original length. The elastic limit, or yield point, as it is sometimes called, must not be less than half the tensile strength. A pull in excess of the elastic limit affects the structure of the metal and weakens it so that the pressure in the boiler should never be permitted to rise to such a point that the pull on the plate exceeds the elastic limit. It is for this reason that the testing pressure of a boiler is restricted to about 25 percent above the working pressure.

73. Elongation.—Elongation is the amount the test piece stretches before it breaks, and is determined by placing the ends together and measuring the distance between two prick-punch marks that were originally 8 inches apart.

The proper amount of elongation insures the flanging of the plates without materially affecting the strength of the metal.

The foregoing definitions do not apply to the special alloy steels that contain nickel, etc., used in some modern boilers.

DEFINITIONS RELATING TO JOINTS

74. Types of Joints.—There are two types of joints used in locomotive boilers, namely, the lap joint and the butt joint. A lap joint is one in which one edge of the plate laps over the other, as in Fig. 27. A butt

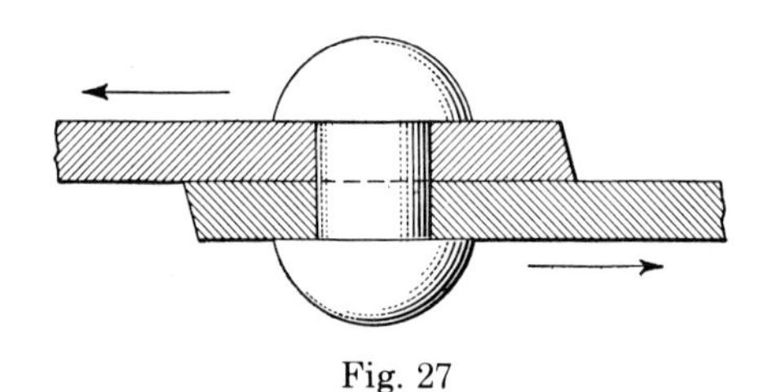

Fig. 27

joint is one in which the edges butt against each other and are held in contact by welts or cover-plates, as in Fig. 28.

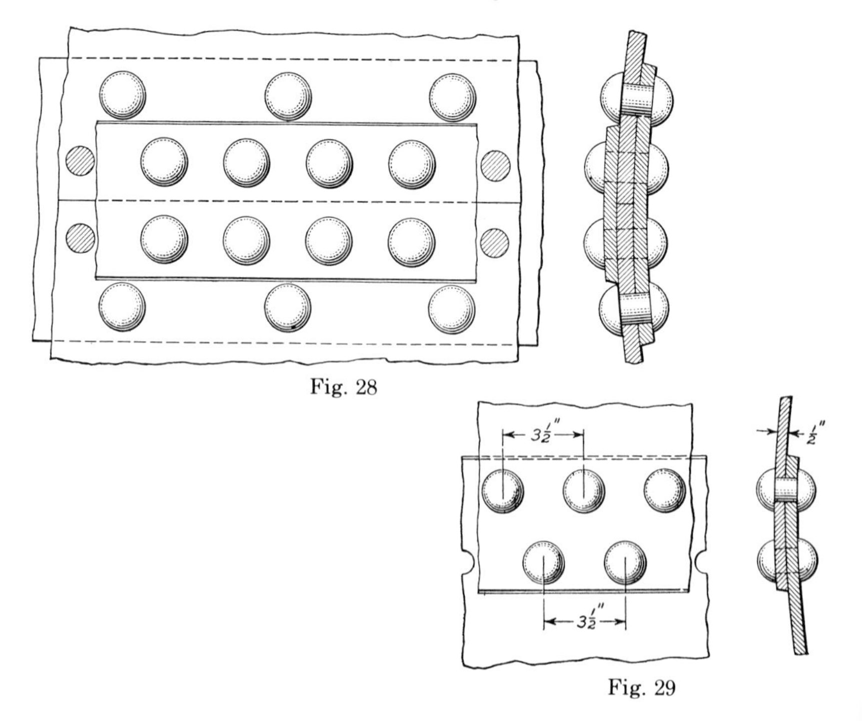

Fig. 28

Fig. 29

75. Single- and Double-Riveted Lap Joints.—A single-riveted lap joint is one in which the edges of the plate are held together by a single row of rivets, whereas a double-riveted lap joint employs two rows of rivets, as in Fig. 29. The various courses of the cylindrical part of a boiler are usually joined together by double-riveted lap joints.

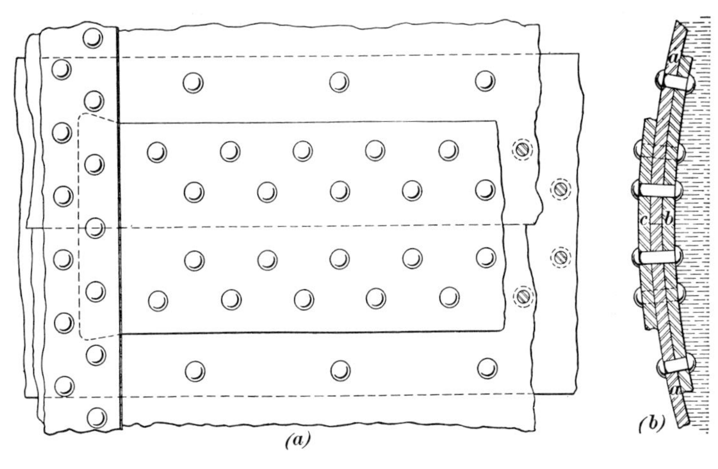

Fig. 30

76. Triple- and Quadruple-Riveted Joints.—A triple-riveted butt joint, Fig. 30, is identified by having three rows of rivets on each side of the seam, and a quadruple-riveted butt joint is one with four rows if rivets on each side of the steam. The latter is generally used in the longitudinal seams of locomotive boilers.

77. Pitch of Rivets.—Pitch of rivets is defined as the distance between the centers of adjacent rivet holes, as in Fig. 31.

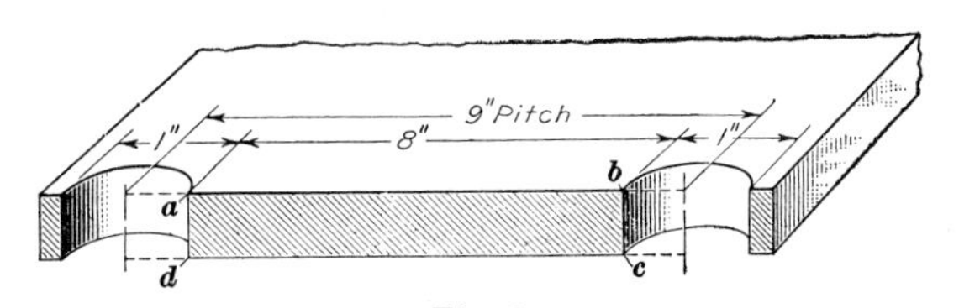

Fig. 31

78. Back Pitch.—Back pitch is the distance between two rows of rivets, one on each side of the joint.

79. Rivet in Single Shear.—A rivet is in single shear when it is possible to shear it in only one place, as in Fig. 32. A pull on the platesas shown by the arrows will cause the rivets to shear, as shown, only along the line *ab*.

80. Rivet in Double Shear.—A rivet is in double shear when it will shear in two places, as at *c* and *d*, Fig. 33. A rivet in single shear involves two plates; one in double shear involves three plates.

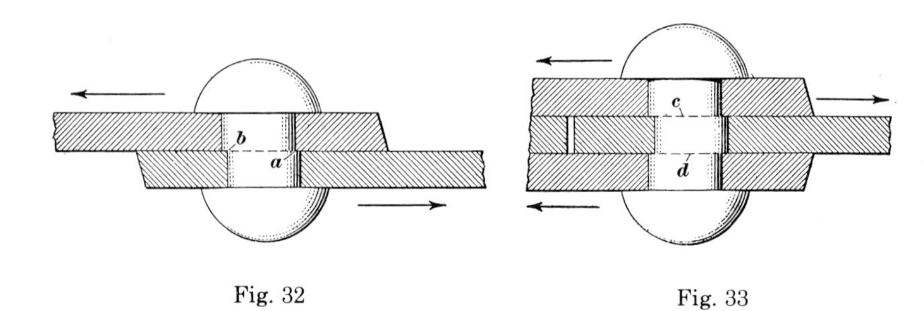

Fig. 32 Fig. 33

81. Relative Strength of Rivets in Single and Double Shear.—A rivet in double shear has twice the strength of a similar one in single shear, because it has to be severed at two points simultaneously. In Canada, however, a rivet in double shear is considered as being only 1¾ times as strong as one in single shear.

82. Rule for Calculating Tensile Strength of Plate.—The rule used to calculate the tensile strength of a piece of boiler plate is as follows:

Rule.—*To find the tensile strength of plate, multiply the cross-sectional area of the plate, in square inches, found by multiplying its width by its thickness, by the unit of tensile strength, here assumed to be 60,000 pounds for each square inch of cross-sectional area.*

EXAMPLE.—What is the tensile strength of the piece of boiler plate shown in Fig. 34, which has a width of 8 inches and a thickness of ⅝ inch, the stress being applied in the direction of the arrows?

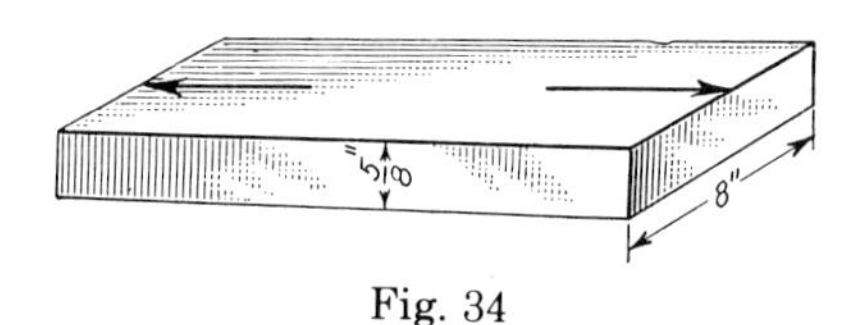

Fig. 34

SOLUTION.—The cross-sectional area of the plate is 8×⅝=5 sq. in. As each square inch has a tensile strength of 60,000 lb., 5 sq. in. will have a tensile strength of 60,000×5=300,000 lb.

83. Calculating Tensile Strength of Plate Between Rivet Holes.—The drilling or punching of rivet holes in a plate of course weakens it most on a line through the centers of the rows of rivets; the strength of the remainder of the plate remains unimpaired. The only difficulty met with when calculating the tensile strength of a plate in which holes are drilled, is to find the area of the section of plate that remains between any two adjacent holes.

The method of finding this area can be more readily understood from Fig. 31, which shows the plates broken off on a line through the centers of the rivet holes and the broken piece removed. Drilling the rivet holes reduces the length of solid plate between the centers of any two adjacent holes by one-half the diameter of two rivet holes, or in this case by 1 inch. The length of any section of plate between two adjacent holes then becomes 9 inches less 1 inch, or 8 inches, and as the plate is ½ inch thick, the cross-sectional area *abcd* between the holes is 8 x ½ = 4 square inches. The unit of tensile strength assumed is 60,000 pounds for each square inch of cross-sectional area, so that the plate has a strength, on a line through the rivet holes, of 4 x 60,000 or 240,000 pounds. It is necessary to calculate only the strength of the plate between two adjacent holes, because as the pitch of all holes in any one row is the same the plate will be weakened the same amount between every two holes. Theoretically, the plate will fail simultaneously on a line through the centers of all holes. The following rule can be deduced from the foregoing:

Rule.—*To find the tensile strength of a plate between rivet holes, subtract one-half the diameter of two rivet holes, or the diameter of one rivet hole, from the pitch and multiply the result by the thickness of the plate and the unit of tensile strength.*

84. Rule for Calculating Efficiency of Plate.—The efficiency of the plate is its tensile strength after drilling the rivet holes, as compared with its strength before, and is found by dividing the former value by the latter. This can be summarized in a rule as follows:

Rule I.—*To calculate, in per cent, the efficiency of a plate in which rivet holes have been drilled, divide the tensile strength of the plate after the holes were drilled, by its tensile strength before the holes were drilled and multiply the result by 100.*

EXAMPLE.—What is the efficiency of the boiler plate shown in Fig. 31?

SOLUTION.—The tensile strength after the holes were drilled has already been shown to be 240,000 pounds. The ultimate tensile strength, according to the rule in Art. 82, is 9 inches multiplied by ½ inch, multiplied by 60,000 pounds. The result is 270,000 pounds. Hence, the efficiency of the plate is $\frac{240{,}000}{270{,}000} \times 100$ or 88.8 per cent. The plate has then been weakened 11.2 per cent.

The efficiency of the plate can also be calculated from the following rule, which involves less work than the one just given:

Rule II.—*To calculate the efficiency of the plate subtract the diameter of one rivet hole from the pitch, then divide by the pitch and multiply by 100.*

For example, the efficiency of the plate shown in Fig. 31 equals $\frac{9-1}{9} \times 100 = 88.8$ per cent.

This rule is derived from the rule just given by canceling out similar quantities.

Thus, efficiency of plate

$$= \frac{(\text{pitch} - \text{diameter of one rivet hole}) \times \text{thickness of plate} \times \text{unit of tensile strength}}{\text{pitch} \times \text{thickness of plate} \times \text{unit of tensile strength}} \times 100$$

It will be noted that thickness of plate and unit of tensile strength cancel out, as each appears above and below the line, leaving the rule as stated.

85. Calculating Shearing Strength of Rivets.—The unit of shearing strength of rivets will be taken as 44,000 pounds for one square inch of cross-sectional area; the driven diameter of the rivet is always taken when calculating the cross-sectional area. The shearing strength of a rivet, like the tensile strength of a piece of plate, depends on the amount of metal in its cross-sectional area, so that the following rule governs:

Rule.—*To calculate the shearing strength of a rivet, multiply its cross-sectional area, by the unit of shearing strength.*

EXAMPLE.—What is the shearing strength of a rivet of a driven diameter of 1 inch?

SOLUTION.—The cross-sectional area of the rivet is found my multiplying the diameter by itself and by .7854. Hence, the area is 1 x 1 x .7854 = .7854 sq. in. Then the shearing strength of the rivet is .7854 x 44,000 = 34,557 lb.

86. Assumption on Which Rules Are Based.—It will be evident from the foregoing that the rules relating to the strength of plates and rivets are based on the assumption that the strength of a plate or a rivet is proportional to the amount of metal in its cross-section. Doubling the cross-sectional area doubles the strength; halving it reduces the strength a corresponding amount.

ANALYSIS OF STRENGTH OF LAP JOINTS

87. Factors on Which Strength of Joint Depends.—The strength of any joint depends on the pitch of the rivets, on the shearing strength of the rivets, and on the tensile strength of the plate. The larger these factors the stronger the joint. However, there is a limit to the pitch, as too great a pitch will impair the tightness of the joint. The joint can never be as strong as the plate itself but, in an ideal joint, the pitch and the diameter of the rivets should be so arranged as to make the strength of the plate between rivet holes of approximately the same strength as the shearing stress of the rivets.

88. Calculating Strength of Plate, Single-Riveted Lap Joint.—The maximum strain that the plate will withstand at the weakest point of the joint shown in Fig. 35, which is through the center of

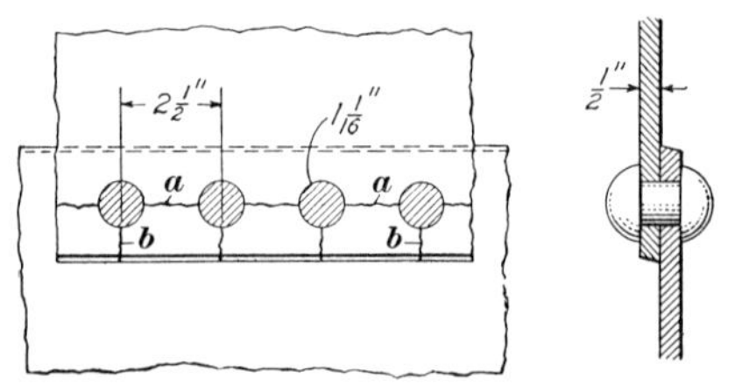

Fig. 35

the rivet holes, can be calculated from the rule in Art. 83. The pitch is 2½ inches, the diameter of rivets is 1$\frac{1}{16}$ inches, and the thickness of the plate is ½ inch, so that

(pitch — diameter of one rivet hole) x thickness of plate x unit of tensile strength = strength of plate, or

$$(2\frac{1}{2} - 1\frac{1}{16}) \times \frac{1}{2} \times 60{,}000 = 43{,}125 \text{ pounds.}$$

89. Calculating Strength of Rivets, Single-Riveted Lap Joint.—The shearing strength of the rivets can be calculated from the rule in Art. 85, it being noted that one-half the diameter of two rivets, or the diameter of one rivet, is to be considered when calculating the cross-sectional area. From this rule,
Cross-sectional area of one rivet x unit of shearing stress = shearing strength of rivet, or

$$1\frac{1}{16} \times 1\frac{1}{16} \times .7854 \times 44{,}000 = 39{,}010 \text{ pounds}$$

As the shearing resistance of the rivets is less than the tensile strength of the plate, the rivets will all shear off first, that is, the rivets constitute the weak part of the joint.

If the rivets are too large for the pitch, the plate will fail along the line *a*, Fig. 35; if the holes are drilled too close to the edge of the plate, the plate will fail in front of the rivets along the lines *b*. But if the distance from the center of a rivet to the edge of the plate is made 1½ times the diameter of the rivet, the latter failure is extremely unlikely to occur.

90. Efficiency of Joint.—The efficiency of the joint is the comparison between the weakest part of the joint and the strength of the solid plate expressed in per cent. The strength of the solid section of plate from the rule in Art. 82 is equal to Cross-sectional area of plate x unit of tensile strength, or

$$2\frac{1}{2} \times \frac{1}{2} \times 60{,}000 = 75{,}000 \text{ pounds}$$

The strength of the weakest part of the joint was shown to be 39,010 pounds. Hence, $\frac{39{,}010}{75{,}000} \times 100 = 52.6$ per cent., or a little over one-half the strength of the plate.

Therefore, the following rule can be used to calculate the efficiency of a joint:

Rule.—*Divide the strength of the weakest part of joint by the ultimate strength of the plate and multiply the result by 100.*

91. Calculating Strength of Plate, Double-Riveted Lap Joint.—The strength of the plate at each row of rivets of the double-riveted lap joint, Fig. 29, is the same because the pitch and the size of the rivets in each row are identical. Hence, it is necessary only to consider the strength at one row. The first row will be taken. From the rule in Art. 83, the strength of the plate is

$$(3\frac{1}{2} - 1\frac{1}{16}) \times \frac{1}{2} \times 60{,}000 = 73{,}125 \text{ pounds}$$

92. Calculating Strength of Rivets, Double-Riveted Lap Joint.—The shearing resistance offered by the rivets can be calculated from the rule in Art. 85, it being remembered that there are two rivets to be sheared off, that is, half of two rivets in one row and a whole rivet in another row. Hence, the shearing strength of the rivets is

$$1\frac{1}{16} \times 1\frac{1}{16} \times .7854 \times 2 \times 44{,}000 = 78{,}020 \text{ pounds.}$$

93. Efficiency of Joint.—The strength of the solid section of plate, from the rule in Art. 82, equals

$$3\frac{1}{2} \times \frac{1}{2} \times 60{,}000 = 105{,}000 \text{ pounds}$$

The strength of the weakest part of the joint was found to be 73,125 pounds. Hence the efficiency of the joint, according to the rule in Art. 90, is equal to $\frac{73{,}125}{105{,}000} \times 100 = 69.6$ per cent of of the strength of the solid plate.

94. Reason for Greater Efficiency of Double-Riveted Lap Joint.—The reason for the greater efficiency of the double-riveted lap joint is the employment of two rows of rivets; this permits of a greater distance between rivet holes, or greater pitch. The greater the pitch the less the strength of the plate is affected along the line of the rivet holes.

The reason for staggering the rivets is to give a tighter joint, but the joint is no stronger than if the rivets of one row were directly in line with those of the other row.

ANALYSIS OF BUTT JOINTS

95. Points of Failure.—On account of its construction the quadruple-riveted butt joint shown in Fig. 36 has many more possible points of failure than a double-riveted lap joint. The joint may fail in the following ways:

(1) The plate may fail between the rivets in the outer row, as at *a*, view (*b*). This failure affects no other part of the joint.

(2) The plate may fail between the rivets in the second row, or at *b*. The plate before it can fail here must also shear off the outer row of rivets.

(3) The plate may fail between the rivets in the third row, or at *c*. Before the plate can fail here, it must also shear off the rivets in the two outer rows.

(4) The two butt straps or coverplates may fail between the rivet holes in the inner row, but this failure will involve no other part of the joint.

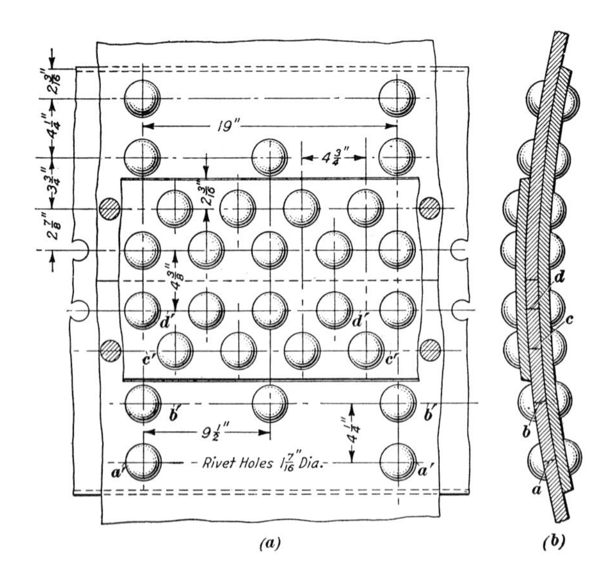

Fig. 36

96. Length of Joint to be Considered.—A longitudinal joint in a locomotive boiler is really about 6 feet in length, but when considering the strength of the seam it is unnecessary to take account of the whole length. To shorten the calculations it is only necessary to take a length of joint equal to the greatest pitch, which is always found at the outer row of rivets in joints of this kind. In this case the greatest pitch is 19 inches, the tensile strength of the plate is taken as 55,000 pounds per square inch of cross-section, and the thickness of the boiler shell is 1 inch.

97. Strength of Joint at Outer Row of Rivets.—From the rule in Art. 83, the strength of the plate at the outer row of rivets, Fig. 36, is

$$(19 - 1\tfrac{7}{16}) \times 1 \times 55{,}000 = \mathbf{965{,}937} \text{ pounds}$$

98. Strength of Joint at Second Row of Rivets.—The resistance of the plate to failure at the second row of rivets, Fig. 36, is also found from the rule in Art. 83, it being noted that the length of the cross-section of plate here is shortened by the diameter of two rivet holes; that is, the two half-diameters of two rivets and the diameter of one. Hence, the strength of the plate at the second row of rivets is

$$19 - 2(1\tfrac{7}{16}) \times 1 \times 55{,}000 = 886{,}875 \text{ pounds}$$

To this must be added the shearing stress of two halves of one rivet, or a whole rivet, in the outer row, which according to the rule in Art. 85 is

$$1\tfrac{7}{16} \times 1\tfrac{7}{16} \times .7854 \times 44{,}000 = 71{,}412 \text{ pounds}$$

The total strength of the joint at the second row of rivets, then, is 86,875 + 71,412 or 958,287 pounds.

99. Strength of Joint at Third Row of Rivets.—The resistance offered by the plate at the third row of rivets, where the plate has four holes, according to the rule in Art. 83, is

$$[\ 19 - 4\ (1\tfrac{7}{16})\ ] \times 1 \times 55{,}000 = 728{,}750 \text{ pounds}$$

To this must be added the resistance of three rivets in single shear. It has already been shown that the shearing stress of one rivet is equal to 71,412 pounds, hence 71,412 x 3 = 214,236 pounds is the shearing stress of three rivets.

Therefore, the total strength of the joint at the third row of rivets is equal to 728,750 + 214,236 = **942,986** pounds.

100. Strength of Coverplates at Inner Row of Rivets.—It can be assumed that the two coverplates each ¾ inch thick are equivalent to one plate 1½ inches thick. The diameters of four rivet holes are to be deducted from the pitch of 19 inches. Therefore, according to the rule in Art. 83, the strength of the coverplates is

$$[\ 19 - 4\ (1\tfrac{7}{16})\ ] \times 1\tfrac{1}{2} \times 55{,}000 = \mathbf{1{,}093{,}125} \text{ pounds}$$

This value, as shown further on, is slightly more than that of the boiler plate.

101. Purpose of Calculations.—The sole purpose of the foregoing calculations is to ascertain where the lowest value comes, as at this point the joint will be the weakest. It is unnecessary to proceed further once it becomes evident that a higher value will be obtained. Consider, for example, a failure of the plate at the inner row of rivets. The rivet holes number the same as in the third row, hence the plate is of the same strength at both points. But before the plate can fail at the inner row, four additional rivets in double shear in the third row must also fail. Hence, there is no need of proceeding further with the calculations.

102. Weakest Part of Joint.—In checking over the values obtained, which are shown in heavy type, it will be noticed that the lowest resistance to rupture is found in the plate between the rivet holes in the third row; the low resistance at this point will also result in the shearing off of the two outer rows of rivets. Hence, the plate throughout its entire length will rupture at that point. However, with locomotive boilers it is apparently assumed that the net section of the plate at the outer row of rivets is the weakest part of the joint. For this to hold true the joint must be designed with this fact in mind, otherwise the weakest part of the joint can be elsewhere.

103. Efficiency of Joint.—The tensile strength of the section of full plate, according to the rule in Art. 82, is

$$19 \times 1 \times 55{,}000 = 1{,}045{,}000 \text{ pounds}$$

Therefore, the efficiency of the joint, according to the rule in Art. 90, is the weakest value divided by the ultimate tensile strength or

$$\frac{942{,}986}{1{,}045{,}000} \times 100 = 90.2 \text{ per cent.}$$

104. Efficiency of Joint at Outer Row of Rivets.—The efficiency of the joint becomes the efficiency of the plate if the weakest part of the joint is assumed to be the net section of plate at the outer row of rivets. Hence, the rule for calculating the efficiency of the plate can in this case be used to calculate the efficiency of the joint. From Rule II, Art. 84, the efficiency equals

$$\frac{19 - 1\frac{7}{16}}{19} \times 100 = 92.4 \text{ per cent.}$$

The foregoing shows that when the efficiency of the joint is also the efficiency of the plate, the efficiency of the joint can be found from Rule II, Art. 84, which is used to calculate the efficiency of the plate. But when the efficiency of the plate is less than the efficiency of the joint, as when the joint is reinforced by rivets in other rows, the efficiency of the joint must be calculated from the rule in Art. 90.

FORCE TENDING TO RUPTURE BOILER SHELL

105. Area To Be Considered.—According to the law of the resolution of forces, the force that tends to rupture a circular boiler shell is the same as the force that tends to rupture a boiler made up of a series of vertical and horizontal surfaces that can be enclosed by a circle of the same diameter as the boiler shell. This is illustrated in Fig. 37. Now the force of the steam that tends to separate one-half of the boiler from the

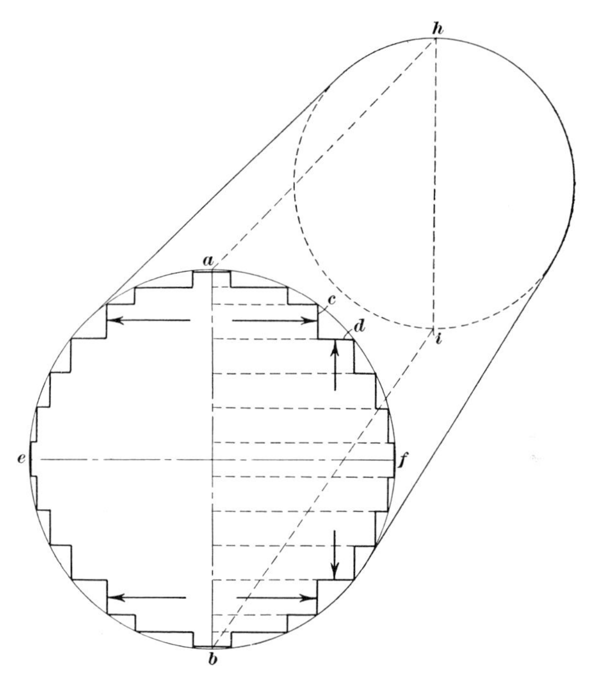

Fig. 37

other half at any two points, such as *a* and *b,* must be exerted at right angles to these points or in the direction of the arrows, and so acts against the vertical surfaces *c* of the boiler shell. The pressure that acts against the horizontal surfaces *d* tends to separate the boiler at *e* and *f* but not at *a* and *b,* and so does not have to be taken into acount when considering the strain at *a* and *b.*

If the vertical surfaces *c* are now assumed to be laid one after the other against an imaginary plate *ahib* in the center of the boiler, it will be evident that the area of these surfaces will be just equal to the complete area of the imaginary plate. It then follows that the area of the boiler against which the pressure is acting to rupture it along the lines *ah* and *bi* is equal to the area of this imaginary plate, this area being found by multiplying the diameter of the boiler by its length.

It then follows that the area to be taken when the strain along one line, such as *ah,* is being considered, is the upper half of the imaginary plate, which is equal to half the diameter of the boiler multiplied by its length. An area equal to one-half the area of the imaginary plate may then be assumed to be concerned with the strain along the line *ah,* and the other half with the strain along the line *bi.*

Although the imaginary plate is shown vertical in the boiler, yet the same reasoning applies if the plate is in any other position. In other words, the boiler may be considered as being made up of an infinite number of halves, each one endeavoring to separate itself from the other.

The steps shown in the boiler shell are purposely made large, but they may be considered as being made infinitely small, in which event they will conform very closely in shape to the circular shell of the boiler.

The following rule can be deduced from the foregoing:

Rule.—*The area to be taken when considering the strain at any lengthwise line along the boiler shell, is equal to the radius of the boiler multiplied by its length.*

EXAMPLE. What area is to be taken when considering the strain on a line along a boiler that has a radius of 36 inches and a length of 120 inches?

SOLUTION.—From the rule, the area to be taken is 36 x 120 = 4,320 square inches. Ans.

106. Calculating Strain Tending to Rupture Boiler.—When the area against which the pressure is acting in tending to rupture the boiler shell is known, it is a simple matter to calculate the strain. To do so it is merely necessary to multiply this area by the boiler pressure. As already explained, the area to be taken when the strain on the boiler shell along any one lengthwise line is being considered, is a plate with a width equal to the radius of the boiler and a length equal to that of the boiler shell. It is then only necessary to multiply the length of the assumed plate by its width and by the boiler pressure to obtain the strain.

The pressure against the vertical surfaces in the left half of the boiler, Fig. 37, does not add anything to the force that tends to rupture the boiler at *a* and *b*. This part of the boiler may be considered as acting merely to resist the effort the right half is making to break away. To make the point clearer, suppose that two men pull in opposite directions on a chain, each exerting a force of 100 pounds. Then the force tending to break the chain is not 200 pounds, but 100 pounds. The effect is the same as if one end of the chain was fastened to a post and one man pulled with a force of 100 pounds.

The following rule can be formulated from the foregoing:

Rule.—*To find the strain that tends to rupture a boiler along any one lengthwise line, multiply the radius of the boiler by its length and by the boiler pressure.*

EXAMPLE.—What is the strain along a lengthwise line of a boiler with a diameter of 72 inches and a length of 120 inches, the steam pressure being 200 pounds per square inch?

SOLUTION.—The radius equals one-half the diameter, or ½ x 72 = 36 in. So from the rule just given, the strain is 36 x 120 x 200 = 864,000 pounds. Ans.

107. Calculating Bursting Pressure.—The bursting pressure of a boiler depends on the strength of the plate and the boiler pressure and is based on the assumption that it will fail at one point only or at the point of greatest weakness. It will therefore be evident that a boiler will burst when the internal pressure becomes equal to the tensile strength of the plate at the weakest point.

The first step in the calculation of the bursting pressure is to find the tensile strength of the plate at its weakest point. Let it be assumed that the boiler has a diameter of 72 inches, and a length of 120 inches, with a thickness of plate of ⅝ inch. The tensile strength of the plate is taken as being equal to 60,000 pounds per square inch of cross-sectional area, which is assumed to be reduced by a joint efficiency of 80 per cent. Expressed as a decimal, 80 per cent becomes .80.

From the rule in Art. 82, the tensile strength of the plate is

$$120 \times 5/8 \times 60{,}000 \times .80 = 3{,}600{,}000 \text{ pounds}$$

The second step is to find the area of the boiler to be taken when the strain along any line is being considered. This, according to the rule in Art. 90, is the area of an imaginary plate, equal to the product of the radius, 36 inches, and the length, 120 inches, or 4,320 square inches.

The pressure by which this area must be multiplied to obtain a pressure at one point of the boiler equal to the tensile strength of the plate, or 3,600,000 pounds, is found by dividing 3,600,000 by 4,320 square inches, or the area of the imaginary plate. The result, or 833 pounds, is then the bursting pressure. In other words, one-half the area of the imaginary plate multiplied by this pressure is equal to the cross-sectional area of the plate multiplied by the tensile strength.

The following rule can be deduced from the foregoing:

Rule.—*To calculate the bursting pressure of a boiler multiply the minimum thickness of the plate by its tensile strength and by the efficiency of the joint expressed as a decimal, and divide the result by the radius of the shell.*

EXAMPLE.—What is the bursting pressure of a boiler with a plate thickness of 3/4 inch (.75), an inside diameter of 90 inches, an efficiency of longitudinal seam of 80 per cent, and a tensile strength of metal of 60,000 pounds per square inch of cross-sectional area?

SOLUTION.—From the rule in Art. 82, the bursting pressure is

$$\frac{.75 \times 60{,}000 \times .80}{45} = 800 \text{ pounds per square inch.} \qquad \text{Ans.}$$

108. Relative Strengths of Girth and Longitudinal Joints.—A girth joint in a boiler is under only one-half the strain of a longitudinal joint of equal length, so that a double-riveted lap joint is usually all that is required for a girth joint.

The foregoing can be easily proved by the rules already given. A boiler diameter of 72 inches will be assumed with a length of joint of 1 inch and a boiler pressure of 100 pounds. With the dimensions kept small as with a joint 1 inch in length, the calculations will be shorter.

From the rule in Art. 106 the strain per inch of the longitudinal joint is

$$36 \times 1 \times 100 = 3{,}600 \text{ pounds}$$

The pull on the girth joint is equal to the product of the boiler pressure and the area of an imaginary tube-sheet in the course, so that it equals

$$72 \times 72 \times .7854 \times 100 = 407{,}151.36 \text{ pounds}$$

This pull is distributed over a length equal to the circumference of the boiler, which is

$$72 \times 3.1416 = 226.1952 \text{ inches}$$

Therefore, the pull on 1 inch of the joint is

$$407{,}151.3600 \div 226.1952 = 1{,}800 \text{ pounds,}$$

or just one-half of the strain per inch of longitudinal joint.

A much briefer calculation can be made by assuming a pressure of 1 pound and a length of joint of 1 inch and also using a letter, as *D*, for the diameter. The strain on the longitudinal joint then becomes

$$\frac{D}{2}, \text{ or the radius of the shell}$$

the pull on the entire length of the girth joint becomes

$$D \times D \times .7854$$

and the pull on 1 inch of the girth joint equals

$$\frac{D \times D \times .7854}{D \times 3.1416} = \frac{D}{4}$$

Therefore, the last result is just one-half of the first.

109. Factor of Safety.—The factor of safety is the number obtained by dividing the bursting pressure of a boiler by the working pressure. Thus, if the bursting pressure is 800 pounds to the square inch and the working pressure is 200 pounds, the factor of safety is $\frac{800}{200} = 4$; that is, the working pressure is only one-quarter of the bursting pressure. The Federal regulations prescribe a minimum factor of safety of 4.

110. Calculation of Working Pressure.—The following rule for the calculation of the working pressure can be deduced from Art. 109:

Rule.—*To find the maximum allowable working pressure, divide the bursting pressure by the factor of safety.*

CALCULATION OF STRESSES ON FIREBOX

111. Nature of Stresses.—The firebox is subjected to two kinds of stresses, one due to the steam pressure and the other to the unequal expansion and contraction of the firebox sheets and the wrapper sheet. The steam pressure acts to bulge the firebox sheets inwards, the crown-sheet downwards, and the wrapper sheet outwards, but as this sheet is somewhat thicker than the firebox sheets the tendency for it to bulge is less.

With the steam pressure acting in one direction to force the wrapper sheet outwards, and the opposite direction to force the firebox sheets inwards, the stress that tends to separate the sheets, aside from the slight tendency to bulge between the stays, comes not on the sheets but wholly on the stays and imposes a heavy tensile or stretching stress on them. Were it not for the stays, the firebox sheets and the wrapper sheet would have to be made very thick, but by using stays the sheets can be made comparatively thin. The circular part, or barrel, of the boiler is, on the contrary, self-supporting and staybolts are unnecessary, so that the pressure of the steam in this case induces a heavy tensile stress entirely on the sheets; hence the necessity for multiple-riveted butt joints in the longitudinal seams of the courses. However, as the firebox sheets and the wrapper sheet are almost entirely supported by the staybolts, single- or double-riveted lap joints or welded joints afford ample strength.

112. The only tensile stress induced in the side sheets of the firebox is that due to the downward pressure of the steam on the mud-ring as well as on the narrow margin of the firebox at the corners not supported by staybolts. This is so small that it may be disregarded; hence, as far as the steam pressure is concerned, the tensile strength of the firebox sheets does not enter greatly into the problem of firebox stresses. However, the stresses induced by unequal expansion and contraction of the sheets often exceed those due to the pressure of the steam, so that firebox steel must possess a high tensile strength so as to be thin enough to transfer heat rapidly, and thick enough to prevent bulging as well as to take in enough threads of the staybolts for support. The thickness of the firebox sheets seldom exceeds ⅜ inch and the thread found to be best adapted to boiler work is twelve to the inch, so that the greatest number of threads that can enter the sheet is four and one-half.

The shape of a locomotive firebox, as far as the water legs are concerned, is due to the necessity for a mud-ring in order that the rear end of the boiler may be secured to the frame. Such a construction also, of course, gives a maximum heating surface.

113. Spacing of Stays.—The stays are spaced close enough to prevent the sheet from bulging and their cross-sectional area is sufficient to prevent them from pulling apart. The staybolts are usually spaced 4 inches apart in each direction and it is assumed that each staybolt supports the sheet halfway to the next staybolt, as shown at *a*, in Fig. 38. With such a spacing, the area supported by one staybolt is a square, 4 inches by 4 inches, or an area of 16 square inches; or the quarters of four staybolts may be considered as supporting the same area, as shown at *b*. The stress on a staybolt with a boiler pressure of 250 pounds to the square inch would then be 16 x 250, or a pressure of 4,000 pounds, although it is permissible to deduct the area of the staybolt at its smallest point from this area.

114. Tensile Strength of Stays.—The maximum allowable tensile strength that can be used in staybolt calculations is 7,500 pounds per square inch of cross-sectional area, although the actual tensile strength

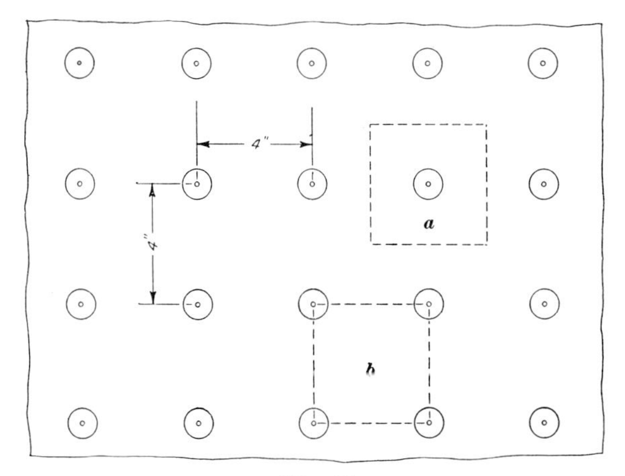

Fig. 38

of staybolt material must fall between 47,000 and 52,000 pounds per square inch of cross-sectional area. Using only 7,500 pounds when the actual tensile strength is as stated insures an ample factor of safety or more than six. Hence the firebox has a higher factor of safety than the shell of the boiler, which may have as low a factor of safety as 4. The cross-sectional area of the smallest staybolt that could be used with a pressure of 250 pounds and a spacing of 4 inches would be $\frac{4{,}000}{7{,}500}$ square inch, or one with an approximate diameter of nearly ⅞ inch. This diameter must be the smallest section of the staybolt, because the body of many staybolts is reduced at the middle to a smaller diameter than at the bottom of the thread. With a straight-body staybolt and a continuous thread the diameter is measured at the root of the thread. Drilling the telltale hole in the staybolt also reduces its strength, hence the area of the hole must be deducted when calculating the tensile strength of the bolt.

The holding power of the thread on the staybolt should be equal to its tensile strength, otherwise it would strip before the bolt would fail. As a means of aiding the thread as well as making the stay steam-tight, the ends are extended through the sheets and riveted over.

115. Maximum Boiler Pressure.—The maximum boiler pressure that can be carried for a certain specified spacing and minimum staybolt diameter can be easily calculated by remembering that the area supported by one staybolt multiplied by the boiler pressure will equal the tensile stress on the staybolt. With a 4-inch spacing each way, thus giving an area support of 16 square inches, and staybolts with a least diameter of 1$\frac{1}{16}$ inches, hence with a cross-sectional area of .886 square inch, less the area of the telltale hole, .028 inch, or a net area of .858 square inch, the maximum boiler pressure would be equal to $\frac{7{,}500 \times .858}{16} =$ 402 pounds. It will be noted that the area supported is equal to the product of the spacing taken each way.

The foregoing shows that the pressure for which a firebox can be stayed depends on the spacing of the stays and their cross-sectional area. Owing to the difficulty in washing out if the spacing were made smaller and the diameter of the stays were increased, the spacing has to be at least about 4 inches and the stay diameter about 1 inch, this giving a distance between stays of 3 inches.

CALCULATING STRESSES DESIGNATED ON SPECIFICATION CARD

116. Stresses on Staybolts and Crown Stays.—The specification Card Form No. 4, Rules and Instructions for Inspection and Testing of Locomotive Boilers and Their Appurtenances, requires the following calculations to be made and the results entered in the proper spaces on the card: The maximum stress on the staybolts at the root of the thread, in pounds per square inch, at the allowed working pressure; the stress on the staybolts at reduced section; the stress on the crown stays or crownbar rivets at root of thread or smallest section, top; the stress on the crown stays or crown-bar rivets at root of thread or smallest section, bottom.

The above stresses can all be calculated from the following rule:

Rule.—*To calculate the stress on a staybolt or a crown stay, in pounds per square inch, of cross-sectional area at either the root of the thread or at the reduced section, multiply the boiler pressure by the area supported by the staybolt and divide by the cross-sectional area at root of thread less the area of telltale hole, or by the cross-sectional area at the reduced section.*

117. It is to be understood that if any part of a crown stay is found to be smaller in area than the section at the root of the thread, then the lesser area is to be used. With a staybolt, the area of the telltale hole must be deducted from the least cross-sectional area.

EXAMPLE.—Find the stress on a staybolt in pounds per square inch of cross-sectional area with a spacing of 4 inches by 4 inches, a boiler pressure of 200 pounds to the square inch, and a diameter of 1 inch.

TABLE I

AREA OF BOILER STAYS AT ROOT OF THREAD

12 Threads Per Inch, **V** *Thread*

Size of stays Inches	Area of Stays at Root of Thread Deduct .028 sq. in. if telltale hole is used.
7/8	.419
15/16	.494
1	.575
1 1/16	.662
1 1/8	.755
1 1/4	.960

SOLUTION.—From Table 1, the area of the staybolt at the root of the thread is .575 sq. in. With a telltale hole 3/16 in. in diameter, an area equal to .028 sq. in. must be subtracted, leaving an area of .547 sq. in. Then, $\frac{4 \times 4 \times 200}{.547} = 5{,}580$ lb. per square inch of cross-sectional area. Ans.

118. Stresses on Round and Rectangular Braces.—The method of calculating the stress on round and rectangular braces that are placed diagonally can be understood from a study of Fig. 39. The area supported by the stay is found in the same manner as for a staybolt, namely, each brace is assumed to support the plate halfway to the next one on all of the

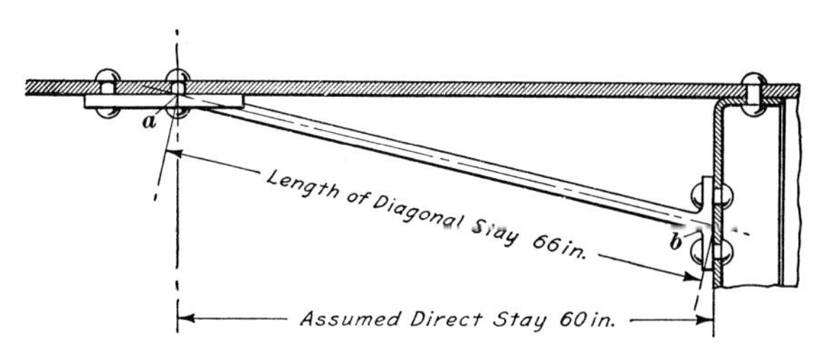

Fig. 39

four sides; or, if an upper brace, the measurement is made to a point 2 inches from the flange of the tube-sheet. An area of 36 square inches will be taken; this multiplied by a boiler pressure of 200 pounds will give a stress of 7,200 pounds on an assumed direct brace. Owing to the diagonal position of the actual brace, the stress on it is more than if the brace were direct, the stress varying as the length. Thus, with a stress of 7,200 pounds on a direct brace of a length of 60 inches, the stress on a diagonal brace 66 inches long will be $\frac{66}{60}$ x 7,200, or 7,920 pounds. Dividing this by the minimum or smallest cross-sectional area of the diagonal brace, here assumed to be 1¼ square inches, gives a stress of $\frac{7{,}920}{1\frac{1}{4}}$, or 6,336 pounds per square inch of cross-sectional area. As this is much less than the 9,000 pounds permitted by the rules, the brace is amply strong.

The solution of this problem is dependent on the length of the assumed direct brace. This length can be found closely enough by suspending a plumb line as nearly as possible at the junction *a* of the center line of the brace with the shell, and then measuring the distance to the tube-sheet. The length of the diagonal brace is measured from *a* to the foot at *b*.

The calculation of the stress on a diagonal brace at the back head will not differ to any great extent from that at the tube-sheet. A slight error will exist owing to the slope or the back head, but this error is on the safe side, and does not have to be considered.

119. Gusset Stays.—Owing to the abrupt angle of the gusset stays, Fig. 40, their cross-sectional areas, according to the A.S.M.E. rules, must be 10 per cent greater than the area of a diagonal stay under the same stress, but the I.C.C. rules do not require this extra allowance. The method of calculating the stress is the same as already given for diagonal braces.

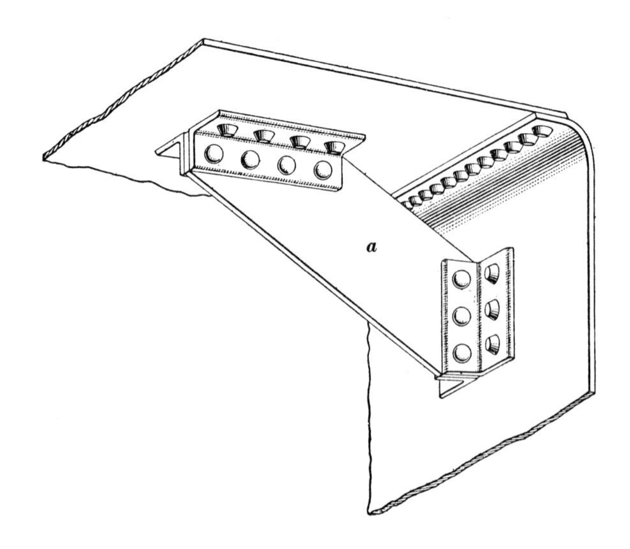

Fig. 40

120. Shearing Stress on Rivets in Diagonal Braces.—Referring to Fig. 39, a pull of 7,200 pounds on the assumed direct brace is resisted by two rivets in single shear in the shell. The shearing stress on the rivets is not equal to the pull on the diagonal brace, because 720 pounds of the pull is exerted as a tensile stress on the rivets, and is so small as compared with common tensile stresses that it does not have to be considered. To determine the shearing stress on these rivets, in pounds per square inch of cross-sectional area, divide the combined cross-sectional area of the two rivets into whatever stress is found to be exerted on the assumed direct brace, here assumed to be 7,200 pounds. The stress on an iron rivet in single shear should not exceed 9,500 pounds per square inch; for a steel rivet a stress of 11,000 pounds per square inch is permitted. These values are obtained by dividing the figures given in Rule 5, Rules and Instructions for Inspection and Testing of Locomotive Boilers and Their Appurtenances, found in the back of this lesson, by the factor of safety, or 4. According to Rule 6, a higher value may be taken when a greater strength of rivet material can be shown.

The tensile stress is also developed on the rivets that secure the brace to the tube-sheet. The combined area of the rivets, in square inches, divided into 7,200 will give the tensile stress per squre inch.

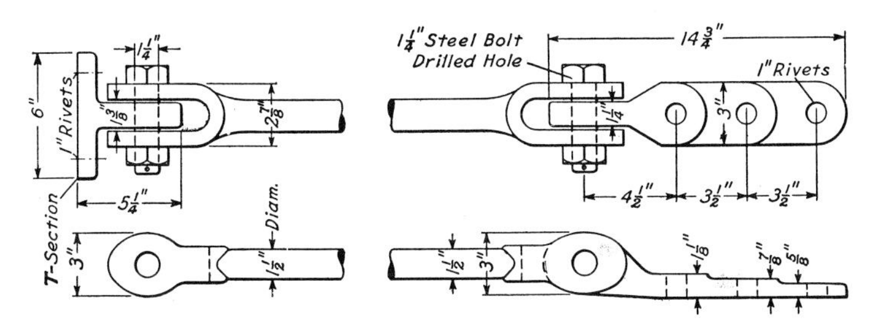

Fig. 41

With diagonal braces of the type shown in Fig. 41, the pin is in double shear. To determine the shearing stress, in pounds per square inch of cross-sectional area, divide the stress on the diagonal brace, which is assumed to be 7,920 pounds, by the cross-sectional area of the steel pin. The reason why the stress on the diagonal stay is taken is that the pin is actually a part of the brace. According to Rule 5 of the foregoing regulations, the result of the calculation must be under 22,000 pounds per square inch.

The area of rivet holes for various sizes of rivets is given in Table II. The area of the drilled hole is always to be taken, and not the cross-sectional area of the rivet before it is driven. An inch rivet is usually driven into a hole 1¹⁄₁₆ inches in diameter; in driving this rivet it is expanded ¹⁄₁₆ inch, hence the area of the hole is taken instead of the cross-sectional area of the rivet.

121. Tension on Net Section of Plate in Longitudinal Seam of Lowest Efficiency.—The analysis of a longitudinal seam, Fig. 36, to determine the point of lowest efficiency was given in Arts. 95 to 100, the lowest resistance to rupture being found in the plate between the rivet holes in the third row.

TABLE II

AREA OF RIVET HOLES

Size of Driven Rivets Inches	Area of Rivet Hole Square Inches
11/16	.371
3/4	.442
13/16	.518
7/8	.601
15/16	.690
1	.785
1 1/16	.887
1 1/8	.994
1 5/32	1.050
1 3/16	1.108
1 1/4	1.227
1 5/16	1.353

122. Calculating Efficiency of Weakest Longitudinal Seam.— If the weakest part of the seam is assumed to be the net section of plate at the outer row of rivets, then its efficiency can be calculated from Rule II, Art. 84. However, if this assumption is not to be made, the weakest part of the seam can be calculated from the rule in Art. 90.

SUMMARY OF RULES

123. The rules given throughout this lesson are here arranged in a convenient form for reference.

Tensile strength of plate = cross-sectional area x unit of tensile strength.

Efficiency of plate (at outer row of rivets)

$$= \frac{\text{pitch} - \text{diameter of rivet hole}}{\text{pitch}} \times 100.$$

Shearing stress of rivets = cross-sectional area x unit of shearing stress.

$$\text{Efficiency of joint} = \frac{\text{weakest part of joint}}{\text{ultimate strength of plate}} \times 100.$$

Efficiency of joint (if equal to efficiency of plate)

$$= \frac{\text{pitch} - \text{diameter of rivet hole}}{\text{pitch}} \times 100.$$

Strain on longitudinal seam = radius of boiler x length x pressure.

Bursting pressure

$$= \frac{\text{minimum thickness of plate x tensile strength x efficiency of joint}}{\text{radius of shell}}$$

Safe working pressure (boiler shell)

$$= \frac{\text{minimum thickness of plate x tensile strength x efficiency of joint}}{\text{radius of shell x factor of safety}}.$$

Safe working pressure (firebox)

$$= \frac{\text{7,500 x net area of staybolt}}{\text{vertical pitch x horizontal pitch}}.$$

Stress on staybolts (root of thread)

$$= \frac{\text{boilure pressure x area supported}}{\text{cross-sectional area, root of thread — area of telltale hole}}.$$

Stress on crown stays (root of thread)

$$= \frac{\text{boilure pressure x area supported}}{\text{cross-sectional area, root of thread}}.$$

Stress on diagonal braces

$$= \frac{\text{boiler pressure x area supported x length of stay}}{\text{length of assumed direct stay}}.$$

RULES AND INSTRUCTIONS FOR INSPECTION AND TESTING OF LOCOMOTIVE BOILERS AND THEIR APPURTENANCES

Approved by orders of the Interstate Commerce Commission, dated June 2, 1911, September 12, 1912, June 9, 1914, April 7, 1919, and July 28, 1925.

RESPONSIBILITY FOR THE GENERAL CONSTRUCTION AND SAFE WORKING PRESSURE.

1. The railroad company will be held responsible for the general design and construction of the locomotive boilers under its control. The safe working pressure for each locomotive boiler shall be fixed by the chief mechanical officer of the company or by a competent mechanical engineer under his supervision, after full consideration has been given to the general design, workmanship, age, and condition of the boiler, and shall be determined from the minimum thickness of the shell plates, the lowest tensile strength of the plates, the efficiency of the longitudinal joint, the inside diameter of the course, and the lowest factor of safety allowed.

FACTOR OF SAFETY.

2. The lowest factor of safety for locomotive boilers which were in service or under construction prior to January 1, 1912, shall be 3.25.

Effective October 1, 1919, the lowest factor shall be 3.5.

Effective January 1, 1921, the lowest factor shall be 3.75.

Effective January 1, 1923, the lowest factor shall be 4.

3. *(a) Maximum allowable stress on stays and braces.*—For locomotives constructed after January 1, 1915, the maximum allowable stress per square inch of net cross-sectional area on firebox and combustion chamber stays shall be 7,500 pounds. The maximum allowable stress per square inch of net cross sectional area on round, rectangular, or gusset braces shall be 9,000 pounds.

(b) For locomotives constructed prior to January 1, 1915, the maximum allowable stress on stays and braces shall meet the requirements of rule No. 2, except that when a new firebox and wrapper sheet are applied to such locomotives they shall be made to meet the requirements of rule No. 3.

TENSILE STRENGTH OF MATERIAL.

4. When the tensile strength of steel or wrought-iron shell plates is not known, it shall be taken at 50,000 pounds for steel and 45,000 pounds for wrought iron.

SHEARING STRENGTH OF RIVETS.

5. The maximum shearing strength of rivets per square inch of cross-sectional area shall be taken as follows:

Iron rivets in single shear	38,000
Iron rivets in double shear	76,000
Steel rivets in single shear	44,000
Steel rivets in double shear	88.000

6. A higher shearing strength may be used for rivets when it can be shown by test that the rivet material used is of such quality as to justify a higher allowable shearing strength.

RULES FOR INSPECTION.

7. The mechanical officer in charge at each point where boiler work is done will be held responsible for the inspection and repair of all locomotive boilers and their appurtenances under his jurisdiction. He must know that all defects disclosed by any inspection are properly repaired before the locomotive is returned to service.

8. The term "inspector" as used in these rules and instructions, unless otherwise specified, will be held to mean the railroad company's inspector.

INSPECTION OF INTERIOR OF BOILER.

9. *Time of inspection.*—The interior of every boiler shall be thoroughly inspected before the boiler is put into service and whenever a sufficient number of flues are removed to allow examination.

10. *Flues to be removed.*—All flues of locomotive boilers in service, except as otherwise provided, shall be removed at least once every four years for the purpose of making a thorough examination of the entire interior of the boiler and its bracing. After the flues are taken out the inside of the boiler must have the scale removed and be thoroughly cleaned and inspected. The removal of flues will be due after 48 calendar months' service, provided such service is performed within five counsecutive years. Portions of calendar months out of service will not be counted. Time out of service must be properly accounted for by out of service reports and notations of months claimed out of service made on the back of each subsequent inspection report and cab card. The period for removal of flues, upon formal application to the chief inspector, may be extended, if investigation shows that conditions warrant it.

11. *Method of inspection.*—The entire interior of the boiler must then be examined for cracks, pitting, grooving, or indications of overheating and for damage where mud has collected or heavy scale formed. The edges of plates, all laps, seams, and points where cracks and defects are likely to develop or which an exterior examination may have indicated, must be given an especially minute examination. It must be seen that braces and stays are taut, that pins are properly secured in place, and that each is in condition to support its proportion of the load.

12. *Repairs.*—Any boiler developing cracks in the barrel shall be taken out of service at once, thoroughly repaired, and reported to be in satisfactory condition before it is returned to service.

13. *Lap-joint seams.*—Every boiler having lap-joint longitudinal seams without reinforcing plates shall be examined with special care to detect grooving or cracks at the edges of the seams.

14. *Fusible plugs.*—If boilers are equipped with fusible plugs they shall be removed and cleaned of scale at least once every month. Their removal must be noted on the report of inspection.

INSPECTION OF EXTERIOR OF BOILER.

15. *Time of inspection.*—The exterior of every boiler shall be thoroughly inspected before the boiler is put into service and whenever the jacket and the lagging are removed.

16. *Lagging to be removed.*—The jacket and lagging shall be removed at least once every five years and a thorough inspection made of the entire exterior of the boiler while under hydrostatic pressure. The jacket and lagging shall also be removed whenever on account of indications of leaks the United States inspector or the railroad company's inspector considers it desirable or necessary. The modification granted in rule 16 in the commission's order of September 20, 1917, on account of the war in which the date for removal of jacket and lagging was advanced for a period equivalent to the duration of the war, such advanced period shall be considered two years.

TESTING BOILERS.

17. *Time of testing.*—Every boiler, before being put into service and at least once every 12 months thereafter, shall be subjected to hydrostatic pressure 25 per cent above the working steam pressure.

18. *Removal of dome cap.*—The dome cap and throttle standpipe must be removed at the time of making the hydrostatic test and the interior surface and connections of the boiler examined as thoroughly as conditions will permit. In case the boiler can be entered and thoroughly inspected without removing the throttle standpipe the inspector may make the inspection by removing the dome cap only, but the variation from the rule must be noted in the report of inspection.

19. *Witness of test.*—When the test is being made by the railroad company's inspector, an authorized representative of the company, thoroughly familiar with boiler construction, must personally witness the test and thoroughly examine the boiler while under hydrostatic pressure.

20. *Repairs and steam test.*—When all necessary repairs have been completed, the boiler shall be fired up and the steam pressure raised to not less than the allowed working pressure, and the boiler and appurtenances carefully examined. All cocks, valves, seams, bolts, and rivets must be tight under this pressure and all defects disclosed must be repaired.

STAYBOLT TESTING.

21. *Time of testing rigid bolts.*—All staybolts shall be tested at least once each month. Staybolts shall also be tested immediately after every hydrostatic test.

22. *Method of testing rigid bolts.*—The inspector must tap each bolt and determine the broken bolts from the sound or the vibration of the sheet. If staybolt tests are made when the boiler is filled with water, there must be not less than 50

pounds pressure on the boiler. Should the boiler not be under pressure, the test may be made after draining all water from the boiler, in which case the vibration of the sheet will indicate any unsoundness. The latter test is preferable.

23. *Method of testing flexible staybolts with caps.*—Except as provided in paragraph *(b)*, all staybolts having caps over the outer ends shall have the caps removed at least once every two (2) years and the bolts and sleeves examined for breakage. Each time the hydrostatic test is applied the hammer test required by rules 21 and 22 shall be made while the boiler is under hydrostatic pressure not less than the allowed working pressure.

(b) When all flexible staybolts with which any boiler is equipped are provided with a telltale hole not less than three-sixteenths (3/16) inch nor more than seven thirty-seconds (7/32) inch in diameter, extending the entire length of the bolt and into the head not less than one-third (1/3) of its diameter, and these holes are protected from becoming closed by rust and corrosion by copper plating or other approved method, and are opened and tested, each time the hydrostatic test is applied, with an electrical or other instrument approved by the Bureau of Locomotive Inspection, that will positively indicate when the telltale holes are open their entire length, the caps will not required to be removed. When this test is completed the hydrostatic test must be applied and all staybolts removed which show leakage through the telltale hole.

The inner ends of the telltale holes must be kept closed with a fireproof porous material that will exclude foreign matter and permit leakage of steam or water, if the bolt is broken or fractured, into the telltale hole. When this test is completed the ends of the telltale holes shall be closed with material of different color than that removed and a record kept of colors used.

(c) The removal of flexible staybolt caps and other tests shall be reported on the report of inspection Form No. 3, and a proper record kept in the office of the railroad company of the inspections and tests made.

(d) Firebox sheets must be carefully examined at least once every month for mud burn, bulging, and indication of broken staybolts.

(e) Staybolt caps shall be removed or any of the above tests made whenever the United States inspector or the railroad company's inspector considers it desirable in order to thoroughly determine the condition of staybolts or staybolt sleeves.

24. *Method of testing flexible staybolts without caps.*—Flexible staybolts which do not have caps shall be tested once each month, the same as rigid bolts.

Each time a hydrostatic test is applied such staybolt test shall be made while the boiler is under hydrostatic pressure not less than the allowed working pressure and proper notation of such test made on Form No. 3.

25. *Broken staybolts.*—No boiler shall be allowed to remain in service when there are two adjacent staybolts broken or plugged in any part of the firebox or combustion chamber, nor when three or more are broken or plugged in a circle 4 feet in diameter, nor when five or more are broken or plugged in the entire boiler.

26. *Telltale holes.*—All staybolts shorter than 8 inches applied after July 1, 1911, except flexible bolts, shall have telltale holes three-sixteenths inch in diameter and not less than 1¼ inches deep in the outer end. These holes must be kept open at all times.

27. All staybolts shorter than 8 inches, except flexible bolts and rigid bolts which are behind frames and braces, shall be drilled when the locomotive is in the shop for heavy repairs, and this work must be completed prior to July 1, 1914.

STEAM GAUGES.

28. *Location of gauges.*—Every boiler shall have at least one steam gauge which will correctly indicate the working pressure. Care must be taken to locate

the gauge so that it will be kept reasonably cool and can be conveniently read by the enginemen.

29. *Syphon.*—Every gauge shall have a syphon of ample capacity to prevent steam entering the gauge. The pipe connection shall enter the boiler direct and shall be maintained steam tight between boiler and gauge. The syphon pipe and its connections to the boiler must be cleaned each time the gauge is tested.

30. *Time of testing.*—Steam gauges shall be tested at least once every three months and also when any irregularity is reported.

31. *Method of testing.*—Steam gauges shall be compared with an accurate test gauge or dead-weight tester and gauges found inaccurate shall be corrected before being put into service.

32. *Badge plates.*—A metal badge plate showing the allowed steam pressure shall be attached to the boiler head in the cab. If the boiler head is lagged, the lagging and jacket shall be cut away so that the plate can be seen.

33. *Boiler number.*—The builder's number of the boiler, if known, shall be stamped on the dome. If the builder's number of the boiler can not be obtained, an assigned number which shall be used in making out specification cards shall be stamped on dome.

SAFETY VALVES.

34. *Number and capacity.*—Every boiler shall be equipped with at least two safety valves, the capacity of which shall be sufficient to prevent, under any conditions of service, an accumulation of pressure more than 5 per cent above the allowed steam pressure.

35. *Setting of safety valves.*—Safety valves shall be set to pop at pressures not exceeding 6 pounds above the working steam pressure. When setting safety valves two steam gauges shall be used, one of which must be so located that it will be in full view of the person engaged in setting such valves; and if the pressure indicated by the gauges varies more than 3 pounds they shall be removed from the boiler, tested, and corrected before the safety valves are set. Gauges shall in all cases be tested immediately before the safety valves are set or any change made in the setting. When setting safety valves the water level in the boiler shall not be above the highest gauge cock.

36. *Time of testing.*—Safety valves shall be tested under steam at least once every three months, and also when any irregularity is reported.

WATER GLASS AND GAUGE COCKS

37. *Number and location.*—Every boiler shall be equipped with at least one water glass and three gauge cocks. The lowest gauge cock and the lowest reading of the water glass shall be not less than 3 inches above the highest part of the crown-sheet. Locomotives which are not now equipped with water glasses shall have them applied on or before July 1, 1912.

38. *Water glass valves.*—All water glasses shall be supplied with two valves or shutoff cocks, one at the upper and one at the lower connection to the boiler, and also a drain cock, so constructed and located that they can be easily opened and closed by hand.

39. *Time of cleaning.*—The spindles of all gauge cocks and water glass cocks shall be removed and cocks thoroughly cleaned of scale and sediment at least once each month.

40. All water glasses must be blown out and gauge cocks tested before each trip and gauge cocks must be maintained in such condition that they can be easily opened and closed by hand without the aid of a wrench or other tool.

41. *Water and lubricator glass shields.*—All tubular water glasses and lubricator glasses must be equipped with a safe and suitable shield which will prevent the glass from flying in case of breakage, and such shield shall be properly maintained.

42. *Water glass lamps.*—All water glasses must be supplied with a suitable lamp properly located to enable the engineer to easily see the water in the glass.

INJECTORS.

43. Injectors must be kept in good condition, free from scale, and must be tested before each trip. Boiler checks, delivery pipes, feed water pipes, tank hose and tank valves must be kept in good condition, free from leaks and from foreign substances that would obstruct the flow of water.

FLUE PLUGS.

44. Flue plugs must be provided with a hole through the center not less than three-fourths inch in diameter. When one or more tubes are plugged at both ends the plugs must be tied together by means of a rod not less than five-eights inch in diameter. Flue plugs must be removed and flues repaired at the first point where such repairs can properly be made.

WASHING BOILERS.

45. *Time of washing.*—All boilers shall be thoroughly washed as often as the water conditions require, but not less frequently than once each month. All boilers shall be considered as having been in continuous service between washouts unless the dates of the days that the boiler was out of service are properly certified on washout reports and the report of inspection.

46. *Plugs to be removed.*—When boilers are washed, all washout, arch, and water bar plugs must be removed.

47. *Water tubes.*—Special attention must be given the arch and water bar tubes to see that they are free from scale and sediment.

48. *Office record.*—An accurate record of all locomotive boiler washouts shall be kept in the office of the railroad company. The following information must be entered on the day that the boiler is washed:

(a) Number of locomotive.
(b) Date of washout.
(c) Signature of boiler washer or inspector.
(d) Statement that spindles of gauge cocks and water-glass cocks were removed and cocks cleaned.
(e) Signature of the boiler inspector or the employee who removed the spindles and cleaned the cocks.

STEAM LEAKS.

49. *Leaks under lagging.*—If a serious leak develops under the lagging, an examination must be made and the leak located. If the leak is found to be due to a crack in the shell or to any other defect which may reduce safety, the boiler must be taken out of service at once, thoroughly repaired, and reported to be in satisfactory condition before it is returned to service.

50. *Leaks in front of enginemen.*—All steam valves, cocks, and joints, studs, bolts, and seams shall be kept in such repair that they will not emit steam in front of the enginemen, so as to obscure their vision.

FILING REPORTS.

51. *Report of inspection.*—Not less than once each month and within 10 days after each inspection a report of inspection, Form No. 1, size 6 by 9 inches, shall be filed with the district inspector of locomotive boilers for each locomotive used by a railroad company, and a copy shall be filed in the office of the chief mechanical officer having charge of the locomotive.

52. A copy of the monthly inspection report, Form No. 1, or annual inspection report, Form No. 3, properly filled out, shall be placed under glass in a conspicuous place in the cab of the locomotive before the boiler inspected is put into service.

53. Not less than once each year and within 10 days after hydrostatic and other required tests have been completed a report of such tests showing general condition of the boiler and repairs made shall be submitted on Form No. 3, size 6 by 9 inches, and filed with the district inspector of locomotive boilers, and a copy shall be filed in the office of the chief mechanical officer having charge of the locomotive. The monthly report will not be required for the month in which this report is filed.

54. *(a) Specification card.*—A specification card, size 8 by 10½ inches, Form No. 4, containing the results of the calculations made in determining the working pressure and other necessary data shall be filed in the office of the chief inspector of locomotive boilers for each locomotive boiler. A copy shall be filed in the office of the chief mechanical officer having charge of the locomotive. Every specification card shall be verified by the oath of the engineer making the calculations, and shall be approved by the chief mechanical officer. These specification cards shall be filed as promptly as thorough examination and accurate calculation will permit. Where accurate drawings of boilers are available, the data for specification card, Form No. 4, may be taken from the drawings, and such specification cards must be completed and forwarded prior to July 1, 1912. Where accurate drawings are not available, the required data must be obtained at the first opportunity when general repairs are made, or when flues are removed. Specification cards must be forwarded within one month after examination has been made, and all examinations must be completed and specification cards filed prior to July 1, 1913, flues being removed if necessary to enable the examination to be made before this date.

(b) When any repairs or changes are made which affect the data shown on the specification card a corrected card or an alteration report on an approved form, size 8 by 10½ inches, properly certified to, giving details of such changes, shall be filed within 30 days from the date of their completion. This report should cover:

A. Application of new barrel sheets or domes.
B. Application of patches to barrels or domes of boilers or to portion of wrapper sheet of crown bar boilers which is not supported by staybolts.
C. Longitudinal seam reinforcements.
D. Changes in size or number of braces, giving maximum stress.
E. Initial application of superheaters, arch or water-bar tubes, giving number and dimensions of tubes.
F. Changes in number or capacity of safety valves.

Report of patches should be accompanied by a drawing or blueprint of the patch, showing its location in regard to the center line of boiler, giving all necessary dimensions, and showing the nature and location of the defect. Patches previously applied should be reported the first time the boiler is stripped to permit an examination.

55. In the case of an accident resulting from failure, from any cause, of a locomotive boiler or any of its appurtenances, resulting in serious injury or death to one or more persons, the carrier owning or operating such locomotive shall immediately transmit by wire to the chief inspector of locomotive boilers, at his office in Washington, D.C., a report of such accident, stating the nature of the accident, the place at which it occurred, as well as where the locomotive may be inspected, which wire shall be immediately confirmed by mail, giving a full detailed report of such accident, stating, as far as may be known, the causes and giving a complete list of the killed or injured.

MONTHLY LOCOMOTIVE INSPECTION AND REPAIR REPORT.

Form No. 1.

..............................19 Locomotive { Number............. / Initial...............

......................................*Company.*

In accordance with the act of Congress approved February 17, 1911, as amended March 4, 1915, and the rules and instructions issued in pursuance thereof and approved by the Interstate Commerce Commission, all parts of locomotive No. ......, including the boiler and appurtenances, were inspected on ...., 19 at ...., and all defects disclosed by said inspection have been repaired, except as noted on the back of this report.

1. Steam gauges tested and left in good condition on....., 19 .
2. Safety valves set to pop at.........pounds, ...pounds, ...pounds on................., 19 .
3. Were both injectors tested and left in good condition? ....
4. Were steam leaks repaired? ....
5. Condition of brake and signal equipment, ...
6. Condition of draft gear and draw gear, ...
7. Condition of driving gear, ...
8. Condition of running gear, ...
9. Condition of tender, ...

I certify that the above report is correct.

......................................,
Inspector.

10. Was boiler washed and gauge cocks and water glass cock spindles removed and cocks cleaned? ....
11. Were steam leaks repaired? ....
12. Condition of staybolts and crown stays, ...
13. Number of staybolts and crown stays renewed, ...
14. Condition of flues and firebox sheets, ...
15. Condition of arch and water bar tubes, if used, ...
16. Were fusible plugs removed and cleaned? ....
17. Date of previous hydrostatic test, ..., 19 .
18. Date of removal of caps from flexible staybolts, ..., 19 .

I certify that the above report is correct.

......................................,
Inspector.

STATE OF.................. }
COUNTY OF................. } *ss:*

Subscribed and sworn to before me this........day of....................19 , by............................, inspectors

of the....................................Company.

....
Notary Public.

The above work has been performed and the report is approved.

....
Officer in Charge.

ANNUAL LOCOMOTIVE INSPECTION AND REPAIR REPORT.

Form No. 3.

...............................19

Locomotive { Number.........
Initial...........

....*Company*.

In accordance with the act of Congress approved February 17, 1911, as amended March 4, 1915, and the rules and instructions issued in pursuance thereof and approved by the Interstate Commerce Commission, all parts of locomotive No. .... including the boiler and appurtenances, were inspected on ...., 19 , at ...., and all defects disclosed by said inspection have been repaired, except as noted on the back of this report.

1. Date of previous hydrostatic test, ..., 19 .
2. Date of previous removal of caps from flexible staybolts , 19 .
3. Date of previous removal of flues, ... 19 .
4. Date of previous removal of all lagging, ..., 19 .
5. Hydrostatic test pressure of.............pounds was applied.
6. Were caps removed from all flexible staybolts? ....
7. Were all flues removed?.......................Number......
8. Condition of interior of barrel, ...
9. Was all lagging removed? ....
10. Condition of exterior of barrel, ...
11. Was boiler entered and inspected? ....
12. Was boiler washed? Water glass cocks and gauge cocks cleaned? ....
13. Condition of crown stays and staybolts, ...
14. Condition of sling stays and crown bars, ...
15. Condition of firebox sheets and flues, ...
16. Condition of arch tubes, ... Water bar tubes,, ...
17. Condition of throat braces, ...
18. Condition of back head braces, ...
19. Condition of front flue sheet braces. ....
20. Were fusible plugs removed and cleaned ....
21. Were steam leaks repaired? ....

I certify that the above report is correct.

............................, *Inspector.*

22. Were steam gauges tested and left in good condition? ....
23. Safety valves set to pop at .... pounds, ... pounds, ... pounds.
24. Were both injectors tested and left in good condition? ....
25. Were steam leaks repaired? ....
26. Hydrostatic test of ...... pounds applied to main reservoirs.
27. Condition of brake and signal equipment, ...
28. Were drawbar and drawbar pins removed and inspected?.....
29. Condition of draft gear and draw gear, ...
30. Condition of driving gear, ...
31. Condition of running gear, ...
32. Condition of tender, ...

I certify that the above report is correct.

............................, *Inspector.*

STATE OF...................} *ss:*
COUNTY OF.................}

Subscribed and sworn to before me this........day of.....................19 , by.............................., inspectors of the.......................................Company.

...................................., *Notary Public.*

The above work has been performed and the report is approved.

...................................., *Officer in Charge.*

Form No. 4.

Specification Card for Locomotive No.__________

Owned by..Railroad Company.

Operated by..Railroad Company.

Builder ....

Builder's No. of Boiler..

When built ....

Where built ....

Type of boiler ....

Material of boiler shell sheets..............................

Material of rivets..

Dome, where located..

Grate area in sq. ft. ....

Height of lowest reading of gauge glass above crown sheet ....

Height of lowest gauge cock above crown sheet ....

Water bar tubes, O. diam......thickness.....

Arch tubes, O. diam..........thickness.........

Fire tubes, number..

" " O. diam............. length...........

Safety valves:

No. Size. Make. Style.

....

....

....

Firebox stay bolts, O. diam....spaced....x...

Combustion chamber stay bolts, O. diam...

" " " " spaced....x...

Crown stays, O. diam., top........bottom.....

" " spaced..................x..................

Crown bar rivets, O. diam, top....bottom....

" " " spaced..............x.............

Water space at firebox ring, sides...........

back.......................... front........................

Width of water space at sides of firebox measured at center line of boiler, front .... back............................

Shell sheets:

Front tube............ thick.

1st course .... " .... I. diam

2d " .... " .... "

3d " .... " .... "

Mem.: When courses are not cylindrical give inside diameter at each end.

Firebox:

Thickness of sheets—

Tube............ Crown............ Side.........

Door.......... Combustion chamber.........

Inside throat (if tube sheet is in two pieces) ....

External firebox:

Thickness of sheets—throat...................

back head ....

Roof.......................... sides.....................

Dome inside diam. ....

Thickness of sheet........base........liner.....

Were you furnished with authentic records of the tests of materials used in boiler ....

Records on file in the office of the............ of the.. Company show that the lowest tensile strength of the sheets in the shell of this boiler is:

1st course ....pounds per sq. in

2d " .... " " " "

3d " .... " " " "

Is boiler shell circular at all points?..........

If shell is flattened state location and amount ....

Are all parts thoroughly stayed?................

Are dome and other openings sufficiently reenforced? ....

Is boiler equipped with fusible plugs?......

Make working sketch here or attach drawing of longitudinal and circumferential seams used in shell of boiler, indicating on which courses used, and give calculated efficiency of weakest longitudinal seam.

The maximum stresses at the allowed working pressure were found by calculation to be as follows:

Stay bolts at root of thread........................ ....lbs. per sq. in

Stay bolts at reduced section........................ .... lbs. per sq. in.

Crown stays or crown bar rivets at root of thread or smallest section, top.......... .... lb. per sq. in.

Crown stays or crown bar rivets at root of thread or smallest section, bottom ....lbs. per sq. in.

Round and rectangular braces.................... .... lbs. per sq. in.

Gusset braces..........................lbs. per sq. in.

Shearing stress on rivets " " " "

Tension on net section of plate in longitudinal seam of lowest efficiency, pounds per sq. in. ....

Dimensions and data taken from locomotive were furnished by..

Data upon which above calculations were made were obtained from drawing No.

....dated.................................. furnished by..................................Company.

STATE OF....................................
COUNTY OF.................................. } *ss:*

....
Mechanical Engineer.

....being duly sworn says that he is the officer who signed the foregoing specification, that he has satisfied himself of the correctness of the drawings and data used, has verified all of the calculations, and has examined the record of present condition of boiler dated..and sworn to by inspector..and believes that the design, construction, and condition of boiler No. .... renders it safe for a working pressure of .... pounds per square inch.

Subscribed and sworn to before me this......... day of 19

....
(name of affiant)

....
Notary Public.

Approved:

....

LOCOMOTIVE BOILERS

(PART 2)

EXAMINATION QUESTIONS

Notice to Students.—*Study the Instruction Paper thoroughly before you attempt to answer these questions.* **Read each question carefully and be sure you understand it;** *then write the best answer you can. When your answers are completed, examine them closely, correct all the effors you can find, and* **see that every question is answered.**

(1) What is the strain on a girth joint as compared with the strain on a longitudinal joint?

(2) The pitch of the rivets in a quadruple-riveted butt joint is 19 inches, the diameter of the rivet holes is $1\frac{7}{16}$ inches. What is the efficiency of the joint, it being assumed that the plate will fail at the outer row of rivets?

(3) Why is the testing pressure of a boiler restricted to about 25 per cent above the working pressure?

(4) Why is the firebox end of a superheater tube smaller in diameter than the remainder of the tube?

(5) What is the disadvantage of having grates with the air openings too large?

(6) Give the progressive action that follows the overheating of a crown-sheet.

(7) What will be the result if the ash-pan has not sufficient air openings?

(8) Upon what does the efficiency of the heating surface depend?

(9) Explain briefly how to make a hydrostatic test of a boiler.

(10) How can the portion of a crown-sheet that has been overheated be identified?

(11) What is the objection to the use of a water-tube firebox when the water is not pure?

(12) Which has the higher factor of safety, the firebox or the shell of the boiler?

(13) Name the two types of joints used in locomotive boilers and explain how they differ.

(14) What is meant by the pitch of rivets?

(15) Why do thermic syphons prevent excessive damage to the crown-sheet in cases of low water?

(16) Why is a firebrick arch applied to a locomotive boiler?

(17) What is meant by the term *circulation* as applied to the water in the boiler?

(18) How can the strain that tends to rupture a boiler along any lengthwise line be calculated?

(19) What is meant by the rate of combustion?

(20) What is meant by the term *factor of safety* and what rule is used to find it?

(21) Explain why the water circulates in a locomotive boiler.

(22) Name the heating surfaces of a boiler in the order of their effectiveness.

(23) Name some of the more important advantages of the firebrick arch.

(24) What is the first action that follows the overheating of the crown-sheet?

(25) What is the combustion chamber and what are its advantages?

(26) What material is used in the construction of a locomotive boiler?

(27) What is the principal purpose of the arch tubes?

(28) How much coal is required to develop one cylinder horsepower for one hour with a modern locomotive?

(29) What is the function of the thermic syphon?

(30) With high steam pressure and a high degree of superheat, how many pounds of steam is required to develop one cylinder horsepower for one hour?

(31) What is the purpose of the tubes and flues?

(32) About how many pounds of water per hour is evaporated for each square foot of heating surface?

(33) Explain how the end of a tube is made to have a tight fit in the back tube sheet.

(34) Define the term *boiler evaporation.*

(35) What is the purpose of the steam dome?

(36) What is meant by the term *direct heating surface?*

(37) How are grates classified?

(38) Name the sheets that comprise the firebox heating surface.

(39) Define the term *heating surface.*

(40) What is the unit of tensile strength?

Locomotive Injectors

By
J. W. Harding

5236 Printed in U.S.A. Edition 1

1946 Edition

LOCOMOTIVE INJECTORS

THEORY, OPERATION, AND DISORDERS

INTRODUCTION

1. Development of Injector.—Among the inventions of the nineteenth century, none were of a more ingenious nature than that of the injector for feeding water to boilers, by Henry J. Gifford, the eminent French engineer and scientist, who obtained his first patent on this device in 1858.

The property of a moving jet of steam to raise water and convey it from one place to another was known and utilized long before this date, so that strictly speaking he was not the original inventor of steam-jet instruments in general. What he really invented or discovered was the detail of the overflow space which for boiler-feeding purposes is invaluable and which is usually located between the discharge end of what is termed the condensing nozzle and the receiving end of what is known as the delivery nozzle.

In starting an injector, more water may enter the instrument under certain conditions than the injector is capable of delivering against the steam pressure in the boiler, and if it were not for the overflow, which permits the surplus water to escape until the jet obtains sufficient velocity to enter the boiler, a back pressure would be produced by the stowing of the water in the nozzles, which would disrupt the continuity of the jet and prevent the prompt starting of the injector. This disruption of the jet, popularly termed the breaking of the injector, may be observed where the overflow pipes are too small or have sharp bends which do not allow a ready outflow of the water, or where the overflow passages of the instrument itself are obstructed by incrustation, or from other causes.

The practical utility of the injector, the convenience of its installation, its simplicity and reliability as a boiler feeder, were quickly recognized. Improvements in the general design and in the constructive details of the nozzles and operating mechanism rapidly developed, so that today the manufacture of injectors forms a conspicuous branch of industrial activity.*

2. Purpose of Injector.—A locomotive injector is a device for supplying feedwater to a locomotive boiler. It is a type of pump in which the actuating steam is condensed and which is capable of placing water under pressure without the employment of any moving parts. Although an injector performs the same function as a pump, it does so in a different manner. With a pump, the water pressure is developed by the action of steam upon two pistons. With an injector, the water in the delivery pipe is placed under pressure by the impact of a jet of water discharging from the injector at a high velocity but at a pressure less than that of the atmosphere. The action of this jet of water may be compared to a piston, pressing forward continuously on the rear of the water in the delivery pipe.

*Nathan Manufacturing Co.

3. Atmospheric Pressure.—The pressure of the atmosphere enters into the operation of a lifting injector, hence an explanation of the term *atmospheric pressure* is necessary. The air that forms the atmosphere surrounding the earth has weight, even though it appears to be so light. The atmosphere is estimated to be about 15 miles high, becoming lighter and rarer on the tops of high mountains. If it were possible to enclose a column of this air measuring 1 square inch and 15 miles high, and weigh it, the weight would be about 14.7 pounds; in other words, the pressure of this column of air on the square inch of surface on which it rests would be 14.7 pounds. Hence, it is customary to say that the atmosphere exerts a pressure of 14.7 pounds per square inch on every square inch of surface of the earth, as well as on all objects at or near the earth's surface. This value, 14.7 pounds per square inch, is known as the atmospheric pressure at sea level. On the tops of mountains, the pressure is less, because there is less air above those points to produce pressure.

4. The fact that the atmosphere has weight and causes pressure on the earth may be demonstrated by using the apparatus shown in Fig. 1. A glass tube about 32 inches long, closed at one end, is filled with quicksilver, or mercury. The tube is then inverted and the end is placed in a cup half-filled with mercury. Part of the mercury will run out of the tube into the cup, but not all of it; and if the distance from the surface of the mercury in the cup to the surface of the mercury in the tube is measured, it will be found to be 30 inches. The mercury stands at this height because of the pressure of the atmosphere on the surface of the mercury in the cup.

The mercury tends to run down out of the tube and thus raise the level in the cup; but the atmosphere exerts a pressure on the surface of the mercury in the cup. This pressure prevents the mercury in the cup from rising, and so balances the downward pressure of the mercury in the tube. Suppose that the tube measures 1 square inch in area, inside; then as the mercury stands at a height of 30 inches, there are $30 \times 1 = 30$ cubic inches of mercury in the tube above the level in the cup. A cubic inch of mercury weighs .49 pound, and so 30 cubic inches weigh $30 \times .49 = 14.7$ pounds. Therefore, the column of mercury in the tube exerts a downward pressure of 14.7 pounds per square inch, which tends to raise the level in the cup. The atmosphere exerts an equal pressure of 14.7 pounds per square inch on the surface of the mercury in the cup and prevents it from rising. The two forces are therefore balanced, and the column of mercury remains in the glass tube at a height of 30 inches.

5. Vacuum—Definition and Measurement.—A vacuum is a space in which there is no pressure, but the term is also commonly used to designate any pressure below that of the atmosphere, or 14.7 pounds per square inch. One method of obtaining a vacuum is to pump the air out of a closed vessel by means of a vacuum pump, but a perfect vacuum can never be obtained by this means.

The amount of vacuum obtained in any vessel may be measured by means of a column of mercury, as will be explained by referring to Fig. 2, which illustrates a vacuum pump. This pump is very similar to a small force pump, except in the arrangement of the valves *a* and *b*. In a force pump, the positions of these valves on the piston and cylinder walls would be interchanged. The pump communicates through the pipe *c* with an air-tight vessel *2* that opens to the atmosphere through the cock *d*. The glass tube *3* is more than 30 inches long and is closed at one end. The tube

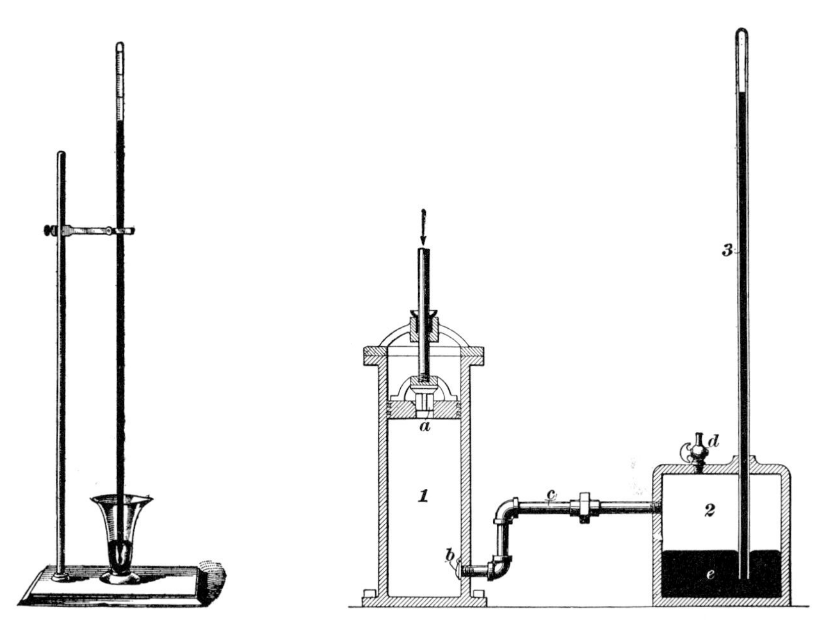

Fig. 1 Fig. 2

has been filled with mercury, then inverted, and after that placed in the mercury *e*. Now if the cock *d* is open so that the vessel *2* is filled with air at atmospheric pressure, the pressure on top of the mercury, will hold the latter up to a height of about 30 inches. If the cock *d* is then closed and a part of the air is pumped out of the vessel, the air that remains will exert less pressure on the mercury, so that the column in the tube will fall from 30 inches to a height that corresponds to the pressure in the vessel. If more air is pumped out, the height of the column will also be lowered more, and, if all the air is removed from the vessel and a perfect vacuum exists therein, the mercury in the tube will drop to a level with the mercury *e*.

In the same way the water is raised from a tank to a lifting injector by the action of atmospheric pressure and a vacuum. A vacuum forms in the injector when it is being started and while it is in operation; the greater pressure of the atmosphere on the surface of the water in the tank then forces the water up into the injector, where the pressure is less.

6. Measuring Air Pressure in a Vacuum.—The amount of the vacuum that exists in any space is usually expressed in the number of inches the mercury falls when connected to such a space. Thus, if the mercury has fallen 20 inches from its original height, the vacuum in the vessel is said to amount to 20 inches of mercury, or to 20 inches. The height at which the column stands merely indicates the extent of reduction in the pressure in the vessel; the actual pressure that exists there can be understood from the following. A vacuum of 20 inches means that the pressure has been reduced enough to support a column of mercury 20 inches high. Since a column of mercury 1 inch high and one square inch in area is equivalent to a pressure of .49 pound per square inch, 20 inches of mercury corresponds to a reduction of pressure of 20 x

.49, or 9.8 pounds below atmospheric pressure. Hence the pressure in the vessel is 14.7 — 9.8, or 4.9, pounds per square inch.

The usual method of measuring a vacuum is by means of a vacuum gage. The dial is usually marked to read in inches of mercury, although some are marked to read in pounds per square inch. Pressure measured above a perfect vacuum or above an absolute zero of pressure is called absolute pressure, hence the pressure just determined is referred to as 4.9 pounds absolute pressure.

7. Absolute Pressure and Gage Pressure.—The ordinary steam gage and air gage register pressures above that of the atmosphere, that is, the starting point on such gages is atmospheric pressure. Hence, to change gage pressure into absolute pressure, add the pressure of the atmosphere to the gage pressure. To change from absolute pressure to gage pressure, subtract the pressure of the atmosphere from the absolute pressure. A gage pressure of 200 pounds to the square inch is equal to an absolute pressure of 214.7 pounds per square inch, and an absolute pressure of 250 pounds per square inch is equal to a gage pressure of 235.3 pounds per square inch.

8. Heat and Work.—The work performed by steam when moving a locomotive or operating an injector should not be ascribed to the pressure of the steam but rather to its heat, which is the term applied to the extremely rapid vibration of the steam particles. It is the movement of these particles that results in pressure; hence any device that employs steam in the performance of work is a heat engine. Steam can be made to do work by two methods: (1) by means of a suitable arrangement of cylinders, pistons, and valves as with a steam locomotive; (2) by means of the steam turbine. By the first method, the steam is admitted to the cylinders and owing to the intense vibration of its particles exerts a pressure on the pistons, which, in the case of a locomotive, causes it to move and develop a pull at the drawbar. With a steam turbine an extremely high velocity but a low pressure is imparted to the steam by passing it through suitably designed nozzles. When directed against the vanes on the turbine wheel, the movement of the steam particles is retarded; hence a pressure is developed against the vanes which causes the wheel to turn. Owing to its compact design and great economy, the turbine engine has almost entirely supplanted the reciprocating engine in large power plants. The steam nozzle of an injector is similar in design to those found in a steam turbine, but with an injector the steam, after leaving the nozzle, condenses and imparts a high velocity to a column of water. Then by retarding the velocity of this water, its pressure increases enough to overcome boiler pressure.

FLOW OF STEAM THROUGH NOZZLES

9. Velocity Affected by Shape of Nozzle.—The steam nozzle may be regarded as the engine of the injector, because it is in this nozzle that the force is developed to place the water under pressure. The steam after passing through the nozzle of an injector must not be considered as being in the same state as when leaving the boiler. The condition of the steam changes and it is this change that is responsible for the action of the injector. Instead of being at boiler pressure, the pressure of the steam on leaving the nozzle is not much above the pressure of the atmosphere, but its velocity is greatly increased and it is on this change in the condition of the steam that the operation of the injector depends. Much of the difficulty in understanding the action of an injector is due to a failure to recognize the real function of the steam nozzle, namely, that its design is such as to convert steam at a high pressure and a low velocity into steam at a low pressure and a high velocity. The other nozzles or tubes in the injector are merely supplementary; the secret of the operation of an injector lies in its steam nozzle.

10. The illustrations in Figs. 3, 4, and 5 are designed to show how the pressure and the velocity of steam are affected by the shape of the nozzle through which it is flowing. The converging tube, Fig. 3, is similar to the earliest form of steam nozzle used with an injector and was patterned after the correct shape of nozzle required to impart a high velocity to a jet of water. It was probably employed with the idea that the shape of nozzle necessary to produce a jet of water at a high velocity would work equally well with steam. However, this is not the case, because a converging nozzle permits the steam to swell out or expand in diameter, as shown, instead of compelling it to expand in the direction of motion so that the steam escapes at the nozzle tip with a pressure much greater than that of the surrounding atmosphere, as indicated by the swelling of the jet just beyond the nozzle. If the steam escaped at the same pressure as the atmosphere, the jet would be straight-sided; and if at a lower pressure, it would become narrower outside the tip, as the atmosphere would press in on it.

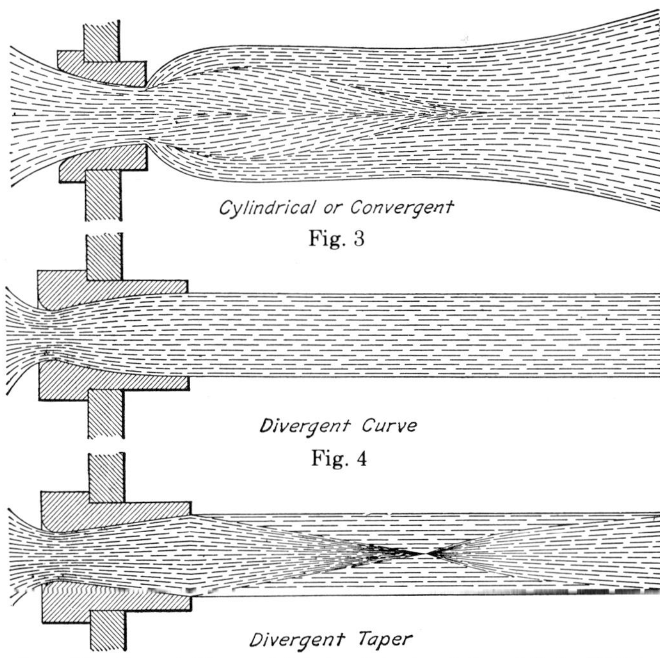

Fig. 3

Fig. 4

From "Practice and Theory of Injectors," by permission of S. L. Kneass.

Fig. 5

It was not until eleven years after the invention of the injector that a nozzle was designed that would expand the steam in the direction of its motion and hence greatly increase its velocity, instead of permitting diametral expansion beyond the nozzle tip as with the converging type. Such a nozzle is shown in Fig. 4, in which the bore diverges or grows larger in a curved line toward the outlet, although ordinarily, owing to the difficulty of manufacturing, the nozzle is made with a straight taper as shown in Fig. 5. However, this type of nozzle does not expand the steam quite so perfectly as the one shown in Fig. 4. For high boiler pressures, the flare of the steam nozzle is lengthened to provide for the required expansion of the steam; the bore of the nozzle is not generally made any larger.

11. Explanation of Increased Velocity.—The action that follows when steam passes through a divergent nozzle with a curved taper can be more easily understood by considering Fig. 6. The nozzle is assumed to be divided into a great number of zones, *a, b, c,* etc., in which, owing to their gradually increasing cross-sectional area, the steam in its passage through the nozzle undergoes a continual expansion; hence the temperature, and consequently the pressure of the steam are progressively lowered. This results in the steam particles passing from zone to zone at an ever-increasing velocity until near the end of the nozzle the steam is moving at a velocity about eight times greater than the velocity at which it enters the nozzle but at a pressure no higher than that of the atmosphere, as indicated by the fact that the broken lines that show the flow of the steam particles are all parallel. This means that a steam gage would indicate no pressure if it were tapped into the side of the nozzle near the tip.

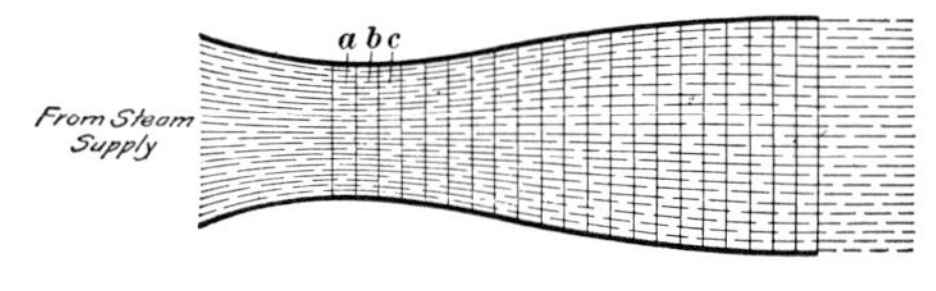

Fig. 6

With a steam pressure of 200 pounds per square inch, the velocity of the steam at the entrance to the nozzle is about 500 feet per second. At the throat or narrowest section, the velocity is about 1,650 feet per second, and as the expansion of the steam continues the velocity increases until, at the outlet, the steam jet is moving at a speed of about 3,850 feet per second and at a pressure equal to about that of the atmosphere.

GENERAL OPERATION OF INJECTOR

12. Conventional Arrangement.—The diagram in Fig. 7 *(a)* is a conventional view of a lifting injector and its related parts; that is, it is a sketch of the essential parts, but not an accurate illustration of the relative sizes and arrangement of those parts. An enlarged view of the injector tubes is shown in *(b).* An injector is made up of the following principal parts: A steam nozzle *a*, a combining tube *b*, and a delivery tube *d*, contained within the body shown, and certain valves to control the flow of steam and water. The opening *e* between the tubes *b* and *d* opens

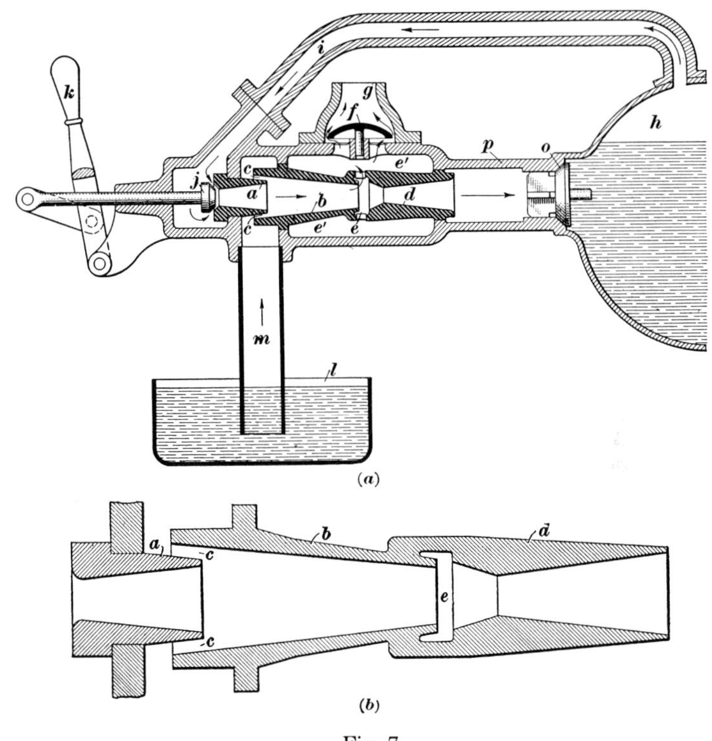

Fig. 7

into a chamber e' surrounding the tubes, and this chamber is fitted with an overflow valve *f* that opens communication with outer air by way of the fitting *g*. Steam is led to the injector from the boiler *h* by the pipe *i*, the flow being controlled by a valve *j* connected to the handle *k*. The water supply is contained in a tank *l* and is drawn into the injector by way of the pipe *m* and the opening *c*. The check-valve *o* allows water discharged from the injector to enter the boiler through the delivery pipe *p*, but prevents any flow in the opposite direction.

13. Operation.—The precise action that occurs in the nozzle *a* has already been fully explained and therefore does not require further mention.

To operate the injector, the handle *k*, Fig. 7 *(a)*, is pulled back so as to open the steam valve *j* only slightly, thus admitting a small quantity of steam to the nozzle *a*. This steam passes through the combining tube *b* into the delivery tube *d*, and the delivery pipe, and meets the closed check-valve *o*, which it cannot move. The steam then passes out through the opening *e* into the chamber e', lifts the overflow valve *f*, and escapes into the atmosphere. In flowing through the combining tube, however, the steam draws with it the air in the tube and in the pipe *m*, thus a vacuum is created inside the pipe. The pressure of the atmosphere on the surface of the water in the tank *l* then forces the water up into the pipe *m* and through the opening *c* into the combining tube. The steam now meets the water and is condensed, giving up part of its velocity to the water; but as the flow of steam is very small, the jet of water has insufficient velocity to build up pressure enough to open the check-valve. As a result, this water flows to the atmosphere by way of the opening *e* and the overflow valve *f*. When water appears at the opening *g*, the injector is said to be *primed*.

14. As soon as the injector has been primed, as indicated by the appearance of water at the overflow *g*, Fig. 7 *(a)*, the valve *j* is opened wide, admitting steam at full boiler pressure to the nozzle *a*. The steam expands in this nozzle to approximately atmospheric pressure, with a great increase in velocity, exactly as already explained. In the combining tube the jet of low-pressure steam at high velocity strikes the water, picks it up and carries it along at increasing speed, the steam being condensed at the same time. The jet of water discharging from the combining tube leaps the gap *e* with no tendency to spill out because the pressure of the jet cannot be more than that of the steam, which is equal to or less than that of the atmosphere. The water next fills the delivery tube and the delivery pipe, where it meets the closed check-valve *o* and stops.

But the column of water in the delivery tube is now being pushed from the rear and compressed by a jet of water from the combining tube, moving at a velocity of about 190 feet per second. This jet of water exerts no pressure against the sides of the tube because the movement of all the particles of the water is forwards. Also, on account of its high velocity, the jet has somewhat the nature of a solid instead of a liquid and may be compared to a piston pressing on the water. The result is that the water in the delivery pipe is compressed and increases sufficiently in pressure to force open the boiler check and enter the boiler. The interval between the priming of an injector and the opening of the boiler check is very brief; the actions described in the foregoing operation of an injector occur almost in an instant. Injectors are designed to compress the water to about 10 per cent higher than the boiler pressure.

15. Summary.—A brief summary of the operation of an injector follows: The steam nozzle is designed to impart an extremely high velocity to the steam, which through condensation imparts considerable velocity to the water in the combining tube; the jet of water from this tube, impinging against the more slowly moving water in the delivery tube, raises the pressure of the water in the delivery pipe above the boiler pressure.

16. Development of Pressure.—The reason why the jet of water, discharging at a high velocity from the combining tube at a low pressure, develops a high pressure is that the jet encounters resistance when directed against the column of water in the delivery tube and delivery pipe; the pressure of this water will then be increased by the impact. If it were not that the water in the delivery pipe offers resistance to movement, there would be but little increase in the pressure. That is, if the end of the delivery pipe were open to the atmosphere, very little pressure would be developed in the pipe. The same would be true of a pump, which would not place water under pressure unless the water in the discharge pipe offered resistance to the movement of the pistons. When starting the injector, the water in the delivery pipe offers resistance to the movement of the jet by stopping momentarily. When the injector is in operation a resistance is set up to the movement of the jet by the water in the delivery pipe moving more slowly than in the injector owing to the boiler check, the opening of which is resisted by boiler pressure. Also, the opening through the check-valve is always smaller than the cross-sectional area of the delivery pipe, thereby further restricting the movement of the water and causing its pressure to increase. Ordinarily the pressure in the delivery tube rises from a pressure of about 13 pounds absolute, or about 2 pounds less than the pressure of the atmosphere at the rear end of the tube, to a pressure of about 220 pounds per square inch at the front end of the tube and in the delivery pipe.

17. Pressure in Combining Tube.—On leaving the steam nozzle and meeting the water in the rear end of the combining tube, the steam has a high velocity but a pressure equal to about that of the atmosphere, which is reduced by condensation to about 4 pounds, absolute, corresponding to 22 inches of vacuum. As the steam moves forward in the combining tube, the pressure of the water rises, until, at the smallest part of the tube, the pressure is about 13 pounds, absolute, corresponding to about 4 inches of vacuum. The reason for this increase in pressure is that the greater part of the condensation occurs in the rear end of the combining tube, with a corresponding increase in the temperature of the water; hence, the remainder of the steam condenses more slowly as the steam moves forward in the tube, and its pressure approaches more nearly to its original or atmospheric pressure.

That the pressure in the combining tube has less lateral or side pressure than that of the atmosphere can be demonstrated by removing the overflow valve from an injector. The water can then be observed flowing past the openings in the combining tube without spilling out, which shows that the atmospheric pressure on the outside of the tube is more than the pressure of the water on the inside. Also, air will be drawn through the overflow valve into the injector, a further proof of a pressure less than that of the atmosphere inside of the tube.

18. Changing from Velocity to Pressure.—The delivery tube has a gradual expanding taper toward the delivery pipe, because if the high-velocity water from the combining tube were permitted to discharge directly against the more slowly moving water in the delivery pipe, the result would be violent swirls and eddies in the water that would react on the entering water and either reduce the efficiency of the injector or prevent it from operating. Such an action can be prevented by changing from velocity to pressure gradually and this is accomplished by expanding the bore of the delivery tube towards the delivery pipe. This construction lessens the velocity of the water more slowly than otherwise and so brings about a gradual equalization of velocities between the fast moving water in the combining tube and the slower moving water in the delivery pipe. The transition from velocity to pressure will then occur less abruptly than if the delivery tube were straight. There is therefore a gradual decrease of velocity in the delivery tube accompanied by a gradual increase in pressure in the delivery pipe until finally the pressure becomes high enough to open the boiler check.

19. Self-Regulation.—The steam and the water must mix in correct proportions to insure efficient operation of the injector. If too much steam is supplied, part of it will not be condensed, and a mixture of steam and water will be discharged at the overflow while the injector is working. If too little steam is used, not all of the water will be forced into the boiler, and the surplus will escape at the overflow. As the steam pressure in a locomotive often varies greatly, fluctuations in the amount of steam admitted to an injector like that just considered would make it impossible to operate without waste at the overflow, except under certain pressures, height of lift, and temperature of water supply. However, it is possible to modify the injector so as to make it self-regulating; that is, the injector automatically reduces the amount of water supplied when the steam pressure falls and increases the amount of water when the steam pressure rises.

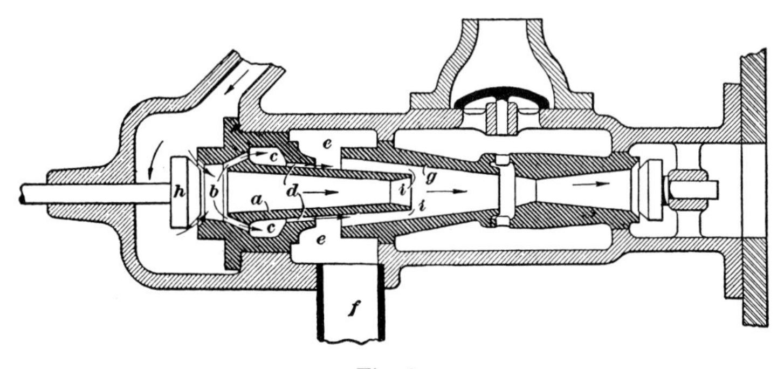

Fig. 8

20. The sectional view in Fig. 8 shows how an injector may be constructed so as to be self-regulating. The only difference between this construction and that just described is in the form and arrangement of the steam nozzle *a*. At the inlet end of the steam nozzle, passages *b* lead into a chamber *c* that surrounds the nozzle; and from the chamber *c*, a passage *d* opens out around the nozzle into the chamber *e*, which is in communication with the suction pipe *f* and the inlet end of the combining tube *g* by the way of passage *i*. The steam nozzle *a* is lengthened, also, and extends well into the combining tube. When the injector is working, some of the steam admitted past the valve *h* flows through the passages *b* into the chamber *c* and thence through the passages *d* and *i*. This rush of steam through the chamber *e* draws the air from that chamber and from the suction pipe *f*, and there is a vacuum formed, the result being that water is lifted into the combining tube of the injector through the passage *i*. In other words, the supplementary jet of steam flowing around the outside of the steam nozzle *a* produces the suction by which water is drawn to the injector.

As the water is lifted and drawn into the combining tube *g*, Fig. 8, through the passage *i* by the action of the supplementary jet from the passage *d*, the jet of steam issuing from the nozzle *a* has no lifting to do, but merely forces the water on into the delivery tube; thus the nozzle *a* becomes simply a forcing nozzle. The self-regulating action is as follows: If the steam pressure in the boiler falls, less steam passes through the forcing nozzle *a;* but at the same time, less steam passes through the passages *b* and *d*, and a smaller quantity of water is drawn in through the suction pipe. Thus the ratio of the amounts of steam and water is kept unchanged. If the steam pressure rises, the supplementary jet raises a greater quantity of water, which is met by an increased quantity of steam flowing through the forcing nozzle *a*.

With the self-regulating feature added, the injector really becomes two injectors combined in one. The passages *b, d,* and *i* form one injector because their action is such as to deliver water to the combining tube. The other injector comprises the forcing nozzle, the combining tube, and the delivery tube.

21. Water Delivered per Pound of Steam.—The maximum weight of water delivered by an injector per pound of steam can be calculated approximately from the following rule:

Rule.—*From the total heat of the steam in British thermal units, subtract the delivery temperature of the water and divide by the delivery temperature of the water less the initial temperature.*

EXAMPLE.—What is the weight of water delivered per pound of steam with a steam pressure of 200 pounds, a delivery temperature of 164 degrees F., and the feedwater at a temperature of 65 degrees F.?

SOLUTION.—One pound of steam at a pressure of 200 pounds contains 1,197.6 B.t.u.'s; hence, the weight of water delivered per pound of steam is $\frac{1{,}197.6-164}{164-65}$ = 10.4 pounds. Ans.

As the steam pressure increases, the temperature of the water increases faster than the B.t.u. content of the steam, hence the water delivered per pound of steam decreases with increasing steam pressures.

PURPOSES OF PARTS

22. General Statement.—The following explanations of the purposes of the various parts of an injector, Fig. 7 *(a)*, applies to injectors of all makes.

23. Overflow Valve.—The overflow valve performs several functions: It permits the water and the steam to escape when the injector is being primed, and thereby prevents a pressure from forming in the overflow-valve chamber and reacting on the water in the suction pipe; it prevents the air from entering and reducing the capacity of the injector when it is working; and it retains the steam in the injector when it is being used as a heater. With a non-lifting injector the overflow valve is also used to prevent the escape of the water from the injector when the feedwater becomes too hot.

24. Overflow Pipe.—The overflow pipe is used to convey to some convenient point the water that discharges from the injector by way of the overflow valve.

25. Water Valve.—The purpose of the water valve is to regulate the amount of water that the injector is delivering to the boiler. If the conditions under which the locomotive is operating is such as to require the delivery of less water to the boiler, the water valve should be partly closed by turning its handle to the right. A reverse movement will increase the amount of water delivered.

26. Steam Nozzle.—The purpose of the steam nozzle is to change steam at a high pressure and a low velocity into steam at a low pressure and a very high velocity. It may be considered as being the power unit of the injector. It is a compact form of heat engine, capable of converting heat into work without the introduction of any moving parts.

27. Combining Tube.—The purpose of the combining tube is to combine the steam and the water and direct or guide the stream of water into the delivery tube. The combining tube is tapered forward so as to conform to the shape of the jet as it decreases in diameter owing to the gradual condensation of the steam.

28. Delivery Tube.—The purpose of the delivery tube is to bring about a gradual equalization of velocities between the fast moving water in the combining tube and the slower moving water in the delivery pipe, thus placing the water in this pipe under pressure with the least loss of efficiency.

29. Line Check-Valve.—The line check-valve, placed in an injector forward of the delivery tube, is to prevent the hot water and the steam from passing back into the injector in the event of a leaky boiler check-valve.

DEFINITION OF TERMS

30. Single-Jet Injector.—A single-jet injector is one through which a single jet of water is passing while the injector is in operation. It may belong to either the lifting or non-lifting class. The Sellers, Nathan, and Ohio injectors are examples of the single-jet type. The conventional injector in Fig. 7 *(a)* is of the single-jet type.

31. Double-Jet Injector.—A double-jet injector is one through which two jets of water are moving when the injector is working. It may be of either the lifting or the non-lifting class. The Hancock inspirator is an example of a double-jet injector.

32. Open-Overflow Injector.—An open-overflow injector is one in which the overflow valve is held to its seat by the pressure of the atmosphere only, when the injector is working. Single-jet injectors are of the open-overflow type.

33. Closed-Overflow Injector.—A closed-overflow injector is one in which the overflow valve is held to its seat by some mechanical means while the injector is working. Double-jet injectors are of the closed-overflow type, as their overflow valves are subject to the pressure in the delivery pipe and must be held shut positively, to prevent the escape of water.

34. Breaking of Injector.—An injector is said to break, when, for any reason, it stops working after having been started. When an injector breaks, there is a violent discharge of steam at the overflow, and the water ceases to be forced into the boiler.

35. Restarting Injector.—A restarting injector is one that will automatically resume operation if the water supply should be interrupted temporarily. The restarting feature is confined to single-jet injectors, and hence to open-overflow injectors, because the overflow valve can lift freely and allow the steam to escape when the injector breaks. Most single-jet injectors are of the restarting type.

36. Priming or Lifting of Injector.—An injector is said to prime, or lift, when the water leaves the delivery tube at too low a pressure to enter the boiler. Priming is indicated by the discharge of water and steam at the overflow when the injector is being started.

37. Highest Operating Temperature.—The highest temperature of the water with which the injector will work without wasting at the overflow with the overflow valve open, or without breaking with the overflow valve closed, is called the highest operating temperature.

38. Maximum Capacity.—The maximum capacity of an injector is the greatest quantity of water the injector is capable of delivering to the boiler in a specified time. It is usually stated in gallons per hour.

CLASSIFICATION OF INJECTORS

39. Types of Injectors.—Injectors may be divided into two classes; namely, lifting injectors and non-lifting injectors. A lifting injector is one that is placed above the level of the feedwater supply. Because of its location, when such an injector is in operation it must raise the water. This is accomplished as follows: When the steam condenses, a vacuum is formed in the combining tube and the suction pipe, and, owing to the pressure of the outside atmosphere, water is forced into the injector.

A non-lifting injector is one that is placed below the level of the feedwater supply, so that the water flows into it by gravity instead of being drawn into it by suction.

40. Lifting Injector.—A lifting injector may be placed in the cab of the locomotive; or, if there is not enough room there, it may be placed just forward of the cab and operated by levers that extend back into the cab. The latter arrangement is shown in Fig. 9. The injector *a* receives steam through a pipe *b* connected to a rectangular turret *c*, to which steam is led from the dome by a dry pipe *d*. The valve *e* serves to shut off steam from the turret. Water enters the injector from the tank through the pipe *f*, first passing through the gooseneck *g*, the hose *h*, and the strainer *i*. A valve operated from the outside of the tank controls the flow of water from the tank to the hose. The water passes from the injector to the boiler through the delivery pipe *j*, which is connected to the body of the check-valve *k*. The check-valve allows water to pass into the boiler, but prevents its return into the pipe *j*. The delivery pipe is usually made of copper, as copper does not corrode as readily as steel. The overflow pipe *l* carries off the water that is wasted while the injector is being started.

A lifting injector is placed about a foot above the highest water level in the tank, and so it has to lift the water only a short distance when the tank is full; but when the tank is nearly empty, the height of lift is considerably greater. The injector should not be placed unnecessarily high, as an increase in the height of lift reduces its capacity.

41. Non-Lifting Injector.—The arrangement of a non-lifting injector on a locomotive is shown in Fig. 10. In this case the injector *a* is placed on the locomotive at a point below the bottom of the tank, consequently the water flows to it by gravity. To the other parts and connections are given reference letters corresponding to the same details in Fig. 9. The principal reason for using the non-lifting injector is that the cabs of modern locomotives do not have available room for a lifting injector, nor is it always convenient to place such an injector forward of the cab. The arrangement in Fig. 10 requires some additional attachments. The starting valve *m* is outside the cab and is operated from the cab by moving the rod *n*. Steam is conveyed to the starting valve through the pipe *b* and from the starting valve to the injector through the pipe *o*. The water valve is operated by the handle *p* and the overflow valve by the handle *q*, these handles being supported by a stand *r* bolted to the floor of the cab. If the injector ceases to operate, the telltale pipe *s* permits a small stream of water to escape from the nozzle *t*, thus warning the engineer.

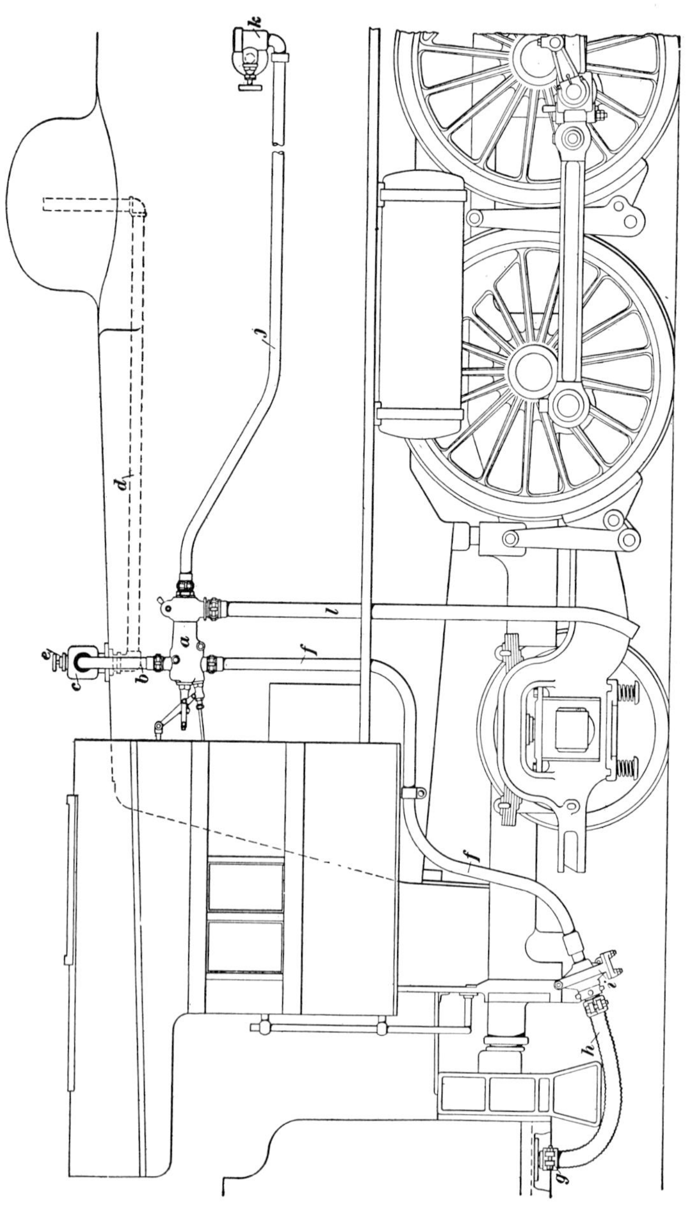

Fig. 9

42. Relative Advantages of Injectors.—The advantages of the non-lifting injector over the lifting injector are as follows: It has a larger capacity for the same size; it can be located outside the cab, in a position easily accessible for repairs, and thus relieve congested cab conditions; it works with less water in the tank and with water of a higher temperature; and it can be graded closer; that is, the ratio of the quantities of steam and water used can be regulated with greater accuracy. On the other hand, it has the disadvantages of being located where it is liable to be

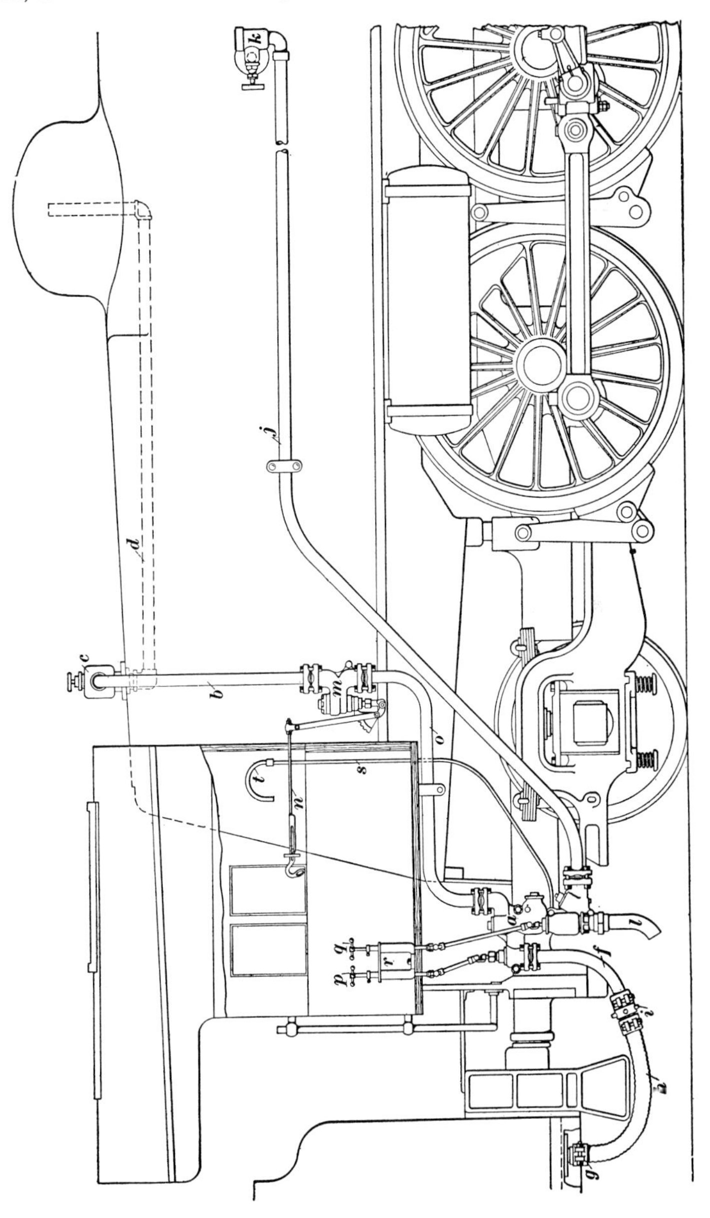

Fig. 10

knocked off, and is subject to freezing in cold weather.

The lifting injector possesses certain advantages, also. Its working can be more readily observed by the engineman; its operating troubles can be more easily remedied on the road; and it has fewer outside operating parts.

SELLERS CLASS N IMPROVED LIFTING INJECTOR

43. Details of Construction.—Every injector is made up of a body into which are screwed a set of tubes, together with the piping for the inlet and discharge of steam and water, and the valves for regulating their flow. In these details, all injectors are similar; but they differ somewhat in the form and arrangement of the parts. An outside view of a Sellers Class N lifting injector is shown in Fig. 11, and a lengthwise sectional

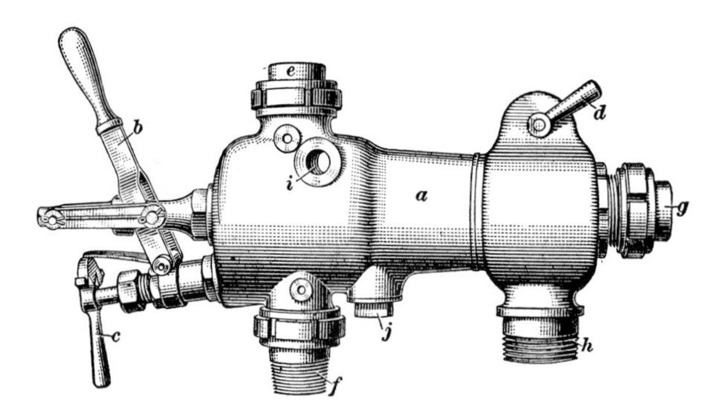

Fig. 11

view of the same injector is shown in Fig. 12. The body *a*, Fig. 11, of the injector carries the starting lever *b*, the water-valve handle *c*, and the overflow-valve handle *d*. Steam is supplied through the connection *e* and water from the tank through the pipe *f*. The delivery pipe *g* conveys the water from the injector to the boiler, and the overflow pipe at *h* carries off the water that wastes from the injector while it is being primed. The pipes are connected to the injector by couplings, and the injector is fixed to the boiler by a stud that passes through the hole *i* in the body. The cage *j* holds the water-inlet valves and may be unscrewed, bringing the valves with it.

44. Sectional View.—A series of holes *m*, Fig. 12, drilled around the rear of the steam nozzle *l*, and annular passages *n* and *p* form a lifting nozzle, the passage of steam through which creates the vacuum that results in the water being lifted from the suction pipe *f* into chamber *o*. The water is then carried through the annular space *p* into the rear end of the combining tube *q*, where it is caught by the steam issuing from the nozzle *k* and forced through the combining tube into the delivery tube *r*. For this reason the nozzle *k* is sometimes referred to as the forcing nozzle. The passages *p* and *n*, together with the holes *m*, may be considered as a separate injector without a delivery tube, the purpose of which is to lift and deliver water at a low pressure to the combining tube *q*.

The projection *j'* on the end of the spindle *v* extends into the steam nozzle when the valve *u* is closed; thus, when the valve is moved slightly from its seat the steam cannot enter the steam nozzle but is forced to flow first through the lifting nozzle.

The combining tube, Fig. 13, has openings or gaps at *x*, *y*, and *z*, and the sections separated by these gaps are held together by the four ribs *a′*. The opening *x* permits the water that is delivered to the combining tube through the lifting nozzle when priming the injector, to escape until the water reduces sufficiently in pressure to pass this opening without

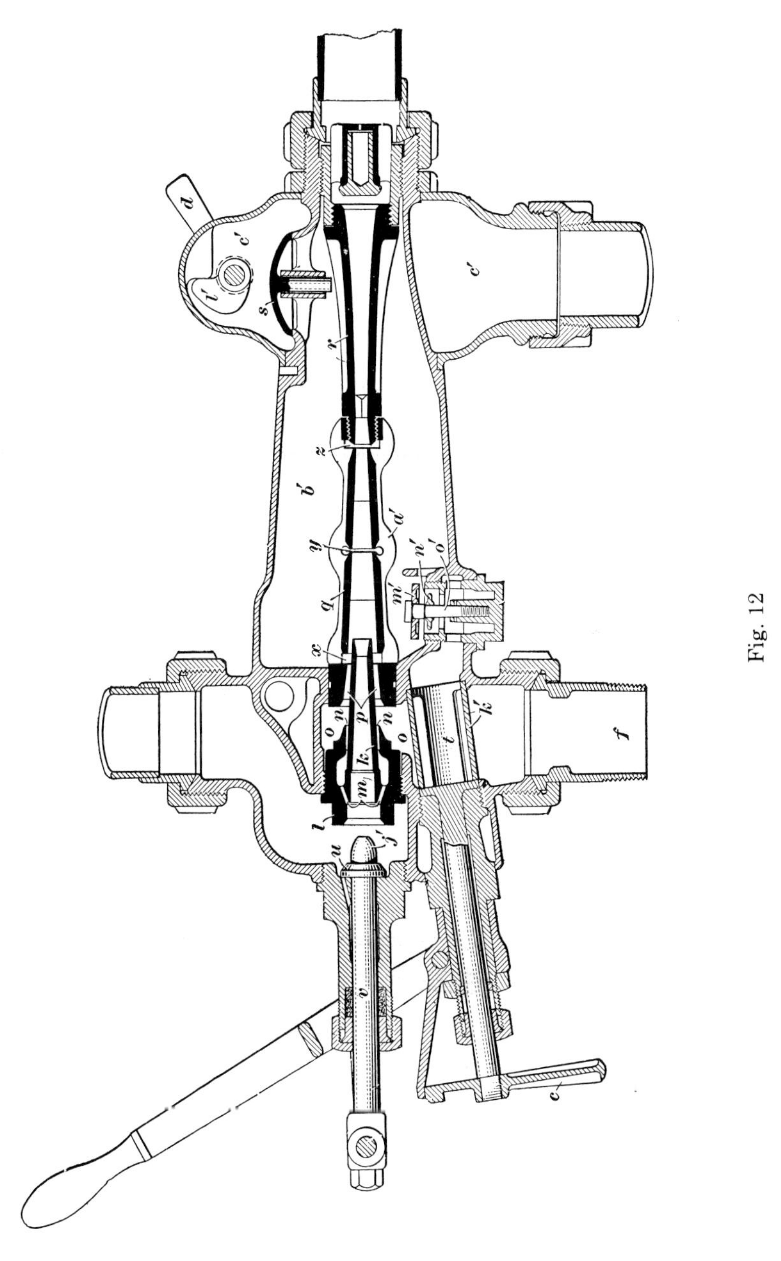

Fig. 12

flowing out. The openings *y* and *z* supplemented by the circular openings *b* permit the water to escape from the combining tube to the overflow pipe when starting the injector. Without them the injector could not be primed, when the overflow valve is located in the combining-tube chamber, because the steam could not get out of this tube and would blow

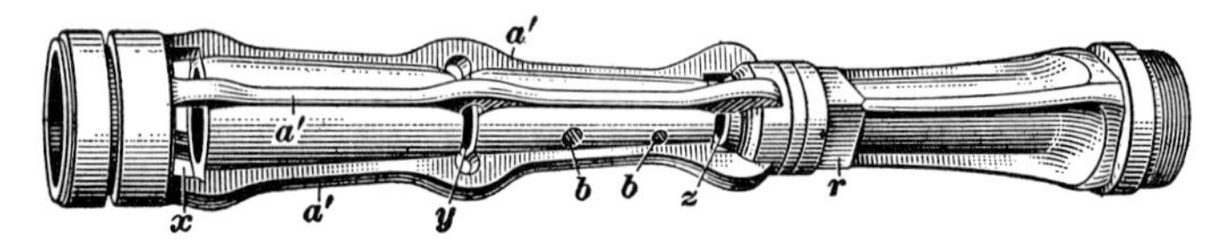

Fig. 13

back into the suction pipe. These openings also permit the injector to restart should it break; also, under certain conditions they serve to increase the capacity of the injector. It was the development of a combining tube with spillways in combination with an overflow-valve chamber that led to the invention of the injector.

45. The openings in the combining tube could be dispensed with and the surplus water, when starting, carried entirely through both tubes. This however, would require an overflow valve forward of the delivery tube, where the water would be above boiler pressure; also, the overflow valve would have to be held forcibly closed after the injector has started and this would complicate the design. It is much simpler with a single-jet injector to cut spillways in the combining tube and have the overflow valve held closed by atmospheric pressure. The design of a double-jet injector, such as the Hancock inspirator, will not permit of spillways in the tubes, hence the overflow valve must be placed forward of the forcer tube and held closed mechanically.

The water valve *t*, Fig. 12, is a hollow plug or cylinder with slots that can be made to register with corresponding slots in the sleeve *k'* by turning the handle *c*, thus allowing water to pass from the suction pipe to the passage *o*; but in the closed position the slots do not register and no water can pass the valve. The overflow valve *s* under normal conditions is free to rise and open communication between the chamber *b'* and the passage *c'*; but, when necessary, it may be held firmly to its seat by turning the handle *d* and forcing the cam *l'* down on top of the valve.

46. Vacuum in Injector.—The vacuum maintained in chamber *o*, Fig. 12, of the injector is equal to an absolute pressure of about 4 pounds per square inch, or about 11 pounds less than atmospheric pressure, and this insures a steady flow of water from the tank. The vacuum in the combining tube varies from 4 pounds absolute in the rear part of the tube to 13 pounds absolute at the front end of the tube. The vacuum in the overflow-valve chamber *b'* is closely related to the vacuum in the combining tube because the two are connected. When the steam pressure is high the vacuum formed in the combining tube and in the overflow-valve chamber may be enough to cause the water in the suction pipe to lift the inlet valves *m'* and *n'* on the stem *o'* and enter the chamber. The water is then drawn through the gaps *y* and *z* and the circular openings in the combining tube, adding about 20 per cent to the quantity of water delivered. Also, the cold water around the delivery tube lessens the temperature inside, hastens condensation and reduces the tendency to form scale on the inside surface of the tube.

47. Self-Regulation and Restarting.—The Sellers lifting injector is self-regulating, since its construction is the same as that shown in Fig. 8 and it is also restarting. The opening *y* and *z*, Fig. 12, in the combining tube, and the use of a large overflow valve and overflow pipe, allow steam to escape freely when the injector breaks, instead of producing a back pressure in the suction pipe. The flow of steam through the lifting nozzle will then reestablish the vacuum in the suction pipe and the water will be lifted and the injector put in operation again.

48. Operation.—The passage of steam and water through an injector has already been explained, hence this description of the operation of the injector will be confined to the correct way to use it. The injector is started by moving the starting lever back slightly until the injector primes as indicated by the discharge of water and the overflow; then the lever is drawn back all the way. The starting lever should never be pulled back suddenly; for, with a boiler pressure of 200 pounds, a sudden opening of the valve may cause a momentary pressure of 300 pounds per square inch in the delivery pipe, which may weaken or rupture the pipe, cause it to creep, or strain the pipe fittings. The injector is stopped by closing the steam valve.

In severe weather, when the injector is not working, it is necessary to heat the water in the suction pipe and the tank hose to prevent freezing. The injector can be converted into a heater by turning down the overflow-valve handle, and drawing the starting lever back slightly. A small amount of steam then enters the injector, and as the overflow valve is held closed, the steam heats any water that remains in the delivery pipe as well as the water in the suction pipe, the tank hose, and the tank. Too much steam must not be admitted in this way, as the water will become so hot that the injector cannot handle it; or the tank hose may be blown off by the pressure.

SELLERS CLASS S NON-LIFTING INJECTOR

49. General Description.—A sectional view of the Sellers Class S non-lifting injector is given in Fig. 14. This injector is operated entirely by one lever, and it can be used as either a lifting or a non-lifting injector. The arrangement for the simultaneous operation of the steam valve *a* and the water valve *b* can be most easily explained by assuming that the rod *c* is pulled upward as when starting the injector. The operating lever *d*, which is pivoted at *e*, then lifts the link *f* and the yoke *g*, the yoke being guided upward in the vertical slot in the guide bar *h* by the pin *i*. The ends of the yoke are flexibly connected to the stems of the water valve and the steam valve, which are accordingly raised off their seats. With this arrangement, a separate rod to operate the water valve from the cab is unnecessary.

The overflow valve is closed automatically when the injector is operating, hence, manual control by means of a handle in the cab is not necessary. When the line check-valve *j* lifts, the pilot valve *k* is unseated and permits the water to flow through port *l* to a passage *m* that leads to the top of the overflow valve *m'*, which is then held to its seat.

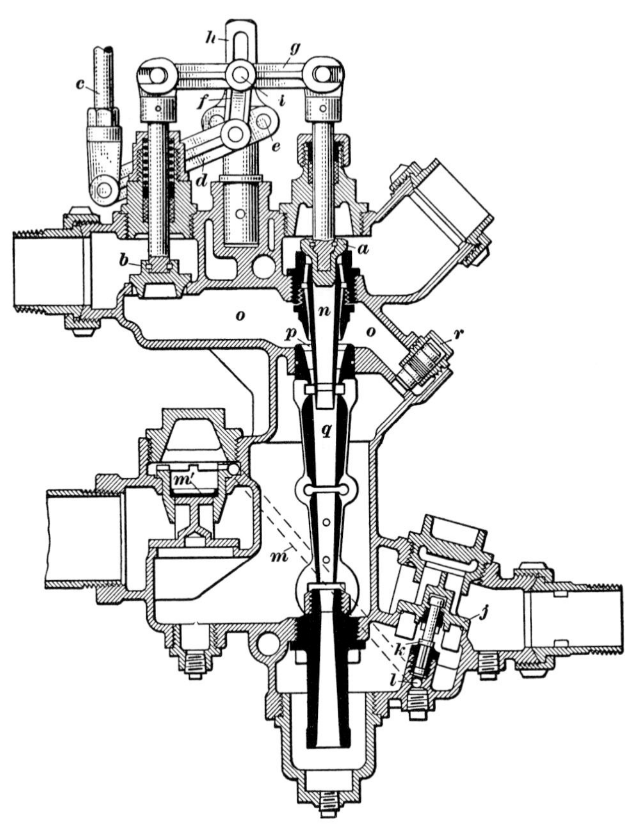

Fig. 14

50. Nozzles.—The construction of the steam nozzle *n*, Fig. 14, does not differ from that of the lifting injector already described. It is made in two parts with one part screwed into the other part as shown. A series of holes in the upper end of the nozzle open into the water passage *o* and these holes in combination with the annular passage *p* between the end of the steam nozzle and the rear end of the combining tube *q* may be considered as constituting a separate injector that regulates the amount of water delivered to this tube to suit the steam supply. The water, after passing into the combining tube as just described, is carried forward to the delivery tube by the high-velocity low-pressure steam that is discharging through the steam nozzle. The openings in the combining tube serve the same purposes as similar openings in the lifting injector.

51. Inlet Valve.—The inlet valve *r*, Fig. 14, acts to increase the capacity of the injector under certain conditions. For example, if the vacuum formed in the chamber surrounding the combining tube and above the valve *r* becomes great enough, it is lifted by the higher pressure in passage *o*, and the water passes into the combining tube through the openings shown. An increase in capacity of about 20 per cent is thus obtained.

52. Cab Stand.—The cab stand used with the Class S injector is shown in Fig. 15. The arrangement is so simple that no description is necessary.

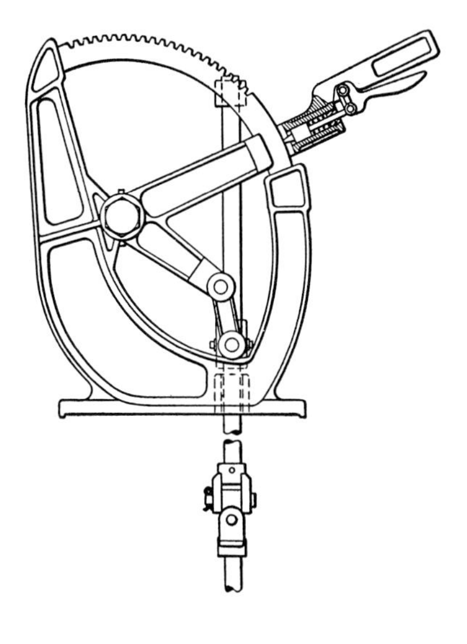

Fig. 15

53. Operation.—To start the injector and operate it at maximum capacity, pull the lever all the way back; to regulate the supply to the boiler, push the lever forward to the desired notch in the quadrant to meet the requirements of the boiler. To stop the injector, push the lever all the way forward; and to use as a heater, draw the lever back slightly.

SELLERS CLASS K NON-LIFTING INJECTOR

54. General Arrangement.—An exterior view of the Sellers Class K non-lifting injector, together with the relative positions of the attachments on the locomotive, is given in Fig. 16. The steam pipe *a* conveys the steam from the cab turret to the starting valve *b*, and steam is led to the injector *c* by the pipe *d*. Water is supplied through the suction pipe *e* and is discharged through the delivery pipe *f*, the water from the overflow being led away by the pipe *g*. The cab stand *h* supports the rods *i* and *j* to which the handles *k* and *l* are attached. The handle *k* is connected by the rod *m* to the spindle *n* of the water valve that controls the flow of water to the injector. The handle *l* controls the overflow valve through the rod *o*. A small copper pipe *p* leads from the body of the injector to the indicator *q* in the cab. When the injector is working properly, the partial vacuum inside it is communicated to the chamber *r* and the piston *s* is drawn up; or, if it overflows, the vacuum is destroyed and the piston *s* drops, thus giving the engineman warning of the trouble. The lever *t* of the starting valve is connected by the rod *u* to the starting handle *v* in the cab.

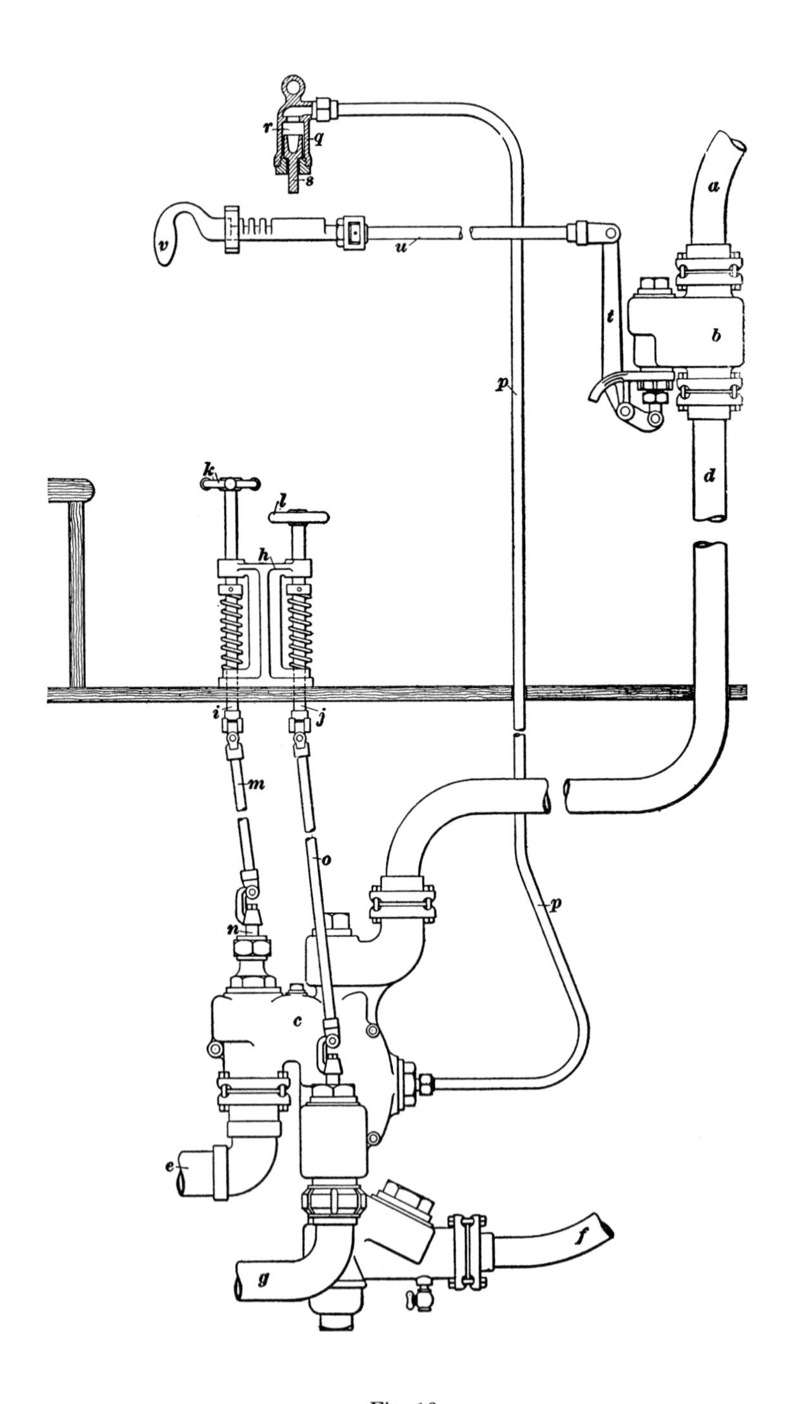

Fig. 16

55. Sectional View.—A sectional view of the Sellers Class K non-lifting injector is given in Fig. 17. The construction of the interior of this injector follows that of the Type S so closely that no description is necessary.

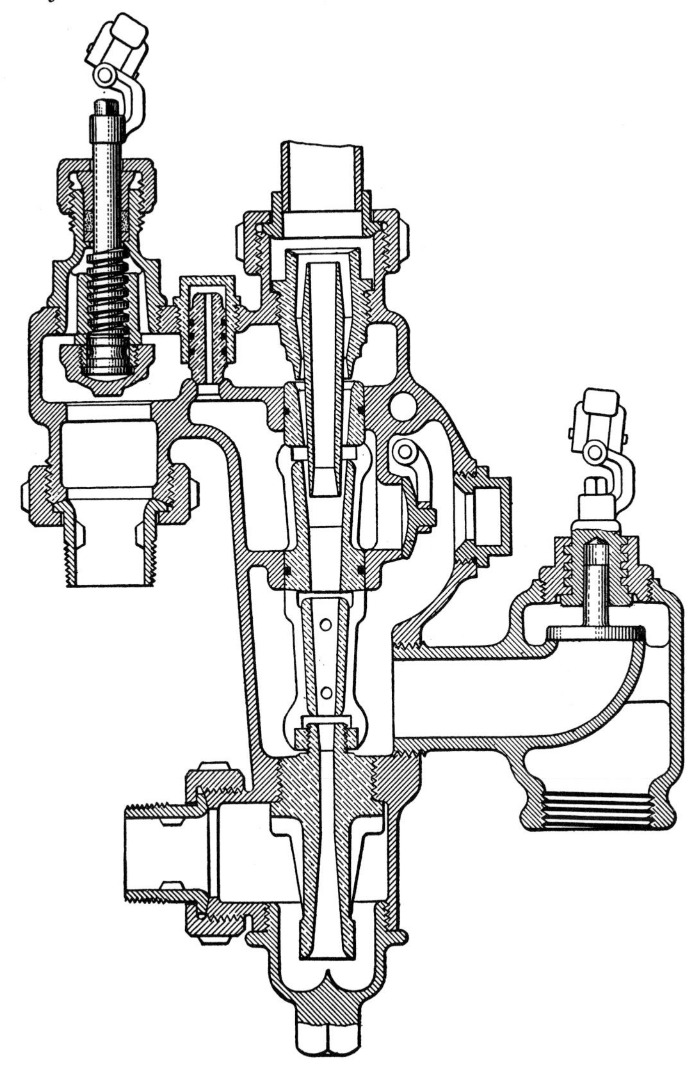

Fig. 17

56. Starting Valve.—A sectional view of the starting valve *b*, Fig. 16, is given in Fig. 18 *(a)*. Steam from the turret enters through the pipe *a* and passes to the injector through the pipe *b*. The pipe *a* communicates with the chamber *c* above the pilot valve *d* and the main valve *e*, and the passage *f* below the valves opens into the steam pipe *b*. The pilot valve seats on the upper end of the main valve, and the latter fits loosely over

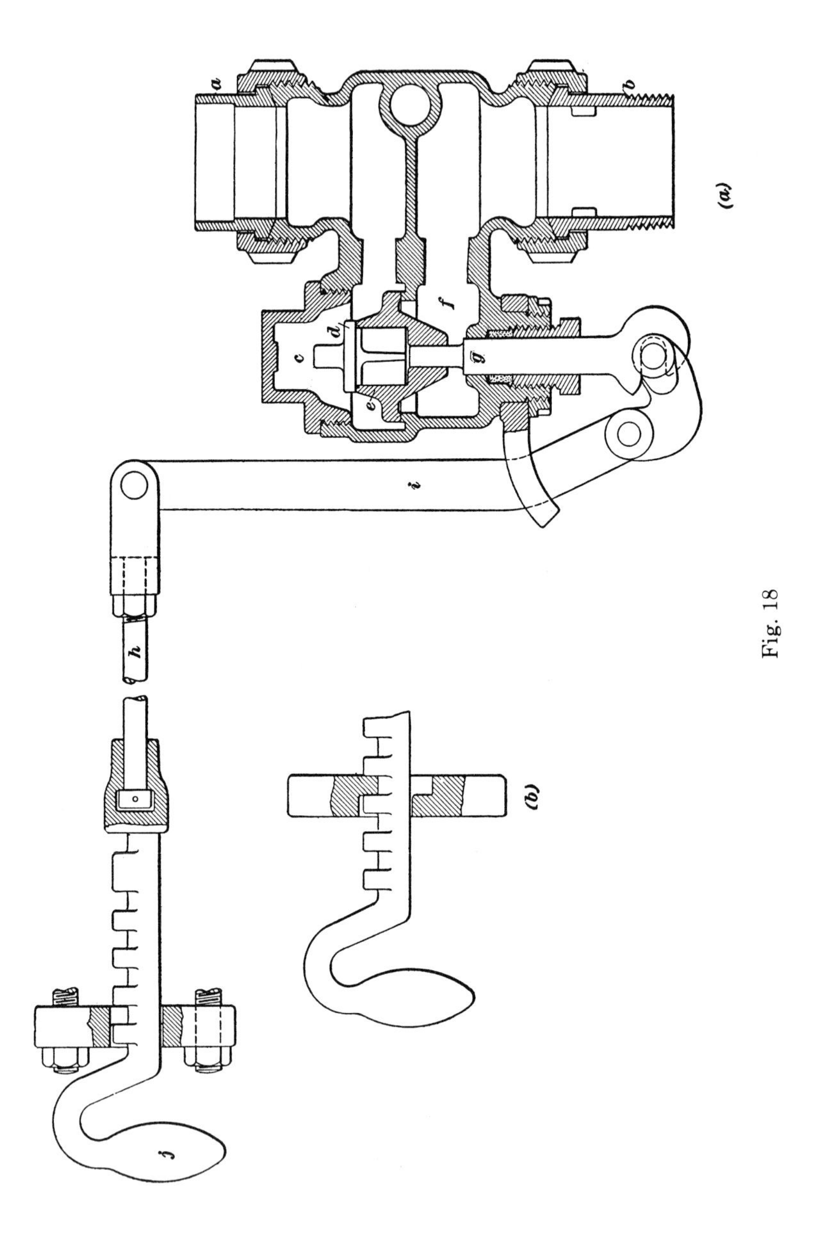

Fig. 18

the upper end of the stem g. When the starting valve is located outside the cab, it is operated by an extension rod h connected to the lever i by which the stem g is raised and lowered. When the rod h is pulled back by moving the handle j, the upper end of the stem g first strikes the pilot valve d and lifts it off its seat. Steam then flows down past the loosely fitting stem in the main valve into the space f and the pipe b. The pressure beneath the main valve thus soon becomes equal to that above it; that is, the valve is balanced. Continued movement of the handle j then brings the shoulder of the stem g against the main valve e, which is easily raised because the pressures above and below it are equal. If the starting valve is located inside the cab, the arrangement shown in Fig. 19 is used. The handle *a* is at the upper end of the operating lever *b* and the spring pawl *c* drops into the notches *d* and *e* to hold the handle in a given position.

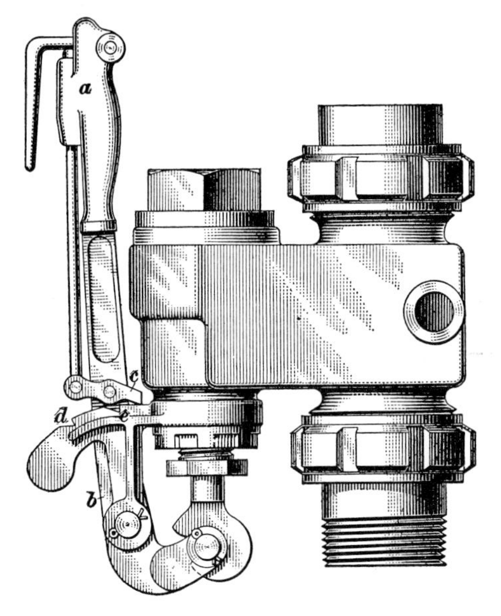

Fig. 19

57. Operation.—To start the injector, open the overflow valve and also the water valve if it has been closed, then move the handle of the starting valve back slowly as far as it will go. To stop the injector, move the handle of the starting valve all the way forward and close either the overflow valve or the water valve. To use as a heater, close the overflow valve, open the water valve, and move the handle of the starting valve back slightly.

58. Temperature of Feedwater.—The non-lifting injector will handle hotter feedwater than will a lifting injector. The reason is that a lifting injector must raise the water, and this action requires the formation of a partial vacuum in the body of the injector. Under a partial vacuum, hot water will give off steam, and this steam will expand in the suction pipe, and destroy the vacuum so that water cannot be lifted. A non-lifting injector does not raise the water, and is therefore not affected by hot feedwater to the same extent as a lifting injector. A non-lifting injector will operate with hot water, provided the water is not too hot to prevent the steam from condensing sufficiently to form the jet.

NATHAN SIMPLEX LIFTING INJECTOR

59. An exterior view of a Nathan lifting injector is shown in Fig. 20, and a sectional view in Fig. 21. The injector tubes with the spillways indicated by the letters *v* and *w* are shown in Fig. 22. The steam pipe is connected at *a*, Fig. 20, the suction pipe at *b*, the overflow pipe at *c*, and

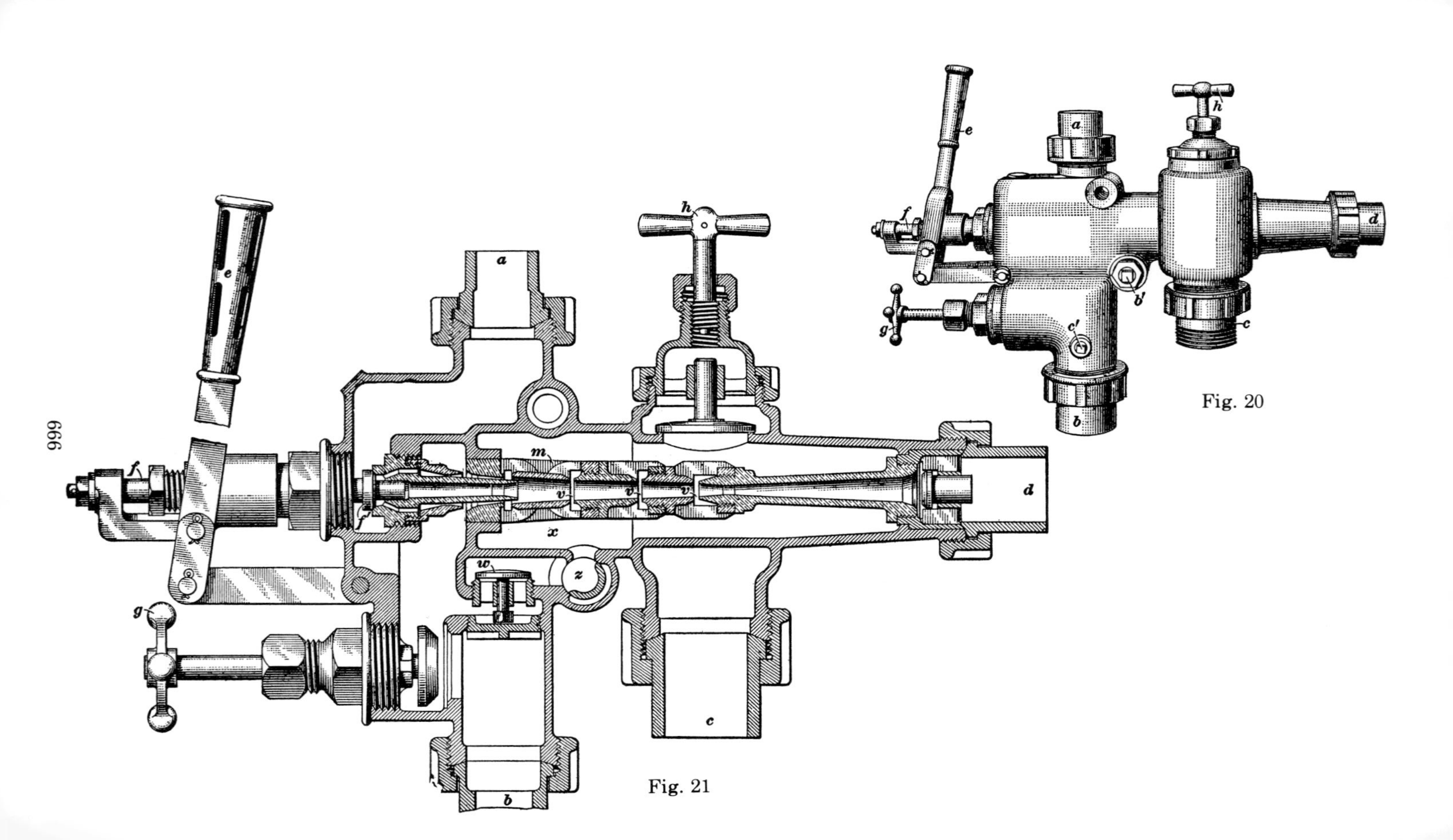

Fig. 20

Fig. 21

the delivery pipe at *d*. The starting lever *e* operates the spindle *f* of the steam valve, and the handles *g* and *h* control the opening and closing of the water valve and the overflow valve, respectively. The plug *c′* permits oil to be introduced into the injector when the feedwater is bad. The construction of the injector in Fig. 20 is so similar to that of the Sellers lifting injector that a detailed description of its construction and operation is unnecessary. However, the arrangement of the inlet valve *w*, Fig. 21, is different. This valve opens when the vacuum in the chamber *x* surrounding the combining tube *m* reaches a certain point, thus admitting water from the chamber *q*. This water enters the combining tube by way of the openings *v* and increases the capacity of the injector.

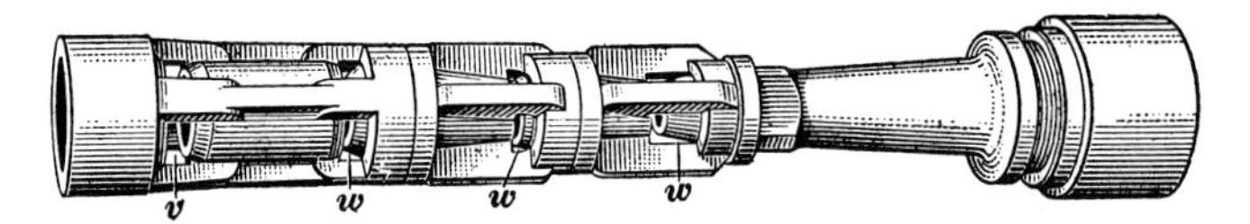

Fig. 22

The purpose of the rotary valve *z* is to close communication between the chamber *q* and the chamber *x* in case the inlet valve leaks; for, in that case, the steam might pass back from the chamber *x* to the suction-pipe chamber *q* in sufficient quantity to prevent the formation of a vacuum and the priming of the injector. It is closed by applying a wrench to the squared end of the stem *b′*, Fig. 20. The end of the valve is marked with the letters *S* and *O*; when *S* is uppermost, the valve is shut, and when *O* is uppermost, the valve is open.

NATHAN NON-LIFTING INJECTOR

60. Arrangement.—An outline drawing of a Nathan non-lifting injector with its control arrangement is shown in Fig. 23. The steam supply to the injector is controlled by the lever shown, the water valve is operated by the handle *a* and the overflow valve by the handle *b*. The telltale pipe *c* permits of a small discharge of water should the injector break. A telltale valve installed in the pipe near where it couples to the injector limits the amount of water that passes through the nozzle in the cab.

The arrangement of the steam nozzle *a*, Fig. 24, the combining tube *b*, and the delivery tube *c* follows that of the non-lifting injector just described, so that no further description is necessary. The steam nozzle is made in two parts, a main part *a* and a part *d* that is screwed on to it. The annular space between these two parts forms a nozzle that serves to deliver water to the combining tube. The connections to the injector for water, steam, overflow, and delivery are indicated on the illustration.

The intermediate check-valve *e* operates only when the feedwater is so warm that it cannot condense all of the steam. In such an event the steam will cause a part of the water to spill out through the opening *f*, lift the check-valve *e*, and escape by way of the overflow valve *g* to the overflow. But if the overflow valve is held closed by screwing down its spindle, the water will enter the combining tube through the openings *h*, *i*, and *j* and will be carried along with the water into the delivery tube. At this time the check-valve prevents the return of the hot water. The water jet is in the process of formation in the nozzle ahead of the opening *f* and if the hot water were permitted to return, it would interfere with the formation of the jet and the injector would break. At the openings *h, i,* and *j* the jet has been fully formed and the entrance of hot water does not affect it.

61. Starting Valve.—The type of starting valve used with Nathan non-lifting injector is shown in section in Fig. 25 *(a)*. The main valve encloses and carries the pilot valve *b*, both valves being moved by the

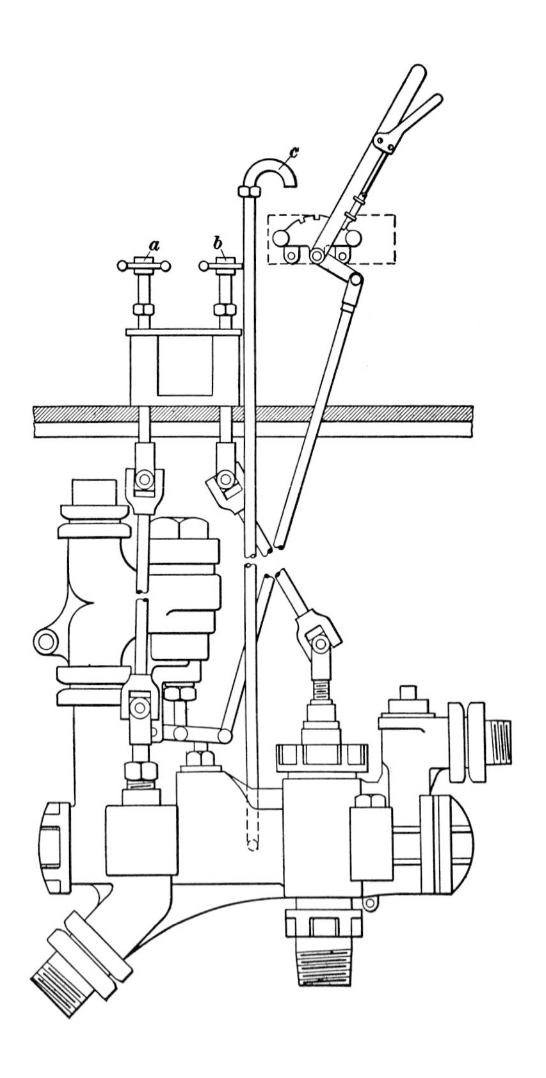

Fig. 23

stem *c*. The construction of the main valve is more clearly shown in *(b)*, which is an outside view, and in *(c)*, which is a part section only, showing the pilot valve raised off its seat in the main valve. With both valves closed, as in *(a)*, steam from the passage *d* surrounds the main valve and fills the space *e* above it. When the stem *c* is forced upward, its end raises the pilot valve *b* off its seat, as shown in *(c)*, and steam flows down the central passage around the stem to the chamber *f* beneath the main valve, as indicated by the arrows, thus balancing the main valve. Then when the shoulder of the stem *c* meets the valve *a*, the main valve is easily

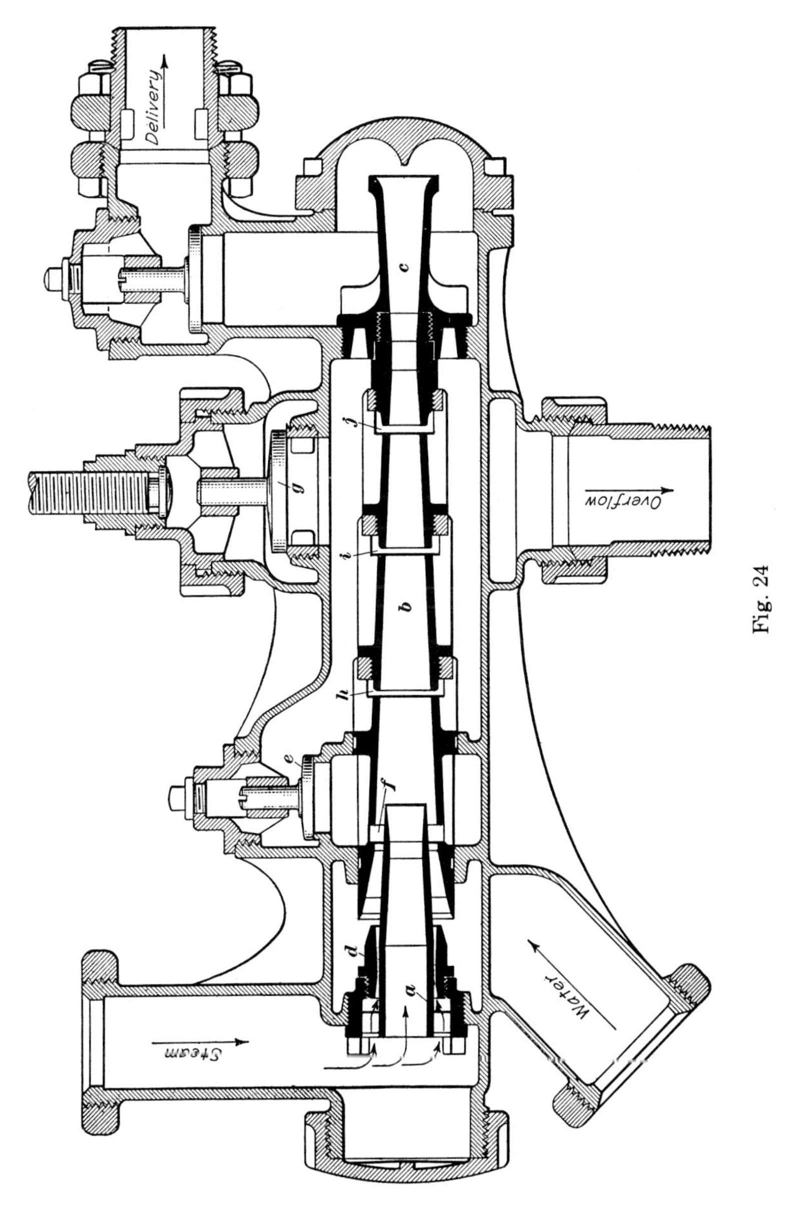

Fig. 24

lifted, and steam flows through to the passage *g* and thence to the injector. The handle *h* is connected to the lever *i* by the rod *j* and has three notches that may be engaged with the plate *k*. The spring-loaded catchbolt *l* holds the slotted handle in place. In the position shown, the starting valve is closed; in the middle notch, the injector will act as a heater; and in the forward notch the starting valve will be fully open.

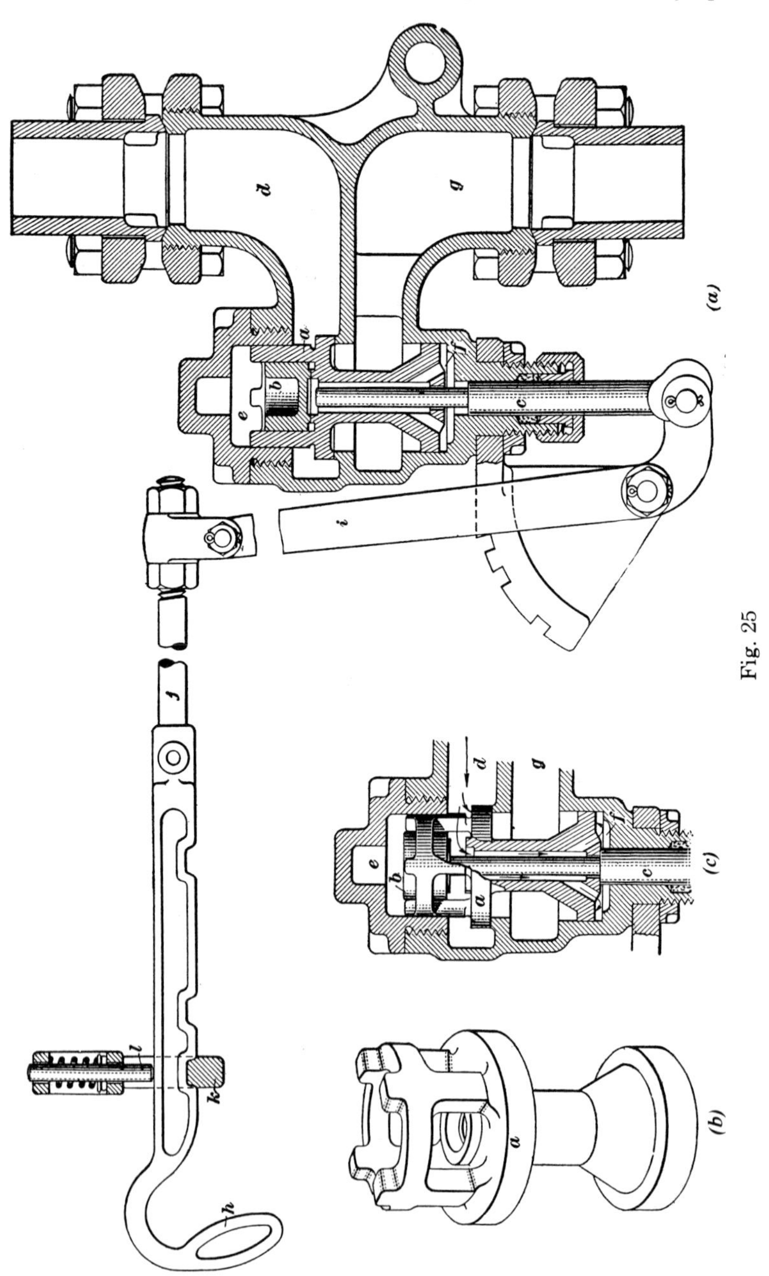

Fig. 25

62. Description.—An exterior view of an Ohio lifting injector, which is of the single-jet type, is shown in Fig. 26. The steam pipe is connected at *a*, the suction pipe at *b*, the overflow pipe at *c*, and the delivery pipe at *d*. The overflow-valve handle is shown at *f*, the water-valve handle and stem at *g*, and the lever, used to start and stop the injector, at *h*. The lever is connected, by the lever links shown, to the valve-stem crosshead *j*, which is screwed on to the end of the steam-valve stem *k*.

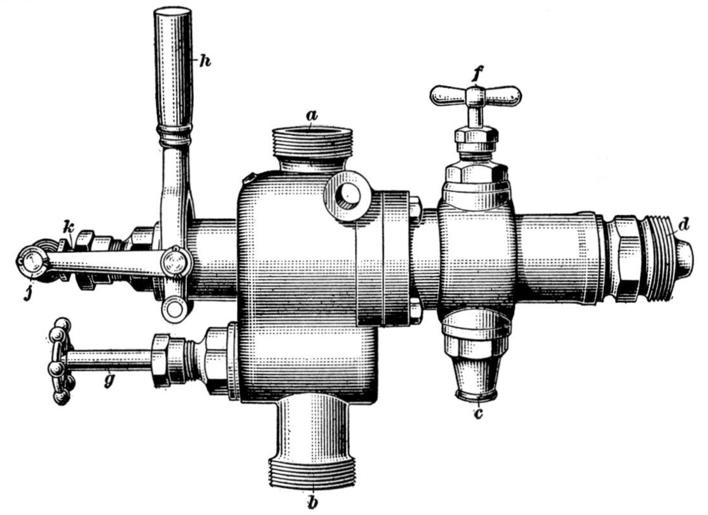

Fig. 26

A sectional view of the injector is shown in Fig. 27 *(a)*. The body of the injector is made in two parts, a back part and a front part, held together by the bolts *e*. The nozzles and tubes in the injector are the steam nozzle *l*, the combining tube, made in two parts *o* and *p*, and the delivery tube *q*. The front combining tube is screwed on to the delivery tube, which in turn is screwed into the delivery-pipe connection *r*. This connection contains the line check-valve *s* and is screwed into the body of the injector as shown. The stem of the line check-valve works in a stop-ring that is screwed into the delivery-pipe connection.

63. Arrangement of Steam Valve.—The arrangement of the steam valve differs from that of the injectors already considered and can be more easily understood by considering its action when starting the injector. A slight movement of the lever *h*, Fig. 27 *(b)*, draws the steam valve *u* to the rear until the part *v* strikes the rib *w* on the primer *m*. Steam from the steam passage *a* view *(a)*, then passes by the unseated steam valve, through the small holes *x* and passage *a'* in the primer into the combining tube *o*, thence through the openings *b'* and *d'* and by the overflow valve *t* to the overflow pipe *c*. This passage of steam expels a part of the air from the injector and the pressure of the air on the water in the tank forces the water to the injector through the suction pipe *b*. The water flows into the rear end of the combining tube through the passage *c'*, passes through the openings *b'* and *d'*, and thence to the overflow, as shown by the arrows, the injector then primes.

A further backward movement of the lever handle causes the steam valve *u* to pull the primer nozzle away from its seat at the entrance of the steam nozzle, as shown in view *(b)*. The gradual withdrawal of the primer nozzle from its seat causes the steam to enter the steam nozzle in an increasing volume, and the injector starts to operate.

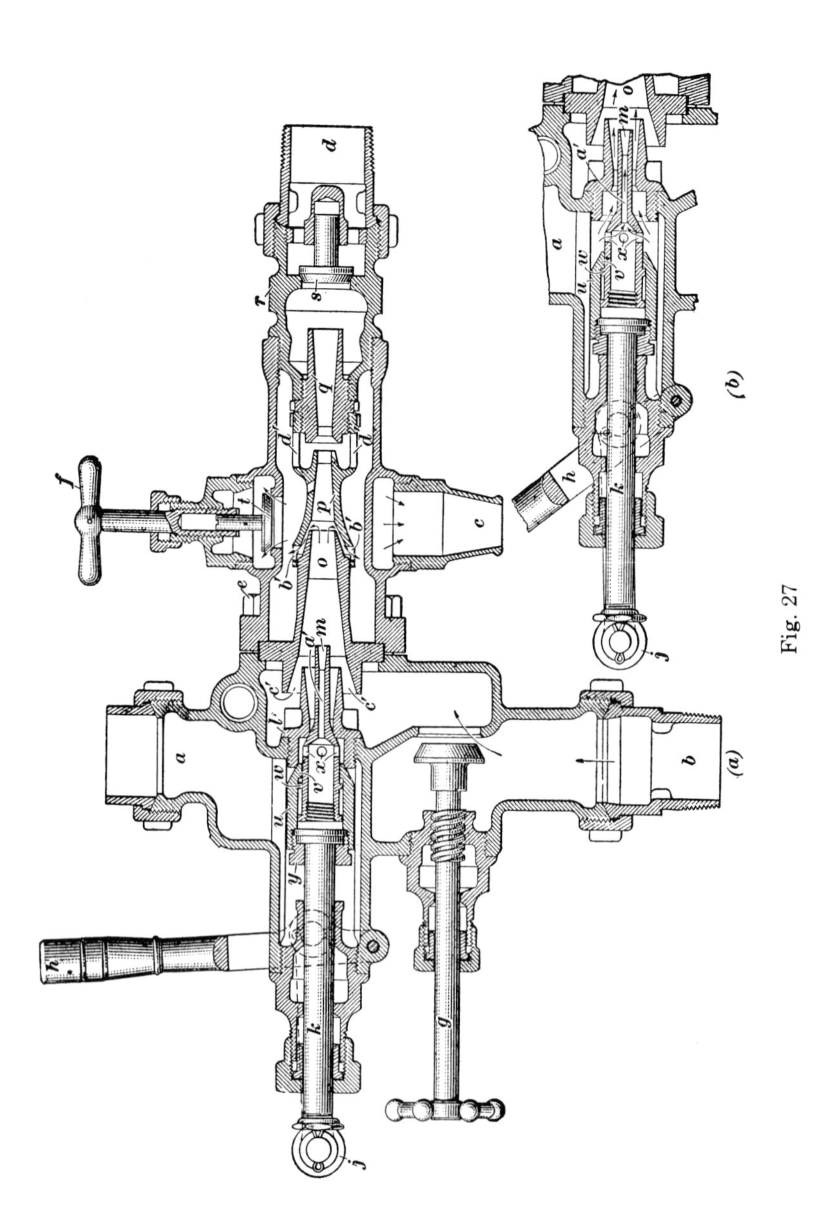

Fig. 27

64. Description.—The exterior view of an Ohio non-lifting injector is shown in Fig. 28, and a sectional view is given in Fig. 29. The steam pipe is coupled to the injector body at *a*, the overflow pipe at *b*, the suction pipe at *c*, and the delivery pipe at *d*. The stem *e* of the overflow valve is connected by a rod to a wheel in the cab, and the stem *f* of the water valve

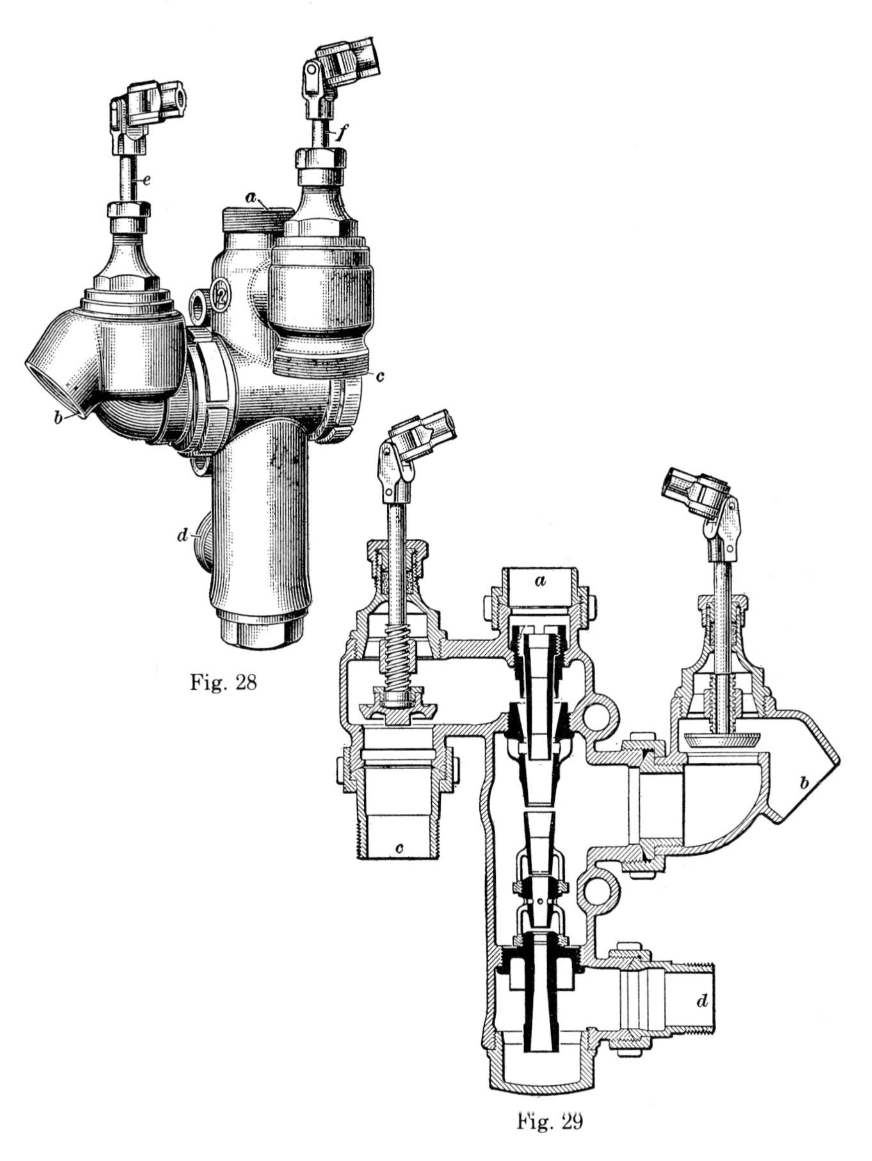

Fig. 28

Fig. 29

is also connected in the same manner to an operating wheel in the cab. The words *water* and *overflow* are cast on these wheels so that the valves can be identified. A sectional view of the lever steam-throttle valve used

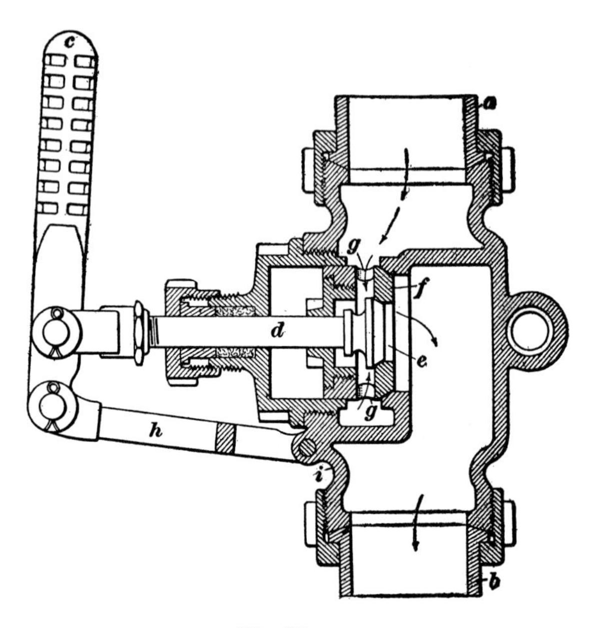

Fig. 30

to control the passage of steam to the injector is shown in Fig. 30. The steam pipe *a* is connected to the cab turret, and the steam pipe *b* leads to the injector. The lever *c* is connected to the spindle *d,* which at the front end controls the passage of steam through the port *e* in the valve *f*. The steam enters the valve *f* through the ports *g*. The stem *d* can be moved back a short distance, thereby opening port *e* before it will come in contact with and open the valve *f;* therefore, the steam is admitted to the injector steam pipe *b* gradually. The lower end of the lever *c* is connected by the links *h* to the body *i*.

The operation of the Ohio non-lifting injector is identical with that of the other types of non-lifting injectors already described.

CONDENSATION OF STEAM WITH HOT WATER

65. Conditions Necessary for Condensation.—If very hot steam is turned into cold water, the steam in heating the water will lose heat and condense, and this process will continue until the water begins to boil. Once boiling begins, the water cannot be made any hotter, hence it will absorb no more heat from the steam and the further cooling and condensing of the steam will stop. The heat now added to the water will be expended in expelling the particles of water in the form of steam.

The boiling temperature of water depends on the pressure to which it is subjected; the higher the pressure, the higher will be the boiling temperature. At atmospheric pressure, water boils at 212° F., hence water at this pressure will condense steam until heated to this temperature, but as the water cannot be made any hotter, the further condensation of the steam will stop. Water under a pressure of 30 pounds must be heated to about 297° F. before boiling will occur. For every pressure there is a certain boiling temperature; this information is given in Steam Tables.

Now should it happen that water under a pressure of 30 pounds is at a temperature of 200° F. instead of 274° F., steam if turned into the water will condense until the temperature is increased to the last named figure. Hence in order to have steam condense with hot water, all that is necessary is to have the water at a temperature below the boiling point for that pressure.

66. Design for High Temperature of Delivery Water. —With the injectors that lift and discharge water to the combining tube at atmospheric pressure or less, the temperature of the water delivered to the boiler with the injector working at or near capacity cannot exceed 212° F., because this is the boiling temperature for this pressure and cannot be exceeded. If it is desired to deliver water at a higher temperature, the injector must be of the double-jet type; that is, one set of tubes must be employed to deliver water under pressure to another set but at a temperature less than that corresponding to this pressure. Then this water, although already heated, is capable of condensing steam when passing through the other set of tubes, thereby raising the temperature of the delivery water. The Hancock inspirator is designed on this principle, hence it will deliver hotter feedwater to the boiler than a single-jet injector.

DOUBLE-JET INJECTORS

HANCOCK INSPIRATORS

67. Origin of Term.—The Hancock inspirator is an injector that is designed to deliver water to the boiler at a higher temperature than the ordinary type of injector.

The term inspirator originated from the fact that John T. Hancock in 1868 made some improvements in an apparatus used to draw in air. He applied the term inspirator to the device, this word being used in the sense of drawing in, or inhaling. In later experiments, the device was used to lift water, and this was later followed by an apparatus that would lift and force water. The term inspirator, which was applied to the original device for drawing in air, continued to be applied to the later devices, with the result that the injector that was finally evolved was called an inspirator.

68. Conventional Design.—The conventional design of the non-lifting double-jet injector shown in Fig. 31 will be used to explain the principle of the Hancock inspirator and why it delivers hotter water than a single-jet injector.

The injector contains two tubes or nozzles with separate steam nozzles *a* and *b* for each. The lower tube is called a lifter tube and is made up of a combining tube *c* and a delivery tube *d* combined in one part. The upper tube is called a forcer tube and it also is made up of a combining tube *e* and a delivery tube *f*. The chamber *g*, with an outlet normally closed by the intermediate overflow valve *h*, connects the front end of the lifter tube to the rear end of the forcer tube. The injector is started by opening the valve *j*, thereby inducing a flow of water through the tubes *c* and *d*, by the valve *h*, and out through the valve *l*. The valve *i* is opened next; this places the water in the tubes *e* and *f* under pressure; the valve *l* is then closed. The valve *h* now prevents the return of water to chamber *g*.

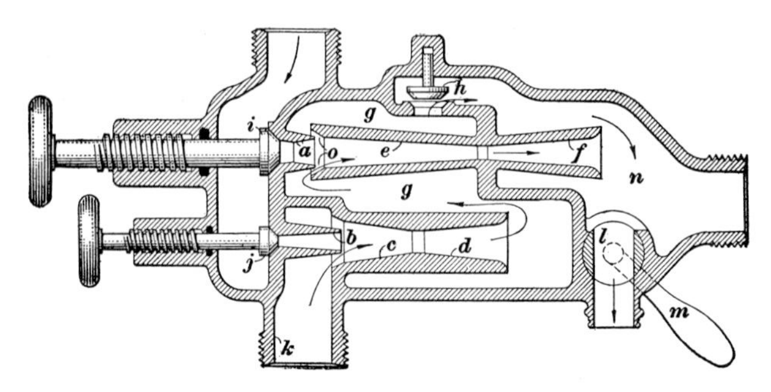

Fig. 31

With steam at a pressure of 200 pounds and water at 75° F., the lifter tube delivers water at a pressure of about 31 pounds and a temperature of about 110° F. to the forcer tube from which the water passes to the boiler. Now the boiling point of water at 31 pounds pressure, or the point up to which it will condense steam, is about 273° F., so that theoretically the steam passing through the forcer tube will condense until the water is raised to this temperature. However, the temperature of the water fed to the boiler depends on the regulation of the injector.

HANCOCK LIFTING INSPIRATOR

69. Description.—An exterior view of a Hancock lifting inspirator is shown in Fig. 32. The steam pipe is coupled at *a*, the suction pipe at *b*, the overflow pipe at *c*, and the delivery pipe at *d*. The lever *e* is connected to the rear end of the heater rod *f* by the stud *g*, and the front end of the rod is connected to the lower end of the overflow crank *h*, the upper arm of which is connected to the overflow-valve stem *j*. The crank *h* is connected by the stud *i* to the crank holder *k*, which is slipped over the intermediate overflow-valve bonnet *l* and is held in position by the capscrew *m*.

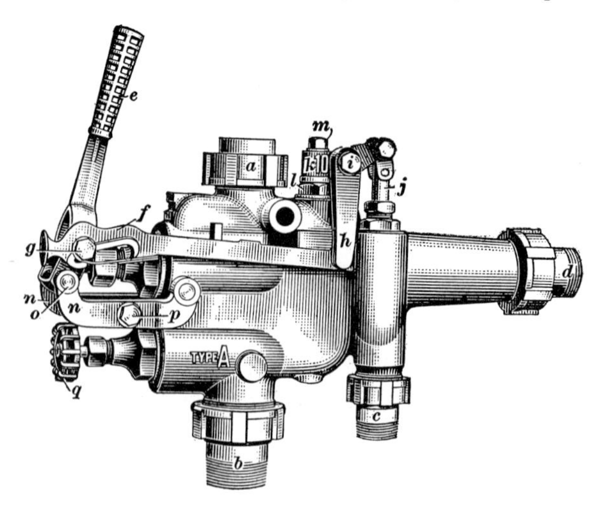

Fig. 32

The lower end of the lever *e* is placed between and is connected to the two side straps *n* by the pin *o*. The front ends of the links are connected to the injector as shown, and the bolt *p* is used to keep the links together. The handle of the regulating valve is shown at *q*.

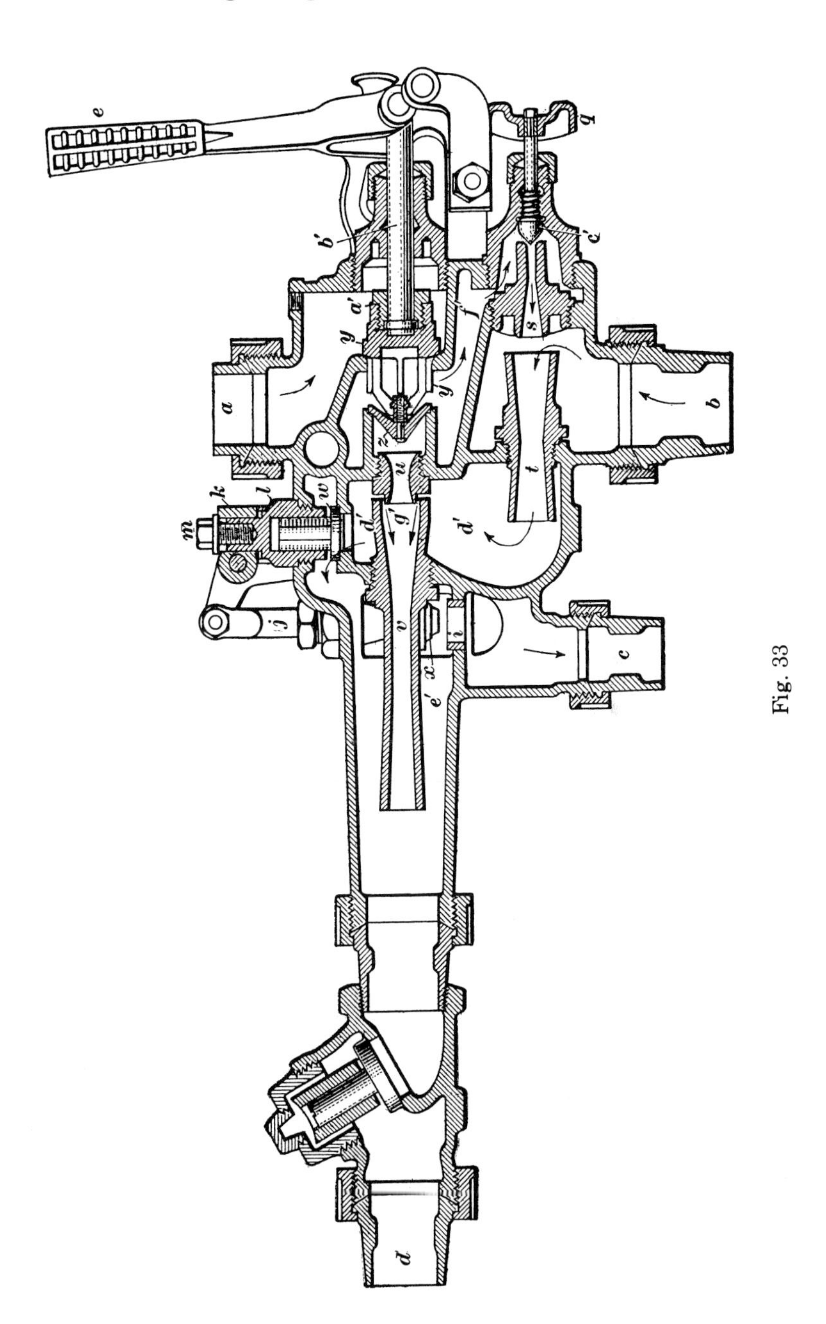

Fig. 33

70. A sectional view of the injector is given in Fig. 33, in which the principal exterior parts are given the same reference letters as in the exterior view. The names of the principal parts in the interior of the injector are the lifter nozzle *s*, the lifter tube *t*, the forcer nozzle *u*, and the forcer tube *v*, all of which are screwed into the body of the injector as shown; the intermediate overflow valve *w*, the final overflow valve *x*, and the compound forcer steam valve that consists of two valves *y* and *z* connected by the coupling nut *a′* to the stem *b′*. The rear end of the stem is connected to the lever *e*. The regulating valve *c′* is operated by the handle *q*.

71. Compound Forcer Steam Valve.—As shown by the detail in Fig. 34, the compound forcer steam valve is made up of a valve *a* and a valve *b*. The valve *a* is connected to the forward ribbed portion of the valve *b* by the sleeve *c* and the bolt *d*. The bolt is threaded through the sleeve and is prevented from backing out by a pin *e*. The sleeve *c* and hence the valve *a* are free to move back and forth in the valve *b*; that is, the valve *a* can move back until the part *f* comes in contact with the part *g* of the valve *b*. The valve can move forward until the collar *h* on the sleeve strikes the part *i* of the valve as shown.

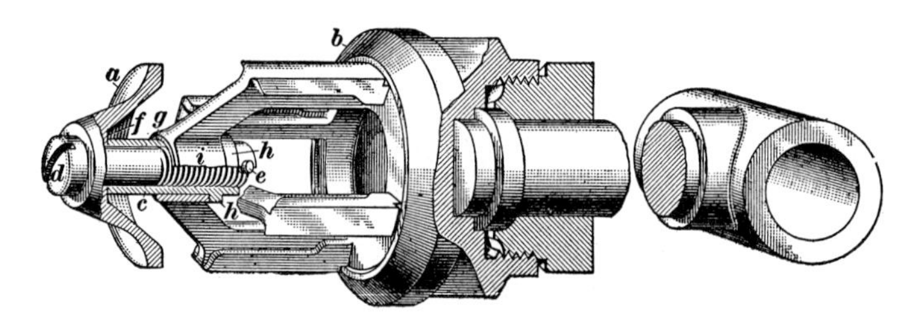

Fig. 34

The valve *a* is merely a sliding valve on the front of the valve *b*. The idea of the design is, when the injector is being started, to open the valve *b* first so as to insure that the water is flowing through the lifter and the forcer tubes before the valve *a* opens and admits steam to the latter tube.

72. Regulating Valve.—The regulating valve *c′*, Fig. 33, operated by the handle *q* is used to govern the amount of steam that passes through the lifter nozzle *s*, and thereby regulates the quantity of water that is being forced through the lifter tube *t* to the forcer tube *v*. This valve can be compared to the water valve of a single-jet injector, because it serves the same purpose, namely, to regulate, when desired, the amount of water that the injector is delivering. However, the regulating valve affects the water supply by regulating the steam supply, whereas the water valve affects the water supply direct. The valve is wide open when a pin in the handle *q* is at the top.

73. Location of Final Overflow Valve.—To insure proper operation the design of a double-tube injector will not permit of spillways being cut in the tubes as with a single-jet injector, hence when priming and starting the injector the surplus water must be carried all the way through the tubes. In this way the water from one set of tubes does not interfere with the passage of the water through the other set. The final overflow valve must then be placed in the forcer-tube chamber, and must be held closed mechanically. The opening normally closed by the intermediate overflow valve in combination with the final overflow valve opening takes care of the surplus water discharged from the lifter tube when starting the injector. The purpose of the intermediate overflow valve is to close this opening when the injector is in operation and thereby prevent the water that is above boiler pressure in the forcer-tube chamber from returning to the lifting-tube chamber and stopping the injector.

74. Operation.—The inspirator is started by pulling back the lever *e*, Fig. 33, slightly, thereby causing the stem *b′* to unseat the valve *y* but not the valve *z*. The steam from the steam pipe *a* then passes by the valve *y* through the passage *f′* and the lifter nozzle *s*, lifts the water from the suction pipe *b*, and carries it through the lifter tube *t* to chamber *d′*. The mixture of steam and water next lifts the intermediate overflow valve *w* and passes by the final overflow valve *x* to the overflow pipe *c*; part also enters the end of the forcer tube *v* by passing through the opening *g′* and after filling the delivery pipe escapes through the overflow pipe.

As soon as the inspirator primes, the lever *e* is drawn back farther, thereby taking up the slack between the valve *y* and the valve *z* and unseating the latter valve. The action that now occurs is similar to that which has already taken place in the other set of tubes. The water that is being delivered through the openings between the forcer nozzle and the forcer tube is carried forward through the rear portion of the tube at a high velocity by the high-velocity low-pressure steam discharging from the forcer nozzle. By this time the starting lever is about all the way back, and the final overflow valve is nearing its seat *i* and is gradually stopping the discharge of water at the overflow. The building-up of pressure in chamber *e′* now forces the intermediate overflow valve *w* to its seat and the discharge of water through it from chamber *d′* stops. With the lever all the way back, the overflow valve is forced to its seat, and by this time the pressure in the forward portion of the forcer tube and in the delivery pipe is sufficient to open the boiler check-valve. The lifter tube delivers the water to the forcer tube at a pressure of about 31 pounds and at a temperature of about 110° F. The temperature of the delivery water is about 220° F. The lifting inspirator does not deliver quite the same amount of water per pound of steam used, which accounts for the slightly higher temperatures than those previously given. The inspirator is self-regulating because if the steam pressure rises, the lifter tube will lift and deliver more water to the forcer tube but, as there is a correspondingly greater amount of steam passing through this tube, proper care is taken of the increased supply of water.

75. Using Inspirator as a Heater.—The inspirator can be used as a heater by lifting the heater rod *f*, Fig. 32, until it is disengaged from the stud *g*, and drawing the rod back until the final overflow valve is closed. The lever *e* should then be drawn back to the lifting position and the amount of steam to the suction pipe should then be regulated by the regulating valve.

76. Types.—The Hancock non-lifting inspirators are known as the type H-N-L, with a capacity of 2,500 to 6,000 gallons per hour, the type L-N-L with a capacity of 6,500 to 8,000 gallons per hour, and the type K-N-L, with a capacity of 8,500 to 12,000 gallons per hour, and the type M-N-L.

77. Exterior View.—In Fig. 35 *(a)* is shown an exterior view of the L-N-L type of inspirator and the operating valve, and in view *(b)* is shown the arrangement for operating the overflow valve. A part of the overflow extension rod *a* is shown connected to the wheel rod *b*, and the lower part of the rod is connected to the overflow valve stem *c*. The steam pipe from the turret is connected to the operating valve at *d*, the suction pipe is connected at *e*, the delivery pipe at *f*, and the overflow pipe at *g*.

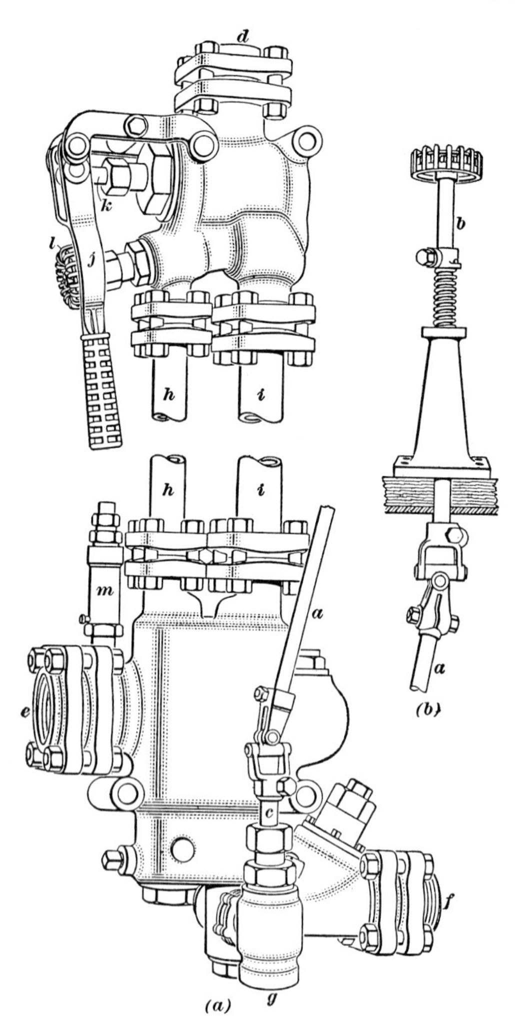

Fig. 35

Two pipes h and i lead from the operating valve to the injector. The pipe h is the regulating steam pipe, and the pipe i is the forcer steam pipe.

The passage of steam from the steam pipe at d to the pipes h and i is controlled by a steam valve in the operating valve body that is operated by the lever j and the valve stem k. The amount of steam that passes

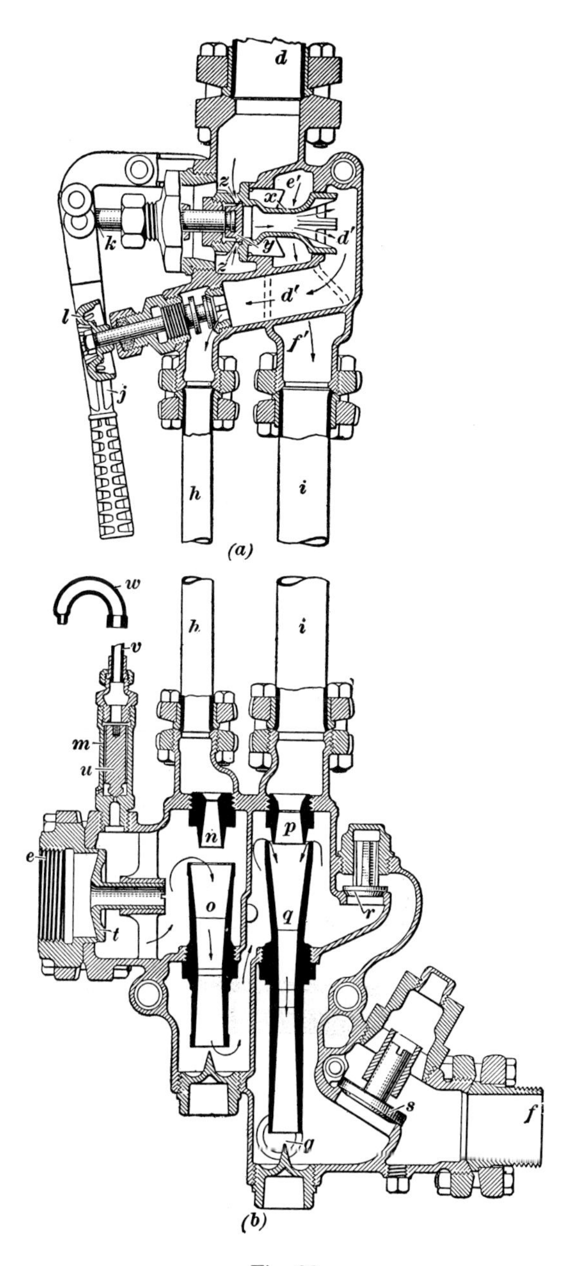

Fig. 36

through the pipe *h* can be regulated by the regulating valve, the handle of which is shown at *l*.

The steam valve and the regulating valve in the operating-valve body serve the same purpose as similar valves in the lifting type of inspirator.

78. Sectional View.—A sectional view of the inspirator is shown in Fig. 36, in which similar parts have the same reference letters as in Fig. 35. The lifter nozzle is marked *n*, the lifter tube *o*, the forcer nozzle *p*, and the forcer tube is *q*. The intermediate overflow valve is shown at *r*, and the line check-valve at *s*. With late types of this injector, the valve *r* is not used. The purpose of the valve *t* here shown seated is to prevent the steam from blowing the tank hose off should the injector break. The final overflow valve is not shown because in order to show a sectional view of the injector the part that contains the overflow valve has to be removed, but the outlet is shown at *g*.

An alarm valve *u*, which is contained within the body *m*, lifts when the injector breaks and permits steam to escape through the pipe *v* connected to the top of the body that leads up into the cab. The upper end of this pipe is fitted with a nozzle *w*.

79. Steam Valve.—The arrangement for admitting steam from the steam passage *d* in the injector to the pipes *h* and *i*, as shown by the sectional view of the steam valve body, Fig. 37, comprises a steam valve *x* and a steam valve auxiliary disk *y*, connected to the steam-valve stem *k*

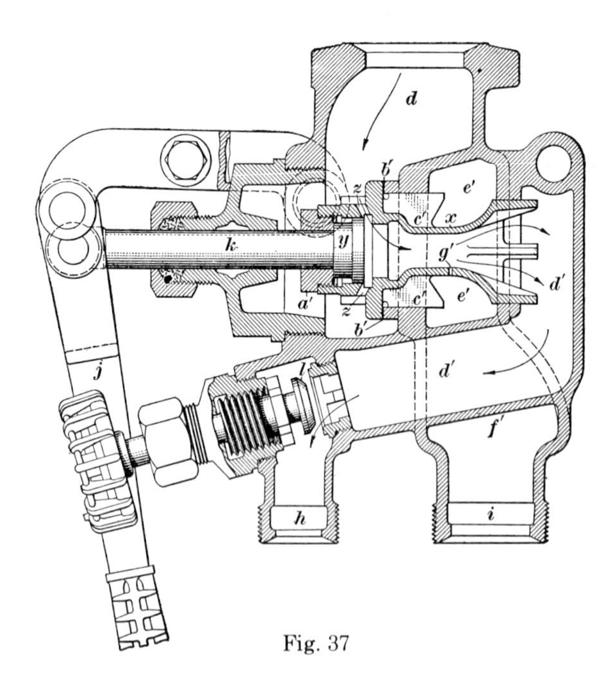

Fig. 37

by the disk nut *a'*. When the valve stem is drawn back by the lever *j*, the auxiliary disk first opens the ports *z* as shown, thereby connecting chamber *d* to the interior of the operating valve *x*. As the lever is pulled back farther, the slack between the auxiliary disk and the disk nut *a'* is taken up and as soon as these parts come in contact a further movement of the lever draws the steam valve x backwards from its seat at b'. The steam valve is guided at the rear by the valve wings c', while at the front the valve is of the piston type and is also wing-guided. The steam that flows through the interior of the steam valve when the auxiliary disk y is opened passes by way of the passage d' and the regulating valve l to the pipe h, while the steam that enters chamber e', when the steam valve is opened, passes through a passage f' around passage d', and enters the pipe i. At this point the steam valve is fully open and the piston at the front end of the valve is withdrawn from its orifice, permitting a full flow of steam from chamber e' to chamber d'. This permits the main steam valve to feed both pipes h and i. The port g' is to allow a little steam down the pipe i to clear out the water and aid the circulation in the forcer tubes when the inspirator is used as a heater.

80. Operation.—To start the inspirator, first open the overflow valve and pull the valve lever back to priming position, thereby establishing the circulation of water through the lifter and forcer tubes. When the inspirator is thus primed, draw the steam-valve lever back slowly to full operating position and close the overflow valve.

The inspirator is stopped by closing the operating valve. The inspirator can be used as a heater by closing the regulating valve and pulling the lever to priming position as shown in Fig. 37. The regulating valve should then be opened the amount required to heat the water, but not enough to open the alarm valve. It is understood that the overflow valve is now closed.

It is unnecessary to trace the flow of water through the injector because this does not differ from the lifting inspirator already described.

81. Type M-N-L.—The principal difference between the L-N-L and M-N-L types of inspirators is that with the latter the separation of the steam for the lifter tubes and the forcer tubes is accomplished in the inspirator. With the former this separation is made in the operating valve, hence two steam pipes are required. The M-N-L inspirator was designed to do away with the operating valve in the cab and one steam pipe.

An exterior view of the type M-N-L Hancock non-lifting inspirator is shown in Fig. 38. The steam pipe is connected to the body of the inspirator at a, the suction pipe at b, the delivery pipe at c, and the overflow pipe at d. The injector steam valve is opened and closed by means of the operating lever e, the connecting-rod f, the operating crank g, and steam valve h. The left end of the operating crank is pinned to the top of the side strap i. The operating lever e is held in any position desired by the notched quadrant j, which is bolted to some convenient point in the cab. The connection between the rod k and the stem l of the water valve and between the rod n and the stem o of the final overflow valve is made by the rods m and p, respectively. The water supply can be regulated either by a water valve, as in Fig. 39, or by a steam valve. With water regulation the water valve w is placed in the water passage. With steam regulation a valve placed in front of the lifter nozzle s regulates the water supply by varying the flow of steam through this nozzle. Either valve is operated by the handle k, Fig. 38.

82. A sectional view of the injector is given in Fig. 39, in which the exterior parts have the same reference letters as in the exterior view. The delivery pipe c, and the final overflow valve r have been moved out of their true positions so as to appear in the section. The arrangement of the nozzles and the tubes do not differ materially from the H-N-L type of injector. The names of the tubes are as follows: s, lifter nozzle; t, lifter tube; u, forcer nozzle; and v, forcer tube. The water or suction valve is marked w, and the intermediate overflow valve is shown at x.

The compound forcer steam valve is similar to that used with the lifting injector. It consists of a valve y, that makes a seat at z and a valve a' that is applied to the valve y, in such a way that this valve can unseat a small amount without unseating the valve a'. The alarm valve b' permits steam to discharge through the pipe shown, should the injector break.

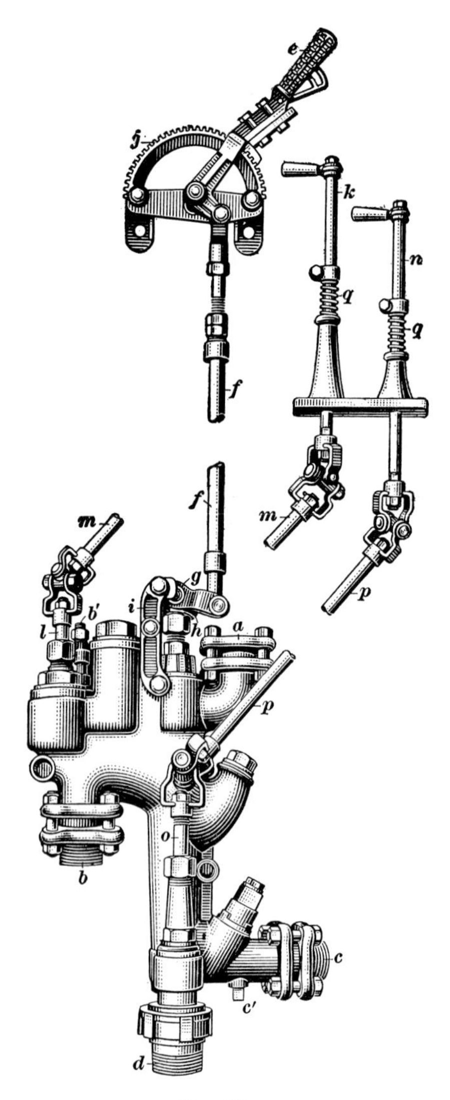

Fig. 38

The intermediate overflow valve *x* and the final overflow valve *r* are used for the same reasons as with the other types of Hancock inspirators. The plug *c'*, permits the delivery pipe to be drained, should it become necessary to do so in severe weather.

83. Operation.—With the suction or water valve *w*, Fig. 39, open, the injector is started by opening the final overflow valve *r*, thereby allowing the water to pass out through the overflow pipe *d*. The operating lever is next drawn back slightly so as to unseat the steam valve *y*, but not the steam valve *a'*. The steam passes by the unseated steam valve *y* through the passage *d'* and the lifter nozzle *s* and starts the circulation of the

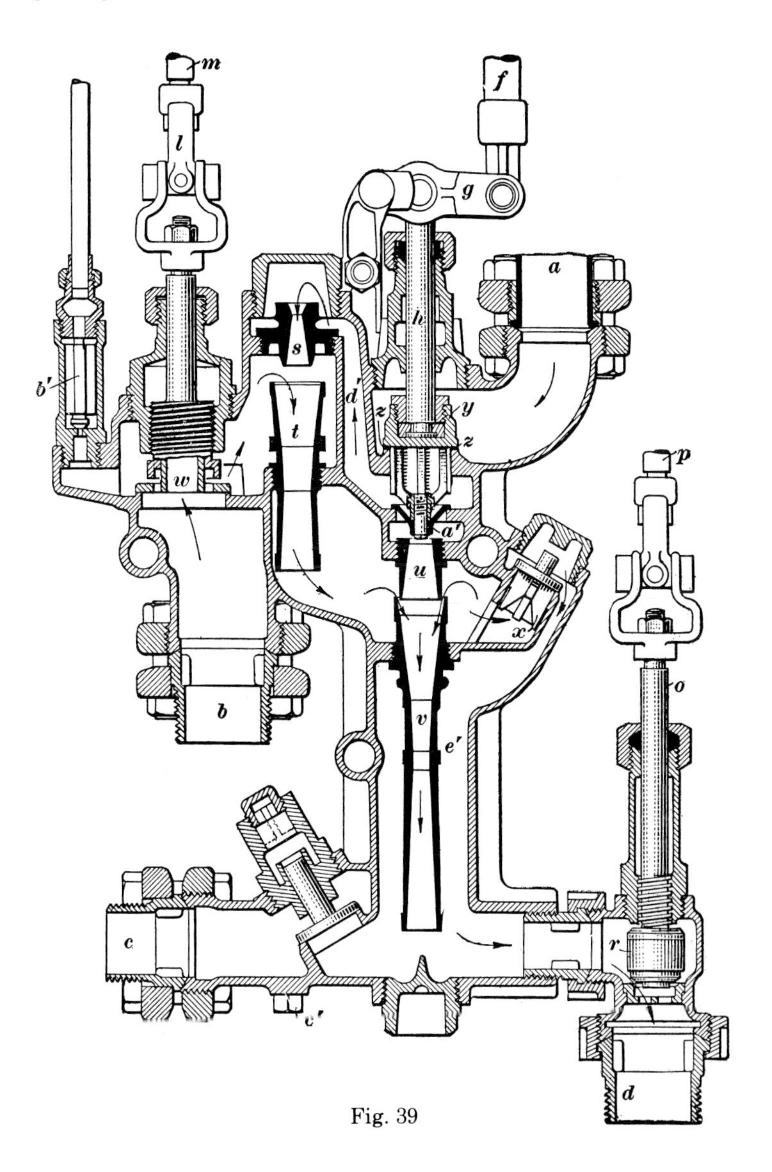

Fig. 39

water through the lifter tube *t* and the forcer tube *v* as shown by the arrows.

Some of the water also passes by the intermediate overflow valve *x* and flows through the passage *e'* to the final overflow valve. When water discharges freely at the final overflow valve, the operating lever is pulled all the way back, thereby opening the steam valve *a'*, and the final overflow valve is closed. The water is then forced into the boiler. The regulation of the water supply is obtained by turning the handle *k*, Fig. 38, in the proper direction.

The injector is stopped by moving the operating lever forward all the way. The injector is used as a heater by closing the final overflow valve, opening the water valve, and then drawing the operating lever back a few notches. Care should be taken not to pull the lever back too far because, in this event, enough steam will be admitted to the injector to lift the annunciator or alarm valve and discharge through the annunciator pipe to the cab.

BOILER CHECK-VALVES

84. Purpose.—The purpose of the boiler check-valve is to close communication between the boiler of the locomotive and the delivery pipe of an injector and thus prevent the return of water from the boiler to the delivery pipe when the injector is not in operation.

Usually a stop-valve is used in combination with a boiler check-valve, so that the check-valve can be removed and ground steam-tight when the boiler is under pressure.

85. Description.—A sectional view of a combination boiler check- and stop-valve is shown in Fig. 40. The connection between the check-valve body *a* and the boiler is made by a flange *b*, screwed on to the body and connected to the boiler by studs and nuts. A steam-tight joint between the boiler and the check-valve body is made by a ring, not shown, which is ball-faced on the side next to the boiler. The steam and water in the boiler can be shut off from the check-valve body *a* and the delivery pipe *d* by closing the stop-valve *e* against the seat *f*. The boiler check-valve *g* can be removed by taking off the check-valve cap *h*.

86. Duplex Top Check and Stop-Valve.—The principal reason for placing the boiler check-valve on top of the boiler is that certain impurities in the feedwater, when it is delivered to the boiler below the water level, will form a hard scale on the sheets, whereas if the water is delivered into the steam space, the impurities will be deposited in the form of a mud that can be blown out.

Also, the water in falling through the steam space becomes heated and does not have such a cooling effect on the sheets as when the water is delivered on the side of the boiler. The two check-valves are combined so as to simplify their application to the boiler.

The arrangement of the check-valve body when applied to the top of the boiler is shown in Fig. 41 *(a)*. The extension *a* with side openings *b* to deflect the water from the dry pipe forms a ball joint with the top sheet *c*, and the flange *d* of the check-valve body is pulled down steam-tight against the extension *a* by the stud and nuts shown. The delivery pipes are marked *e* and *f*. Either of the boiler check-valves can be removed by unscrewing the caps *h* and *i*. The stems of the stop-valves used to shut off the steam from the boiler when necessary to remove the check-valves are marked *j* and *k*.

A part sectional view of the check-valve body *g* is shown in Fig. 41 *(b)*, in which the stop-valve *k* is shown turned around out of its true position in order that the arrangement of the passages in the check-valve body may be more easily seen. The water from the right-hand delivery pipe passes, as shown by the arrows, under the boiler check-valve *l* and thence through passage *m* and by way of the stop-valve *k* to the boiler. The

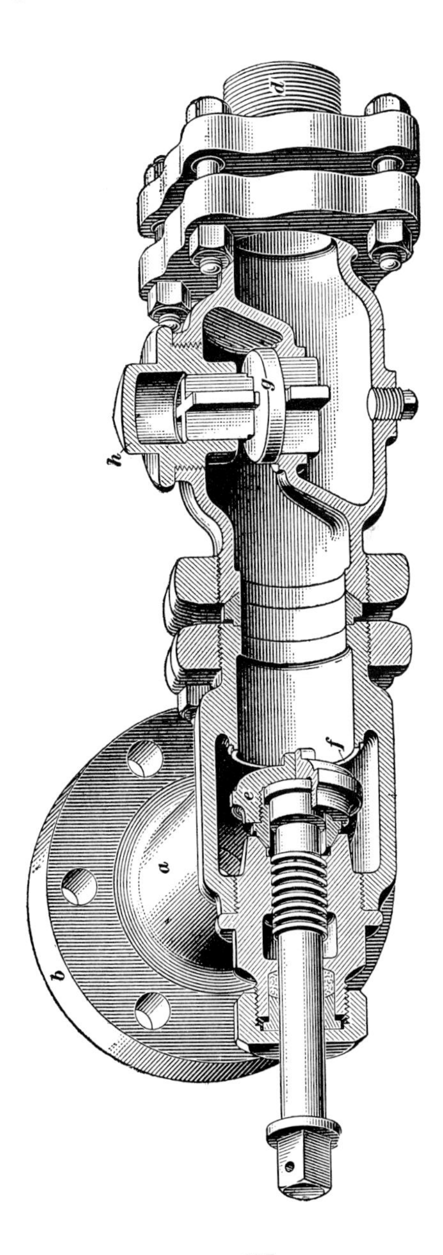

Fig. 40

pressure in the boiler can be shut off from the check-valve *l* by closing the stop-valve *k*. However, before the cap *i* is removed and the check-valve taken out, the relief plug *n* must be unscrewed far enough, as shown, to permit the steam in passage *m* to escape through passage *o*. A similar arrangement of passages, check-valve, relief plug, and stop-valve is contained in the other half of the check-valve body.

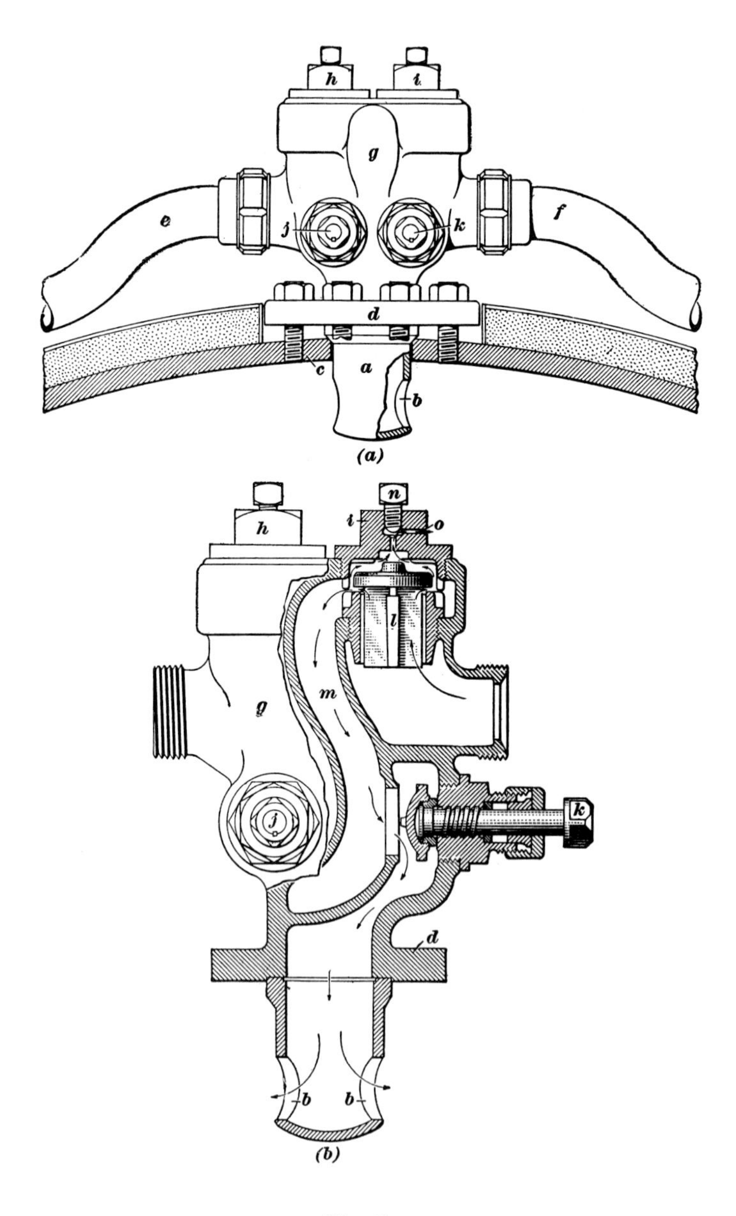

Fig. 41

87. The purpose of the coal sprinkler, or squirt, is to wet down the coal in the tender or flush the deck of the engine. The sprinkler was formerly connected to the delivery pipe of the injector but the high pressure of the water frequently blew off the hose, and caused accidents. Modern coal sprinklers take the water from the suction pipe and discharge it through the sprinkler pipe and hose at a low temperature and velocity.

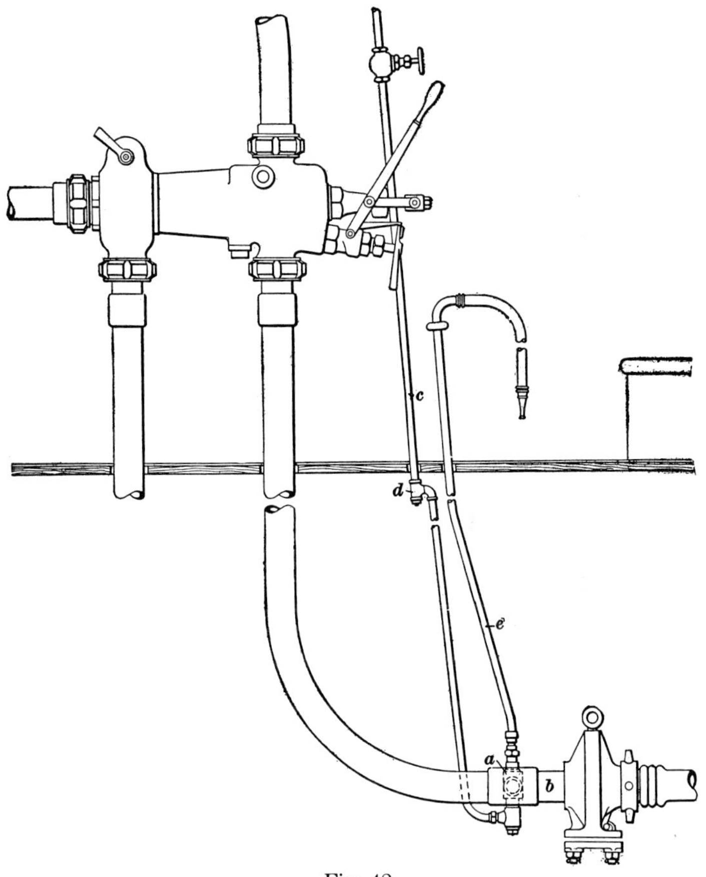

Fig. 42

The arrangement of the Sellers sprinkler is shown in Fig. 42, in which *a* is the sprinkler body connected to the side of the suction pipe *b*, *c* the steam pipe with a globe steam valve and a vent valve *d*, and *e* the sprinkler pipe with a hose. A sectional view of the sprinkler body and the interior parts is shown in Fig. 43. The steam enters the sprinkler, when the steam valve is opened, through the pipe *c* and closes the vent valve *d* and the drain valve *f*. The steam then seats the check-valve *g* and begins to move the piston *h* and the water valve *i* upwards. As soon as the water valve opens, the water that is always present in the suction pipe *b* and in chamber *j* passes to the delivery tube *k*. The piston *h* continues to move

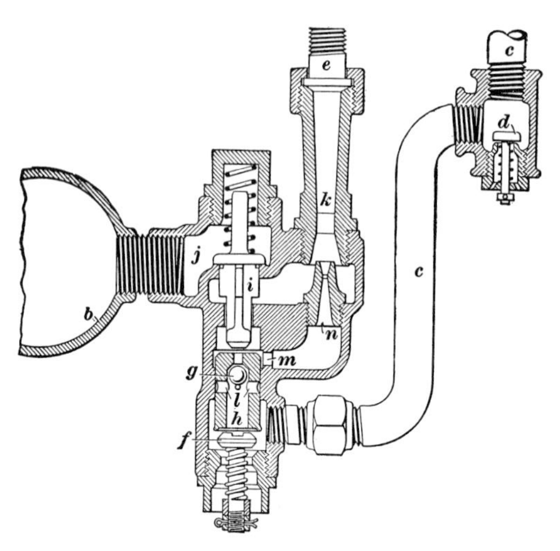

Fig. 43

upwards until the ports l register with the ports m. The steam then passes through the steam nozzle n and a jet of water forms in the delivery tube k. As soon as the jet forms, the water is drawn from the suction pipe and delivered to the pipe e. The sprinkler is really an injector that is designed to deliver water at a low velocity and at a low temperature. When the steam valve is closed, the water valve i is closed by its spring and the pressure of the water in the suction pipe. The check-valve g unseats and permits the water to drain from the pipe e and the tube k past the drain valve f, which is unseated by the spring beneath it. The vent valve d also opens and allows the air to enter and thereby force the water and the condensed steam from the steam pipe.

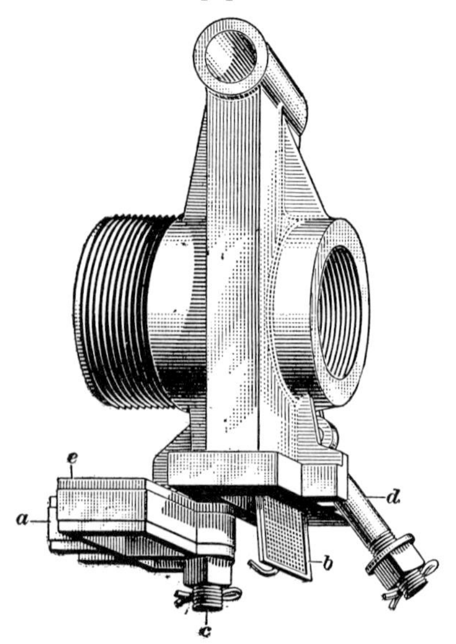

Fig. 44

LOCOMOTIVE FEEDWATER STRAINER

88. The purpose of the feedwater strainer is to prevent dirt or foreign particles that may pass through the strainer in the tank that encloses the tank valve from entering the injector. The feedwater strainer is placed between the end of the suction pipe and the tank hose as shown in Fig. 42. An exterior view of the Sellers strainer is shown in Fig. 44 with a cap *a* turned out of position so that the strainer *b*, here shown partly withdrawn, can be removed for cleaning.

The strainer can be removed by slackening the nut on the fixed stud *c* and the nut on the **T**-head bolt *d*, which is pivoted at the upper end. The bolt *d* is then pulled out of a slot in the cap and the cap turned to the position shown. The gasket *e* makes a water-tight joint between the cap and the body of the strainer.

TANK VALVE

89. The purpose of the tank valve, two of which are applied to a tank, is to admit water to the gooseneck and thence to the suction pipe of an injector. A perspective view of one type of tank valve with one side as well as the wall of the tank broken away is shown in Fig. 45. The tank valve *a*

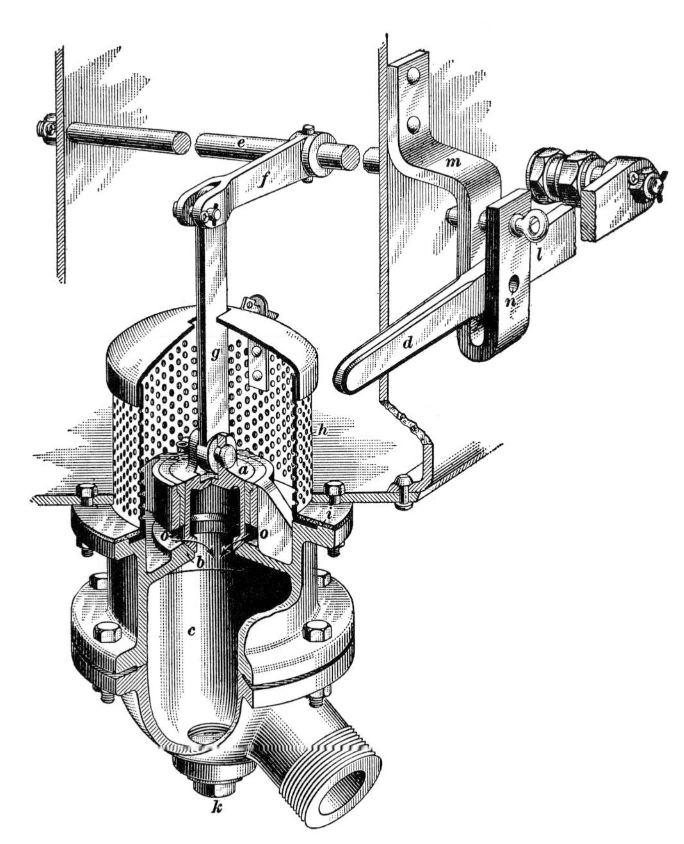

Fig. 45

that normally rests on its seat *b* in the tank-valve chamber *c* is operated by the lever *d*, through the medium of the rod *e*, the arm *f*, and the link *g*. The rod *e* is carried at the ends in the outer and inner walls of the tank. The tank valve is covered by a perforated brass plate *h* so as to prevent dirt from getting into the suction hose. A rubber gasket *i* makes a water-tight joint where the tank-valve chamber is bolted to the bottom of the tank. The gooseneck, a part of which is threaded for the suction hose, is bolted to the bottom of the tank-valve chamber *c*. A plug *k* in the gooseneck permits the tank to be drained when necessary, and it also permits the tank valve to be opened should the operating parts become disconnected. The tank valve is shown seated, but it can be kept in open position by removing the pin *l*, from the bracket *m* and inserting it in the hole *n* after the handle has been raised. When the valve is unseated, the water passes from the chamber enclosed by the plate *h*, through the opening *o* to the chamber *c* as shown by the arrows.

DISORDERS OF SINGLE-JET INJECTORS

GENERAL CONSIDERATIONS

90. Conditions Assumed.—In considering injector disorders it is necessary to regard the tank valve, the suction pipe, the hose, the hose strainer, the delivery pipe, and the boiler check-valve as if they were parts of the injector because any disorder in these parts will affect its operation. The causes for a failure of the injector to operate properly are usually found in the above parts, as about the only disorders in the injector that would lead to trouble are worn out or loose tubes, or clogged openings in the combining tube.

The length of time that the tubes will wear depends largely on the condition of the feedwater; if the water carries solid matter in minute particles, such as sand, the tubes will wear out very rapidly, owing to the wearing action of the solid matter in the water. The most wear occurs where the velocity of the water is the greatest or at the smallest diameter of the delivery tube. When sufficient wear occurs, the velocity of the water that enters the tube will not be transformed into sufficient pressure in the delivery pipe to open the boiler check-valve, and the injector will break.

Another cause for the failure of the delivery tube to convert the velocity of the water into pressure is due to the pitting action of the uncondensed particles of steam that are carried with the water into the delivery tube. The steam particles are forced out against the bore of the tube and pit or roughen it just beyond its smallest diameter. The surface of the tube finally becomes so rough as to decrease the velocity of the water to such an extent that enough pressure is not developed to open the check-valve, and the injector breaks.

The pitting action of the steam is greater when the injector is worked at or near its minimum capacity than when it is worked at its maximum capacity. The reason is that there is more steam in the water when the injector is worked at less than its capacity than when worked at its full capacity.

Wet steam has a cutting effect on the steam nozzle and will interfere with the function of this nozzle in transforming steam at a high pressure and a comparatively low velocity into steam at a low pressure and an extremely high velocity.

Worn tubes will be indicated by a gradual failure of the injector to start properly as well as by the discharge of more or less water at the overflow when the injector is working. One injector should not be operated continuously but both should be worked in turns.

The vibration of the locomotive sometimes causes the tubes to unscrew and loosen in the injector body and result in a failure of the injector.

INJECTOR WILL NOT PRIME

91. Causes.—The principal causes for the failure of an injector to prime are as follows: No water in the tank, strainer stopped up or loose hose lining, air leaks in suction pipe, water in the suction pipe or in the tank too hot, tank valve closed or disconnected and shut, obstruction in injector tubes, leaky inlet valves, the tank cover air-tight, or obstructed overflow pipe.

92. Strainer Stopped Up or Loose Hose Lining.—The strainer, if stopped up, can be blown out by converting the injector into a heater. The liability of the strainer's becoming stopped up is lessened by taking care to prevent coal or any other foreign substance from getting into the manhole of the tank in taking water or at any other time. For this reason the back of the tank should always be kept clean. A loose hose lining obstructs the passage of water through the hose.

93. Air Leaks in Suction Pipe.—A leak of air into the suction pipe above the water level in the pipe, prevents the injector from forming the required vacuum to raise the water. The leaks can be located by closing the tank valve, and opening the steam valve of the injector enough to cause the leaks to show. If the hose and suction pipe are moved from side to side, leaks may show that otherwise would not be located. Slight leaks in the suction pipe cause the injector to work with a rumbling sound.

94. Water Too Hot in Suction Pipe or Tank.—Feed-water that is heated above normal temperature may have one of two effects upon the operation of the injector, depending upon the temperature to which the water is heated. The temperature at which water boils and gives off a steam vapor depends on the pressure on the water. As the pressure lessens, the water will boil at a lower temperature.

If the temperature of the water is sufficient to cause the water to boil and generate steam under the partial vacuum required to raise the water to the injector, the steam will so reduce the vacuum as to prevent the water from being lifted to the injector. If, on the other hand, the temperature of the water is not such as to generate steam under the vacuum established, but is too hot to condense all of the steam issuing from the steam nozzle, then a portion of the steam will pass uncondensed through the injector and cause it to break, because not sufficient velocity is given to the water in the combining tube to develop enough pressure in the delivery tube to open the boiler check-valve.

95. If the water is merely hot in the suction pipe, the hot water can be blown back into the tank by using the injector as a heater. Cold water must be added, if the water in the tank is too hot, although sometimes the injector can be made to work by reducing the steam pressure by partly closing the valve in the cab turret.

Hot water in the tank or in the suction pipe is caused by the improper use of the injector as a heater, by leaks in the steam valve of the injector, or by a leaky boiler check-valve or line check-valve.

To ascertain whether the boiler check-valve or the injector steam valve is leaking, the valve at the cab turret should be closed. If the injector steam valve leaks, the discharge of water and steam at the overflow will stop; but if the check-valve leaks, the discharge will continue. With a combination boiler check and stop-valve, the check-valve can be reseated by closing the stop-valve and removing the cap nut.

Where no stop-valve is used, the check-valve may be seated by pouring cold water on the case, and lightly tapping it, or by starting the injector and then causing it to break. The injector may be started by opening the frost cock in the delivery pipe, if there is one, or by slacking off carefully on the spanner nut, thereby reducing the pressure in the delivery pipe and permitting the injector to prime. When the injector is primed, it is easily started, and in most cases, if the feedwater supply is then shut off, the injector will break and in so doing will cause the check-valve to seat.

96. Tank Valve Disconnected and Shut.—If the tank valve becomes disconnected from the valve rod, and the valve is closed, it can be held open by inserting a piece of wood of the required length through the opening made by the removal of the plug *k*, Fig. 45. The strainer above the check-valve prevents it from being blown off its seat by converting the injector into a heater.

97. Obstruction in Combining Tubes.—The openings in the combining tubes of single-jet injectors may become incrusted, and restrict the discharge of air and steam when the injector steam valve is opened. The injector cannot be started because some of the steam blows back and forms a pressure in the suction pipe. The remedy is to remove the tubes at the first opportunity and place them in a bath consisting of one part of muriatic acid to ten parts of water. The tubes should be removed as soon as the gas bubbles cease to be given off, otherwise the acid will attack the nozzles and will pit and rough them.

98. Leaky Inlet Valves.—Leaky inlet valves with the Sellers and Simplex injectors permit steam to pass back into the suction pipe and prevent the injector from priming. A leaky inlet valve with a Simplex injector can be cut out by closing the plug cock. Leaky hinge check-valves with the non-lifting types of Sellers and Nathan injectors will not affect the operation of the injector unless an effort is made to make the injector work with hot water.

99. Tank Cover Air-Tight.—A failure of the injector to prime may be due to a tank cover that is air-tight; in any event if the air cannot enter the tank freely, the injector will finally break, and cannot be started again. When the tank cover freezes air-tight, the air entrapped in the tank expands as the water is withdrawn and the pressure exerted by the air lessens until it is not sufficient to force the water into the injector.

100. Obstructed Overflow Pipe.—If the overflow pipe is badly obstructed by scale or other causes, the steam and the water cannot escape freely when the injector is being started, and a back pressure will form in it that will prevent it from operating.

INJECTOR PRIMES BUT BREAKS OR SPILLS WATER

101. Causes.—The causes for an injector's priming but breaking or spilling part of the water at the overflow as soon as the starting lever is pulled all the way back are as follows: Boiler check-valve sticks and fails to lift properly, delivery tube worn or obstructed, feedwater too warm, or loose tubes.

102. Boiler Check-Valve Sticks.—If the boiler check-valve sticks shut, or will only lift slightly, the injector will prime but will break as soon as the injector steam valve is opened wide. In this event, the other injector should be operated.

A boiler check with a reduced lift is also sometimes indicated by the fact that the injector will work satisfactorily at a low steam pressure and will discharge part of the water at the overflow with a high steam pressure. The explanation is that less water is delivered with a low steam pressure and the boiler check with a reduced lift can accommodate the supply while with a higher steam pressure the boiler check cannot take care of the additional water and a part spills at the overflow.

103. Delivery Tube Worn or Obstructed.—If the delivery tube is badly worn, the water in the delivery pipe cannot be brought up to the required pressure by the jet in the combining tube, to open the boiler check-valve. The injector therefore breaks.

The obstruction in the tube may be due to a piece of coal or waste. The other injector should be worked until repairs can be made.

INJECTOR PRIMES AND FORCES BUT BREAKS FREQUENTLY

104. Causes.—If the injector primes and forces, and after working for a short or a long time, suddenly breaks, and this action is repeated frequently, the trouble may be due to one of the following causes: *(a)* Leaks in the suction pipe above the water level that are caused by a sudden jarring of the locomotive; loose connections at the hose or suction pipe couplings; or by kinks in the hose; *(b)* water low in the tank so that the motion of the water leaves the tank dry at times, *(c)* tank cover air-tight.

DISORDERS OF DOUBLE-JET INJECTORS

105. Inspirator Will Not Prime.—In addition to the causes already given for failure of an injector to prime, the Hancock lifting inspirator will not prime if the intermediate overflow valve w, Fig. 33, sticks and will not lift or if the valve z is not steam-tight. With a sticky intermediate overflow valve all the water from the lifter tube t has to pass through the openings g' to the forcer tube v, and the water cannot do this without exerting a back pressure on the water that is discharging through the front end of the lifter tube into chamber d'. The intermediate overflow valve can be examined by removing the cap nut l.

If the valve z leaks considerably, the steam will pass by it through the opening g' and back through the lifter tube t to the suction pipe and to the overflow pipe as soon as the valve y is moved back to priming position. The valve z can be tested by closing the regulating valve c' and moving the valve y to priming position. Steam will discharge at the overflow pipe, if the valve z leaks, because the steam that passes the valve y is trapped in chamber f' when the valve c' is closed. The overflow pipe can be tested to ascertain whether it is clear by seeing whether the injector will prime when the pipe is uncoupled.

106. Inspirator Primes But Will Not Force.—In addition to the causes already given for an injector failing to force water, the inspirator will prime but will not force if the intermediate overflow valve w, Fig. 33, leaks or if there is a leak between its removable seat and the injector body, or if an improper relation exists between the opening of the valve z and the closing of the final overflow valve x.

If the intermediate overflow valve leaks or if the valve or its removable seat leaks, the water which is under pressure in the delivery pipe and in chamber e' will return by the valve to chamber d' and interfere with the delivery of water through the lifter tube t. A leaky intermediate overflow valve is indicated when the injector gradually works weaker and finally breaks after being started.

In order that the inspirator may work properly, too much lost motion must not be permitted in the connections between the lever e and the final overflow valve x; otherwise the proper relation will not be maintained between the opening of the valve z and the closing of the final overflow valve. The valve z must be opened wide enough for sufficient steam to pass by to force the water into the boiler before the overflow valve x closes. If the final overflow valve x closes before the valve z opens far enough, the inspirator will break, after being primed. The reason is that the premature closing of the final overflow valve before the jet of water forms in the forcer tube causes the water in the tube to exert a back pressure on the water that is passing through the lifter tube.

A leaky final overflow valve will cause water to waste at the overflow when the inspirator is working, because this valve is then subjected to the pressure of the water in the delivery pipe, or to a pressure in excess of the boiler pressure.

THE STEAM LOCOMOTIVE

EXAMINATION QUESTIONS

Notice to Students.—*Study the Instruction Paper thoroughly before you attempt to answer these questions.* **Read each question carefully and be sure you understand it;** *then write the best answer you can. When your answers are completed, examine them closely, correct all the errors you can find, and* **see that every question is answered.**

(1) Name the two classes of injectors.

(2) What is meant by the term self-regulating as applied to an injector?

(3) What is a double-jet injector?

(4) Give a brief summary of the operation of an injector.

(5) What is the purpose of the injector?

(6) What is meant by a vacuum?

(7) On what nozzle does the action of an injector primarily depend?

(8) What term is applied to pressure measured above an absolute zero of pressure?

(9) What change occurs in the condition of the steam in its passage through the steam nozzle?

(10) What is the purpose of the combining tube?

(11) What is a single-jet injector?

(12) What is the pressure in the rear end and in the front end of the combining tube when the injector is working?

(13) What is the purpose of the delivery tube?

(14) What is the purpose of the openings or gaps in the combining tube of a single-jet injector?

(15) What is the pressure in the rear end and in the front end of the delivery tube when the injector is operating with a steam pressure of 200 pounds?

(16) What is the pressure of the atmosphere at sea level?

(17) How is the amount of vacuum usually expressed?

(18) Explain why the velocity of the steam increases and the pressure reduces when steam is passing through the steam nozzle.

(19) What must be the shape of the bore of the steam nozzle in order that a high velocity of the steam may be obtained?

(20) Why should the starting lever never be pulled back sudenly?

(21) Give some of the causes for failure of an injector to prime.

(22) Explain how to use the Hancock inspirator as a heater.

(23) Name the nozzles and tubes in a single-jet injector and in a double-jet injector.

(24) Explain how to start and stop a lifting injector and how to use it as a heater.

(25) What is the purpose of the regulating valve used with the Hancock inspirator?

(26) What type of injector must be used when it is desired to deliver feedwater above a temperture of 212° F.?

(27) What tube in an injector is subject to the most wear and where does the wear occur?

(28) How will worn tubes in an injector be indicated?

(29) What is the purpose of the intermediate overflow valve used with a double-jet injector?

(30) What is the reason for placing the boiler check-valve on top of the boiler?

(31) What condition is necessary in order to have steam condense with hot water?

(32) Explain how to seat a boiler check-valve.

(33) Where must the final overflow valve be placed with a double-jet injector?

(34) At what pressure does the lifter tube deliver water to the forcer tube with a Hancock inspirator?

(35)What disorders will cause an injector to break after being primed?

(36) What pressure forces the water from the tank to a lifting injector?

(37) How is a leaky intermediate overflow valve indicated with a double-jet injector?

(38) What is the effect of a leak in the suction pipe?

(39) How can it be ascertained whether the boiler check-valve or the injector steam valve is leaking?

(40) What will be the effect if the final overflow valve of the Hancock inspirator closes before the valve z, Fig. 33, opens far enough?

NOTES

NOTES

NOTES

NOTES

NOTES

NOTES